12th International Symposium on Continuum Models and Discrete Systems 2011

(CMDS-12)

Journal of Physics: Conference Series Volume 319

Kolkata, India
21-25 February 2011

Editors:

Abhik Basu
Anjan Kumar Chandra
Bikas K. Chakrabarti

ISBN: 978-1-61839-289-3
ISSN: 1742-6588

12th International Symposium on Continuum Models and Discrete Systems 2011

(CMDS-12)

Journal of Physics: Conference Series Volume 319

Kolkata, India
21-25 February 2011

ISBN: 978-1-61839-289-3
ISSN: 1742-6588

IOP Institute of Physics

Printed from e-media with permission by:

Curran Associates, Inc.
57 Morehouse Lane
Red Hook, NY 12571

Some format issues inherent in the e-media version may also appear in this print version.

Printed by Curran Associates, Inc. (2011)

For permission requests, please contact the Institute of Physics
at the address below.

Institute of Physics
Dirac House, Temple Back
Bristol BS1 6BE UK

Phone: 44 1 17 929 7481
Fax: 44 1 17 920 0979

techtracking@iop.org

Additional copies of this publication are available from:

Curran Associates, Inc.
57 Morehouse Lane
Red Hook, NY 12571 USA
Phone: 845-758-0400
Fax: 845-758-2634
Email: curran@proceedings.com
Web: www.proceedings.com

TABLE OF CONTENTS

PREFACE

The 12th International Symposium on Continuum Models and Discrete Systems (CMDS-12) (http://www.saha.ac.in/cmp/cmds.12/) took place at the Saha Institute of Nuclear Physics in Kolkata from 21–25 February 2011. Previous CMDS symposia were held in Kielce (Poland, 1975), Mont Gabriel (Canada, 1977), Freudenstadt (Federal Republic of Germany, 1979), Stockholm (Sweden, 1981), Nottingham (United Kingdom, 1985), Dijon (France, 1989), Paderborn (Germany, 1992), Varna (Bulgaria, 1995), Istanbul (Turkey, 1998), Shoresh (Israel, 2003) and Paris (France, 2007). The broad interdisciplinary character, limited number of participants (not exceeding 100) and informal and friendly atmosphere of these meetings has made them a well-acknowledged place to make highly fruitful contacts and exchange ideas, methods and results.

The purpose of CMDS is to bring together scientists with different backgrounds who work on continuum theories of discrete mechanical and thermodynamical systems in the fields of mathematics, theoretical and applied mechanics, physics, material science, and engineering. The spirit of the CMDS meetings is to stimulate extensive and active interdisciplinary research. The International Scientific Committee members of this conference were:

- David J Bergman (Chairman CMDS 10), Tel Aviv University, Israel;
- Bikas K Chakrabarti (Chairman CMDS 12), Saha Institute of Nuclear Physics, India;
- Alex Hansen, Norwegian University of Science and Technology, Norway;
- Hans Jürgen Herrmann, Institute for Building Materials, ETH, Switzerland;
- Esin Inan (Chairman CMDS 9), Istanbul Technical University, Turkey;
- Dominique Jeulin (Chairman CMDS 11), Ecole des Mines de Paris, France;
- Frank Juelicher, Max-Planck-Institute for the Physics of Complex Systems, Germany;
- Hikaru Kawamura, University of Osaka, Japan;
- Graeme Milton, University of Utah, USA;
- Natalia Movchan, University of Liverpool, UK; and
- Ping Sheng, The Hong Kong University of Science and Technology, Hong Kong.

At CMDS-12 the topics of interest were:

- Thermodynamics, transport theory and statistical mechanics in the context of continuum modeling discrete systems;
- Statistical mechanics and dynamics of fluid flows;
- Continuum mechanics of complex fluids and deformable solids with microstructure;
- Fundamentals of fracture, defect dynamics, fatigue, and fracture dynamics;
- Statics and dynamics of dislocations, dislocation mediated phase transitions and plasticity;
- Granular materials: statics and dynamics;
- Physics and mechanics of earthquakes;

- Transport in composite materials; and
- Continuum theory of soft matter systems of biological motivation and living structures.

The scientific program consisted of General Lectures and Research Communications. The four General Lectures (40 min + 5 min) introduced the most recent ideas and advances in the fields covered. The 25 Research Communications reported new results and methods in these fields and were presented orally (30 min + 5 min). In addition there were 15 posters which also presented interesting results and five of these posters were selected by a special committee chaired by David J Bergman to contribute to this proceedings. The selected five posters were presented by Moutushi D Choudhury, Morten Grova, Arya Paul, Arnab Saha and Amartya Sarkar. We were happy to note that 32 scientists from ten countries and a large number of students (both from India and other countries) participated in this symposium. The talks and poster sessions generated a lot of discussions, arguments and collaborations. The articles that appear in this proceedings are based on the invited talks and selected poster presentations.

We would like to thank the *Journal of Physics Conference Series* (IOP Publishing) for publishing the proceedings of the conference. We would also like to thank all the contributors to this proceedings for their kind co-operation and the referees for their prompt and active support. A number of invited participants (Amit Acharya, Alexander Altland, Kamal K Bardhan, Davide Bigoni, Andrej Cherkaev, Hikaru Kawamura, Subhasish Mazumder, Srutarshi Pradhan, Jacques Prost, Sriram Ramaswamy, Ping Sheng, Asok K Sen) were unable to contribute to this proceedings volume; on the other hand, although Kamal could not participate in the symposium, he contributed to the proceedings volume. We are grateful to Lucilla de Arcangelis, Kamal K Bardhan, Abhik Basu, Jayanta K Bhattacharjee, Andrej Cherkaev, Yilong Han, Alex Hansen, Takahiro Hatano, Esin Inan, Hikaru Kawamura, Pradeep K Mohanty, Akio Nakahara, Surajit Sengupta, Ping Sheng, Sujata Tarafdar and Uwe Tauber for reviewing the papers submitted to this proceedings volume.

We would like to take this opportunity to thank all the members of the organizing committee and acknowledge the financial support received from the Centre for Applied Mathematics and Computational Science, Saha Institute of Nuclear Physics, Kolkata.

We look forward to meeting you all at the next event in this series, CMDS 13, to be held in 2014/2015 in Utah, USA.

Bikas K Chakrabarti
Chairman of CMDS 12
Kolkata, 8 August 2011

Editors of the proceedings

Abhik Basu *abhik.basu@saha.ac.in*
Bikas K Chakrabarti *bikask.chakrabarti@saha.ac.in*
Anjan Kumar Chandra *anjan.chandra@saha.ac.in*
Saha Institute of Nuclear Physics, Sector-I, Block - AF, Bidhannagar, Kolkata 700064, India

Micromechanics and statistics of slipping events in a granular seismic fault model

L. de Arcangelis

Department of Information Engineering and CNISM, Second University of Naples, Aversa (CE), Italy

M. Pica Ciamarra

CNR–SPIN, Dipartimento di Scienze Fisiche, Universitá di Napoli Federico II, Italy

E. Lippiello

Department of Environmental Sciences and CNISM, Second University of Naples, Caserta, Italy

C. Godano

Department of Environmental Sciences and CNISM, Second University of Naples, Caserta, Italy

E-mail: dearcangelis@na.infn.it

Abstract. The stick-slip is investigated in a seismic fault model made of a confined granular system under shear stress via three dimensional Molecular Dynamics simulations. We study the statistics of slipping events and, in particular, the dependence of the distribution on model parameters. The distribution consistently exhibits two regimes: an initial power law and a bump at large slips. The initial power law decay is in agreement with the the Gutenberg-Richter law characterizing real seismic occurrence. The exponent of the initial regime is quite independent of model parameters and its value is in agreement with experimental results. Conversely, the position of the bump is solely controlled by the ratio of the drive elastic constant and the system size. Large slips also become less probable in absence of fault gouge and tend to disappear for stiff drives. A two–time force–force correlation function, and a susceptibility related to the system response to pressure changes, characterize the micromechanics of slipping events. The correlation function unveils the micromechanical changes occurring both during microslips and slips. The mechanical susceptibility encodes the magnitude of the incoming microslip. Numerical results for the cellular-automaton version of the spring block model confirm the parameter dependence observed for size distribution in the granular model.

1. Introduction

The understanding of the earthquake phenomenology represents a challenge also due to the impossibility of simulating seismic events in laboratory. It is therefore extremely important to develop models for earthquakes which can be investigated, at the same time, theoretically, experimentally and numerically. An earthquake is a mechanical process characterized by two

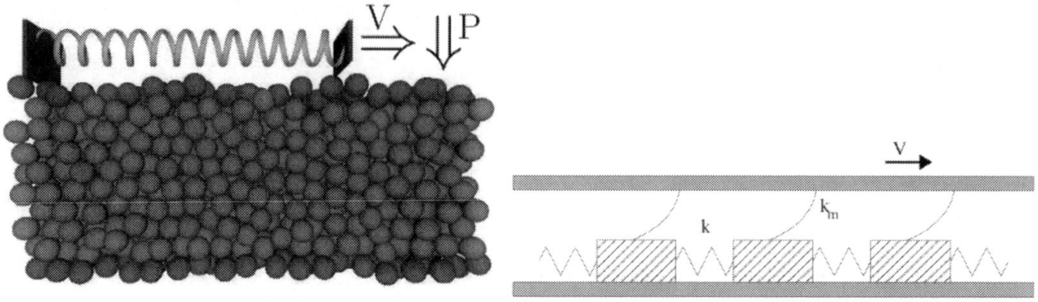

Figure 1. The two considered models of a seismic fault. Left panel: granular particles (gray grains) are enclosed between two rigid plates (red grains) under constant pressure, and represent the fault gouge. The bottom plate is kept fixed. The top plate is driven via a spring: one extreme of the spring is attached to the plate, while the free one moves with constant velocity V. Right panel: the usual spring block model where the top plate is driven at constant velocity.

phases: one in which the fault is locked and elastic energy is stored and one in which the elastic energy is suddenly released as soon as the system becomes unable to sustain the applied stress. In this mechanical transition, the fault gouge appears to play an important role. Gouge is the granular material present between the planes of a fault, being the product of past wearing, and it controls the macroscopic sliding friction and the frictional stability of the fault. This observation has triggered the experimental study of the stick–slip dynamics of granular models of faults [1, 2, 3, 4, 5, 6, 7, 8, 9, 10, 11], as the one shown in Fig. 1 (left panel), in which the shear stress is applied via a spring mechanism, at constant pressure P. A particular interest has received the study of the role of the granular medium on the mechanical properties, such as strain hardening, crystallization, velocity dependent friction and shear localization [1, 2, 3, 4, 8]. Experimental results have also shown evidence [9, 10] of novel features of the non linear elastic relaxation in a granular medium perturbed by an acoustic seismic wave. This has provided a novel interpretation for some highly non linear behaviour of seismic occurrence, as for instance triggering of earthquakes at very large distances [12]. In fact, the non linear behaviour could be explained considering that small perturbations applied to a granular system modify the contact force network, setting up a memory mechanisms which could be responsible of the spatio–temporal correlations between earthquakes.

Numerical simulations of granular models for seismic faults allow to investigate the mechanical properties of the system at a level of spatial and temporal resolution not accessible experimentally [13, 19, 14, 15, 16, 17, 18, 20]. Some studies investigate the effect of fragmentation of the gouge [19, 14, 15], others focus on the mechanisms at the basis of the unjamming transition [16, 17, 18, 20]. An important open question concerns the ability of granular fault models to reproduce fundamental phenomenological laws for seismic occurrence, as for instance the Gutenberg-Richter (GR) law [21]. This states that earthquake magnitudes are exponentially distributed. The logarithmic relation between earthquake magnitude and fault slip implies that slips follow a power law distribution, which is found to be characterized by an exponent $a \simeq 1.7$ independent of the geographic region and quite stable in time [22, 23]. Such a power law distribution is reproduced by the cellular-automaton version of the spring–block model [24, 25, 26, 27] (Fig. 1, right panel) which is a model of an elastic fault, where a slip may involve the displacement of a few as well as of many blocks. Conversely, in the granular model of fault shown in Fig. 1 (left panel), slips of different size typically involve the entire top plate. It could be therefore surprising that this and similar granular models of faults have been shown to reproduce the GR distribution experimentally [5, 6, 7] and numerically [20, 13]. However, it is not clear what is the role of the granular media in the overall dynamics, and how the slip size probability distribution is affected by the model parameters.

In this paper, we review results from three dimensional Molecular Dynamics simulations of a seismic fault model, where granular media play the role of fault gouge. Moreover we compare the statistics for the size of slipping events with numerical simulations of the cellular-automaton version of the spring-block model. The paper is organized as follows. We first present the granular seismic fault model in Sec. 2 and its dynamical behaviour in Sec. 3, then discuss the slip size distribution in Sec. 4. We analyze the micromechanical properties of slipping events in Section 5. We present the spring–block results in Sec. 6, where we compare the size distribution with the granular model results. Conclusions are drawn in the last section.

2. The granular model

We perform Molecular Dynamics simulations of a fault model, where the fault gouge is represented by spherical granular particles, as illustrated in Fig. 1. The fault gouge is enclosed between two rigid rough plates of dimension $L_x \times L_y$. Each plate is made of $L_x \times L_y/d^2$ spheres of diameter d placed in random position in the xy plane. Spheres are shifted by a random $\delta z \in [0, d/2]$ in the perpendicular direction and then glued to each other. In order to make the plates rigid, the particles keep their relative positions. While the bottom plate is kept fixed, the top one is subject to a constant pressure P and attached to a spring of elastic constant k_m (see Fig. 1). The free end of the spring moves with constant velocity V in the x direction. Accordingly, in the jammed phase, when the top plate does not move, the shear stress $\sigma = k_m \Delta l / L_x L_y$ increases linearly in time, Δl being the deformation of the spring.

We consider a system of N monodisperse spheres of diameter d and mass m, enclosed between the rough plates. We keep the transverse length L_y fixed and explore both the effect of changing the system length L_x, as well as the number of particles N. Since each layer contains roughly $L_x L_y/d^2$ particles, in the following we will focus on the effect of changing the number of layers, defined as $h = Nd^2/L_x L_y$ (note that N does not include the particles forming the confining plates). We use periodic boundary conditions along x and y. We employ a contact force model that captures the major features of granular interactions, known as spring-dashpot model, also taking into account the presence of static friction [28].

Static friction is implemented by recording the shear displacement of contacting grains. Let's consider two grains $\{i, j\}$, at positions $\{\mathbf{r}_i, \mathbf{r}_j\}$, with velocities $\{\mathbf{v}_i, \mathbf{v}_j\}$ and angular velocities $\{\boldsymbol{\omega}_i, \boldsymbol{\omega}_j\}$. To describe the force acting on particle i we introduce the normal compression δ_{ij}, the relative normal velocity $\mathbf{v}_{n_{ij}}$, and the relative tangential velocity $\mathbf{v}_{tg_{ij}}$, given by

$$\delta_{ij} = d - r_{ij}, \tag{1}$$

$$\mathbf{v}_{n_{ij}} = (\mathbf{v}_{ij} \cdot \mathbf{n}_{ij})\mathbf{n}_{ij}, \tag{2}$$

$$\mathbf{v}_{tg_{ij}} = \mathbf{v}_{ij} - \mathbf{v}_{n_{ij}} - \frac{1}{2}(\boldsymbol{\omega}_i + \boldsymbol{\omega}_j) \times \mathbf{r}_{ij}, \tag{3}$$

where $\mathbf{r}_{ij} = \mathbf{r}_i - \mathbf{r}_j$, $\mathbf{n}_{ij} = \mathbf{r}_{ij}/r_{ij}$, with $r_{ij} = |\mathbf{r}_{ij}|$, and $\mathbf{v}_{ij} = \mathbf{v}_i - \mathbf{v}_j$. The elastic tangential displacements $\mathbf{u}_{tg_{ij}}$, set to zero at the initiation of a contact, evolve according to

$$\frac{d\mathbf{u}_{tg_{ij}}}{dt} = \mathbf{v}_{tg_{ij}} - \frac{(\mathbf{u}_{tg_{ij}} \cdot \mathbf{v}_{ij})\mathbf{r}_{ij}}{r_{ij}^2}, \tag{4}$$

where the second term in Eq.4 arises from the rigid body rotation around the contact point and ensures that $\mathbf{u}_{t_{ij}}$ always lies in the local tangent plane of contact. Normal and tangential forces acting on particle i are given by

$$\mathbf{F}_{n_{ij}} = k_n \delta_{ij} \mathbf{n}_{ij} - \gamma_n m_{eff} \mathbf{v}_{n_{ij}}, \tag{5}$$

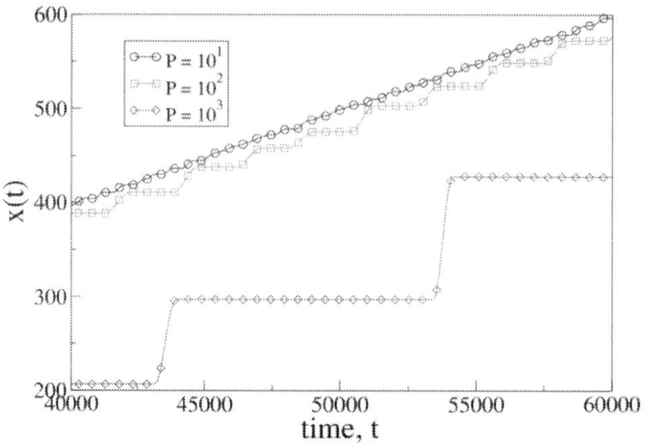

Figure 2. Time evolution of the top plate position, for a system loaded at $V = 1$, and subject to different values of the pressure. Increasing the pressure, the system shows a transient from a flowing configuration to a stick-slip dynamics.

$$\mathbf{F}_{tg_{ij}} = -k_t \mathbf{u}_{tg_{ij}} - \gamma_{tg} m_{eff} \mathbf{v}_{tg_{ij}}, \tag{6}$$

where $k_{n,tg}$ and $\gamma_{n,tg}$ are elastic [29] and viscoelastic constants, respectively, and $m_{eff} = m_i m_j/(m_i + m_j)$. Here the subscripts n and tg indicate the normal and tangential components. We use $\gamma_{tg} = 0$ and γ_n chosen in such a way that the restitution coefficient is $e = |\mathbf{v}_{in}/\mathbf{v}_{out}| = 0.8$, where $\mathbf{v}_{in}$ ($\mathbf{v}_{out}$) is the relative normal velocity of two particles before (after) their collision. Coulomb friction is taken into account via the introduction of a friction coefficient μ, chosen to be $\mu = 0.1$, so that $|\mu \mathbf{F}_{n_{ij}}| \geq |\mathbf{F}_{tg_{ij}}|$.

In this manuscript, we measure the length in units of d and time in units of $\sqrt{m/k_n}$. The fixed bottom plate is treated as an object with infinite mass, while the mass of the top plate is $L_x L_y/d^2$. When not stated otherwise, we fix $L_y = 5$, $m = 1$, $k_n = 2\ 10^5$ and $k_t = 2k_n/7$, the spring constant of the drive mechanisms to $k_m = k_0 = 10^2$, and the driving velocity to $V = 5\ 10^{-2}$.

3. Phase diagram

The rate at which the shear stress increases in absence of plate motion, $kV/L_x L_y$, and the confining pressure, P, are the two parameters which most greatly affect the dynamics of the system, which either flows continuously, or exhibits sick–slip motion. The crossover between these two regimes is shown Fig. 2, where we plot the time evolution of the top plate position $x(t)$ for $V = 1$, and three different values of the confining pressure, $P = 10^1, 10^2, 10^3$. At high pressure, the system is characterized by a stick-slip dynamics: when the system is jammed the applied stress increases linearly in time, and the flow begins when the system becomes unable to sustain the applied stress. In the flow regime the stress decreases, and the system eventually jams again. Conversely, at small values of the pressure the system flows continuously, as it is never able to sustain the applied shear stress. The role of the two control parameters P and V is summarized in Fig. 3. In agreement with experimental results of a similar system [11], as well as with numerical results found in two dimensions [18], we find either stick–slip motion at high pressure and small loading velocity, or continuous flow at low pressure and high loading velocity.

In this paper, we choose the parameter values in order to observe stick-slip behaviour for the system, more precisely we set $V = 510^{-2}$ and $P = 100$. In the stick phase, the system evolves as

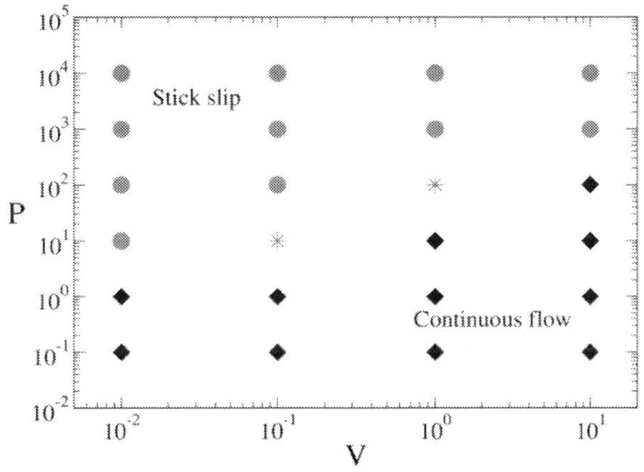

Figure 3. Different dynamical regimes are observed depending on the value of the control parameters, the imposed pressure P, and the velocity of the free extreme of the spring V. At low pressure and high velocity, the system continuously flows (diamonds), while at high pressure and small velocity stick-slip motion is observed (circles). In the transition region, an irregular motion is observed (stars).

it adapts to the increasing stress. Its evolution consists of large slips and small slips, in which the top plate moves by a fraction of a particle diameter d. These microslips are difficult to identify from the top plate displacement. A more accurate identification is based on the evaluation of the top plate velocity, which exhibits peaks in correspondence of each slips, as illustrated in Fig. 4 (lower panel). We define the starting time $t_s^{(i)}$ of the i–th slip as the time when v becomes larger than a given threshold $v_t = 10^{-5}$ (other threshold values give similar results, if small enough). Similarly, the ending time $t_e^{(i)}$ is the time when v goes back to values smaller than v_t. Having identified the starting and the ending time of each slip, we define, as customary, the slip size as $S_i = x(t_e^{(i)}) - x(t_s^{(i)})$.

4. Slip size distribution

The slip size distribution, $p(S)$, in the case of real earthquakes can be derived from the Gutenberg–Richter law for the magnitude distribution, $P(M) \propto 10^{-bM}$, where $b \simeq 1$. In fact, the slip size S is proportional to the seismic moment, $\mathcal{M} = \mu AS$, where A is the slipping area and μ the shear modulus. The seismic moment is related to the magnitude M by a logarithmic relation $M = 2/3\log(\mathcal{M}) + const$ [30]. Accordingly, since in our case the slipping area is constant, one finds $p(S) \propto S^{-a}$, with $a = 1 + (2/3)b \simeq 1.7$, for $b \simeq 1$. Here we examine the dependence of this probability distribution on the number of layers, on the system length and on the spring constant of the driving mechanism.

4.1. Effect of the number of layers

Fig. 5 shows the slip size distribution for a number of layers ranging from $h = 0$ to $h = 8$. In all cases we identify two different regimes in the distribution: a power law behaviour at small slip sizes, named microlips, with $S < 1$, and a bump at a characteristic size $S \sim L_x$. These events are named slips. The power law behaviour, $p(S) \propto S^{-a}$, holds over about four decades with an exponent a in good agreement with the one predicted by the Gutenberg–Richter law, $a \simeq 1.7$, except for the cases $h = 1$ and 2. Note that the exponent $a \simeq 1.7$ is also observed for $h = 0$, which corresponds to the case of solid–on–solid friction. In the case of one or two granular layers,

Journal of Physics: Conference Series **319** (2011) 012001 doi:10.1088/1742-6596/319/1/012001

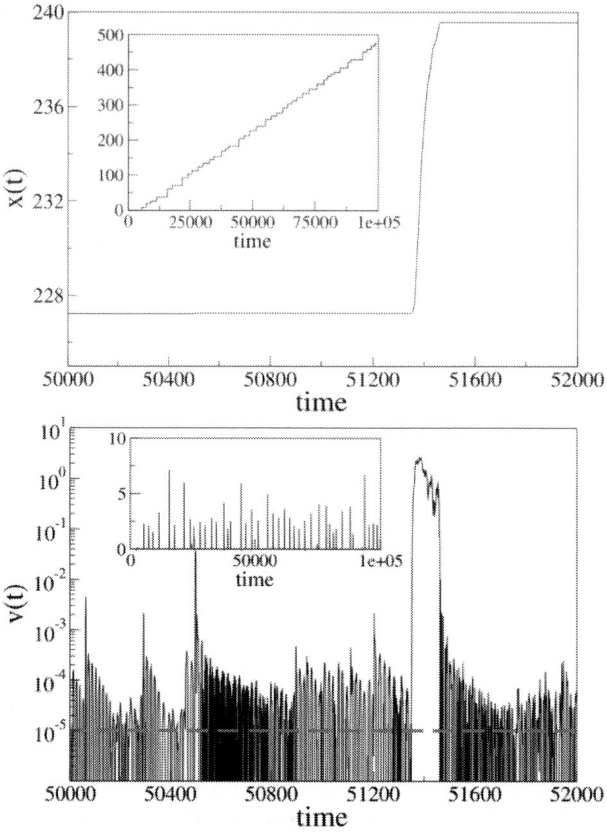

Figure 4. Evolution of the top plate position (upper panel) and velocity (lower panel) as a function of time. The insets illustrates the typical stick slip motion. The main panels correspond to a zoom on a small time interval across a single large slip. The velocity evolution has a peaked structure, which signals the presence of small slips, so small that they are difficult to identify from the behavior of $x(t)$ plotted in the upper panel. The dashed horizontal line in the lower panel marks the threshold value of the velocity used to identify the slips. Here $L_x = 20$, and $h = 1$.

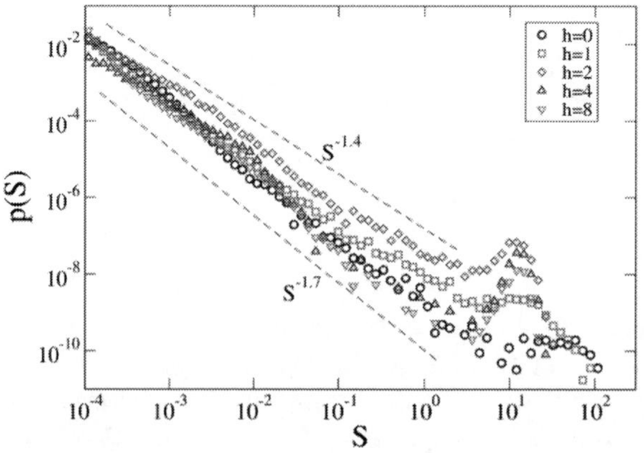

Figure 5. Slip size distribution for $L_x = 20$, $L_y = 5$ and different numbers of granular layers h.

the distribution shows a larger probability for intermediate slip sizes, and therefore a smaller scaling exponent. For $h > 2$ the scaling behaviour is independent of the number of layers. This result can be understood measuring the average displacement, Δx, of particles at height z from the bottom plate during a slip event. The displacement configuration indicates that particles in the two uppermost granular layers respond more intensively to the external shear and therefore the system reorganization during a slip event involves mostly few uppermost layers, not the entire system. Accordingly, only when $h < 2$ the finite thickness of the system interferes with the dynamical process, changing the scaling behaviour of the probability distribution $p(S)$. This is also consistent with the fact that, for $h > 2$, the bump is centered at about the same slip size, regardless of h.

This analysis clarifies that the thickness h of the fault gouge does not affect much the slip size probability distribution, as long as $h > 2$. This is consistent with the finding of a universal exponent a for the slip size of real earthquakes, which occur on faults certainly having different gouge thickness.

4.2. Role of different L and k_m

In Fig. 6 and 7 we show the effects of the different system size L_x and the elastic constant of the driving spring, k_m, on the slip size distribution for $h = 8$ and $L_y = 5$. We first discuss the behaviour of the distribution for microslips. We observe that, keeping k_m fixed, the initial power law regime is independent of L_x (Fig. 6, upper panel). Moreover, Fig. 7 indicates that at small S the distribution is neither affected by the k_m values. On the other hand, k_m introduces a large size cut-off. The maximum slip, indeed, decreases by increasing k_m, being roughly inversely proportional to k_m. Therefore, for very stiff external drive, large slips are not observed. In general for large k_m, separation of length scales between microslips and slips is no longer observed, and for instance for $k_m = 10k_0$ the largest events merge into the tail of the power law behaviour. The above results suggest that the self-similar organization of microslips is independent of parameters with a stable exponent $a \simeq 1.7$, in agreement with experimental values for earthquakes. Note that k_m could be related to the shear modulus of the earth crust close to the fault, which is different on different faults as it depends on the particular rock composition. The initial power-law regime being independent of k_m is therefore in agreement with the experimental finding of a fault–independent a exponent [22, 23].

Conversely, the position of the bump depends on both L_x and k_m. Since in a large slip most of the accumulated stress is released, in order to understand the L_x and k_m dependence we consider, as a simple example, a toy model made of a single block of dimension $L_x \times L_y$, pulled on a rough surface by a spring of elastic constant k_m. Assuming that a slip, occurring when the stress reaches a critical threshold σ_c, releases the whole accumulated stress, the corresponding slip size is $S = \sigma_c L_x L_y / k_m$, and scales as L_x/k_m. We will show in the following that the large slips of our granular model present this scaling behaviour.

First, we investigate the dependence on the system size L_x at fixed values of k_m. Fig. 6 (upper panel), shows that L_x affects the position of the bump which moves towards larger slip sizes as L_x increases. The position of the bump, in agreement with the simple model, linearly scales with L_x, as illustrated in Fig. 6 (lower panel), where the bump collapse is obtained by plotting $p(S)$ versus S/L_x. Second, in order to verify the scaling of the bump position with k_m, as suggested by the single block model, we perform simulations changing at the same time L_x and k_m, keeping their ratio k_m/L_x fixed. Results, plotted in Fig. 8 clearly show that the dependence on k_m and L_x enters via the variable k_m/L_x. Indeed, data corresponding to different L_x collapse onto the same curve, as predicted by the single block calculation. Notice that not only the bumps but the whole distributions collapse onto a unique curve if the ratio k_m/L_x is kept fixed.

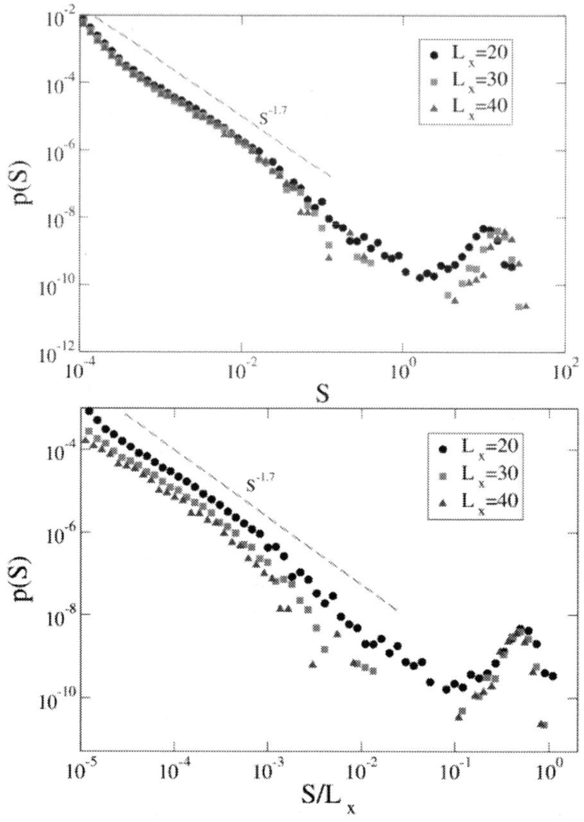

Figure 6. Slip size distribution $p(S)$ versus S for $L_y = 5$, $k_m = 100$, $h = 8$ and different system size L_x (upper panel). Same distributions for slip size rescaled by system size S/L_x (lower panel).

5. Micromechanics of an earthquake

5.1. Onset of a slip

To understand the mechanisms acting at the onset of a slip, we need to identify its precise starting time t_u. Here we describe the analysis performed on the particular slip occurring at time $t \simeq 1600$. Similar results are obtained for other slips. We consider a replica of the system at time t, and follow its time evolution at zero driving velocity $V = 0$. If the replica made at time t resists to the applied stress, then $t \leq t_u$. Conversely, if a slip is observed, $t > t_u$. We define t_u as the largest time where no slip occurs, and we identify it (one for each slip) with an accuracy equal to the time step of integration of the equation of motion δt. This procedure is equivalent to a quasi-static simulation [31] around the un-jamming time t_u and gives the value of the shear stress above which the system starts to flow. The identification of t_u is possible for both slips and microslips. We here concentrate our analysis on slips, where major rearrangements of the system occur.

We have performed a number of checks which suggest that no structural changes occur at t_u. For instance, the comparison of the state of the system at time t_u with the one at shortly earlier

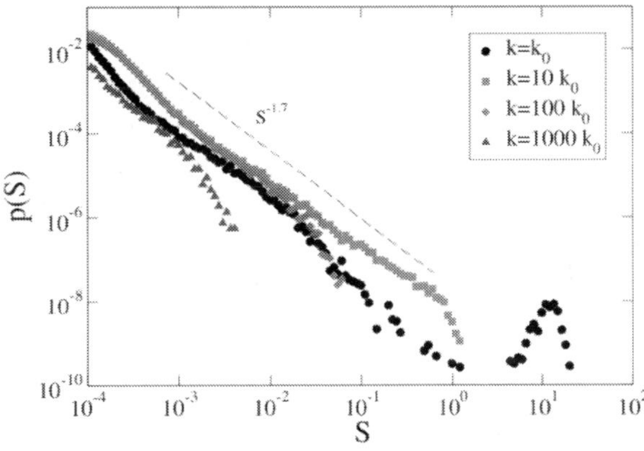

Figure 7. Slip size distribution $p(S)$ versus S for$L_x = 20$, $L_y = 5$, $h = 8$ and different elastic spring constants k_m, for $k_0 = 100$.

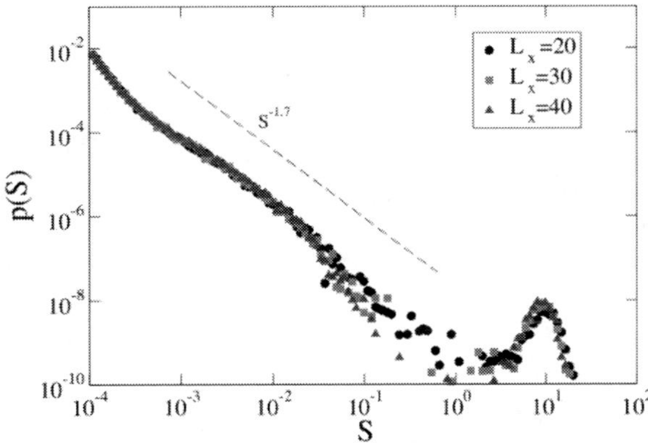

Figure 8. Slip size distribution $p(S)$ versus S for $L_y = 5$, $h = 8$ and different L_x and k_m, with fixed ratio $k_m/L_x = 5$.

and later times, shows that no contact breaks at t_u. We have also considered the distribution of the parameter $\lambda = |\mathbf{f}_t|/\mu|\mathbf{f}_n|$ where $\mathbf{f}_t$ and $\mathbf{f}_n$ are the tangential and the normal forces. When $\lambda > 1$ a contact breaks as the Coulomb condition is violated. The maximum of the probability distribution $P(\lambda)$ gradually moves toward 1 as t_u is approached, indicating the weakening of the solid [18]. However, neither at t_u the number of contacts with $\lambda \simeq 1$ overcomes a given fraction, nor they appear to be spatially organized. The absence of structural changes at t_u supports a scenario in which the system is located in an energy minimum which slowly becomes an inflection point at time t_u, and therefore the smallest eigenvalue of the dynamical matrix continuously decreases to zero [32, 33].

5.2. Evolution in the force space
When the system sticks and the shear stress increases, no macroscopic motion is observed. However, the system microscopically changes since it sustains an increasing shear stress. The time evolution of the top plate position $X(t)$ (Fig. 9a), consists in an elastic deformation, where $X(t)$ increases very slowly in time, interrupted by sudden microslips. To characterize the evolution of the system in the force space, we introduce a two time force-force correlation function for the normal forces, defined as

$$C^n(t_0,t) = \frac{\sum_{ij} |\mathbf{f}^n_{ij}(t_0)||\mathbf{f}^n_{ij}(t)|}{\sum_{ij} |\mathbf{f}^n_{ij}(t_0)|^2}, \tag{7}$$

where the sum running over all couples of particles (i,j) corresponds to a spatial average. An equivalent definition holds for the correlation of tangential forces, C^{tg}. This function could be used to monitor the temporal evolution of the force network approaching any characteristic time t_0. Being interested in the unjamming transition of the slip event found previously, here we fix $t_0 = t_u$, and consider the evolution of the correlation function $C^n(t_u,t)$ for earlier times, $t < t_u$ (Fig. 9b). Similar results are found for other slips. The force correlation function C^n (and C^{tg}, not shown) exhibits small jumps in correspondence of microslips (Fig. 9a), revealing the unusual occurrence of bursts in the reorganization of the force network. During these bursts, the energy due to the tangential interaction decreases, whereas the one due to the normal interaction increases. A possible interpretation is in terms of a two force network scenario, in which the applied stress σ is supported by a stress σ_n due to the normal force network, and by a stress σ_t due to the tangential forces, $\sigma = \sigma_n + \sigma_t$. In a burst, few contacts break, leading to a decrease of σ_t, $\sigma_t \to \sigma_t - \delta\sigma$. A microscopic slips is observed since the normal forces quickly adapt and succeed in sustaining the applied stress, $\sigma_n \to \sigma_n + \delta\sigma$. This physical interpretation is supported by the inset of Fig. 9b, which shows both C^n and C^{tg} across a microslip, with t_0 its starting time. C^n slightly increases and overcomes 1, while C^{tg} exhibits a sharp drop due to the breaking of several contacts.

The force correlation function (Eq. 7) gives also insights into the system evolution during a slip. In Fig. 10 we plot $C^{n,tg}(t_u,t)$ for $t > t_u$, and for comparison the scaled top plate position $(X(t) - X(t_u))/(X(t_\infty) - X(t_u))$, where t_∞ is a time following the slip event, whose precise value does not influence our results. We first notice that the forces evolve on a timescale much shorter than the plate motion. For instance the correlation functions reach the value 0.1, denoting an almost complete relaxation, when the top plate moved only by 10% of its total displacement. The presence of different time scales in the relaxation process is evidenced by the self-scattering correlation function $F(q,t) = \frac{1}{N}\left|\sum_j \exp[i\mathbf{q}\cdot(\mathbf{r}_j(t) - \mathbf{r}_j(t_u))]\right|^2$, where $\mathbf{r}_j$ represents the position of the jth particle. Since the system is sheared along x and confined along z, we have considered wave vectors along y, $\mathbf{q} = (0,q,0)$. For large q, $F(q,t)$ probes small scale relaxation, and coincides with $C^n(t_u,t)$, as shown in Fig. 10. Conversely, at small q, $F(q,t)$ relaxes on a time scale comparable to that of the upper plate motion. The relaxation time τ_q, $F(q,\tau_q) = 1/e$, indeed increases as q decreases. This scenario also holds for the approach to jamming. Moreover, tangential forces decorrelate before normal ones. This can be explained considering the unjamming transition as a buckling-like instability of the chains of large normal forces, which are sustained by weaker tangential contacts. When the weaker sustaining contacts break, either the normal forces adapt to sustain the extra load, leading to a microslips, or a buckling-like instability occurs, giving rise to a slip. The same quantities can be used to investigate the subsequent jamming transition. To this end, the force network at time t is compared with the force network after the slip event studying $C^{n,tg}(t,t_\infty)$. Normal forces

Journal of Physics: Conference Series **319** (2011) 012001 doi:10.1088/1742-6596/319/1/012001

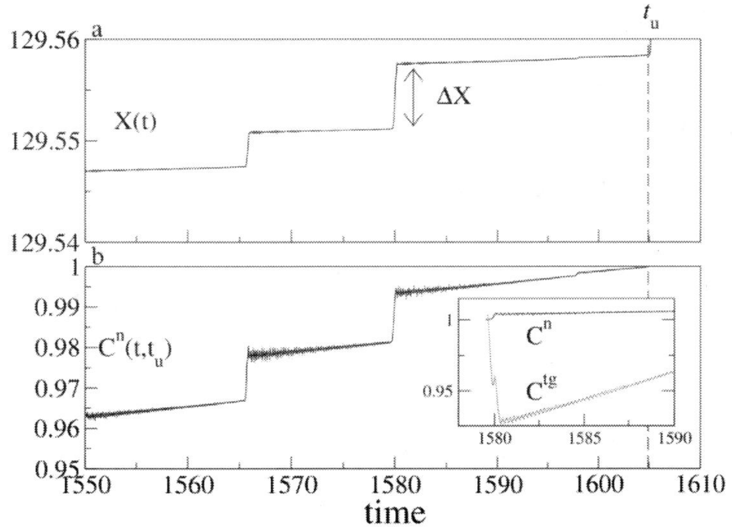

Figure 9. Time evolution of the top plate position (a) and of the correlation function of the normal forces $C^n(t,t_u)$ (b). Dashed lines identify the unjamming time t_u, i.e. the time slip begins. The inset show C^n and C^{tg} across a microslip at $t = 1579.5$.

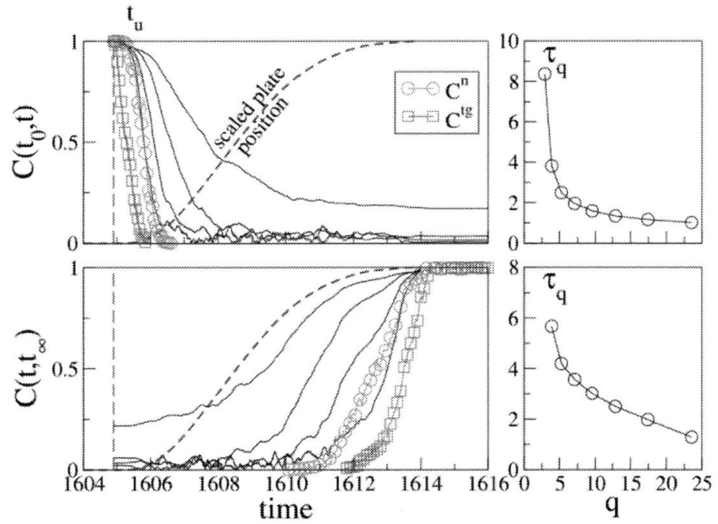

Figure 10. Correlation functions at the unjamming (upper panels) and at the subsequent jamming transitions (lower panels). The left panels show the normal and tangential force correlation functions $C^{n,tg}(t_0,t)$ and $C^{n,tg}(t,t_\infty)$ (symbols) during a slip event. The dashed blue lines indicate the time evolution of the plate position scaled between 0 and 1. The plain lines show the self-scattering function $F(q,t)$ for $q = 3, 5, 9, 17$. The corresponding τ_q are shown in the right panels. t_0 is fixed equal to t_u for the unjamming transition, whereas $t_\infty = 1614$ for the subsequent jamming.

correlate before tangential ones, confirming the stabilizing role of tangential forces in the force network.

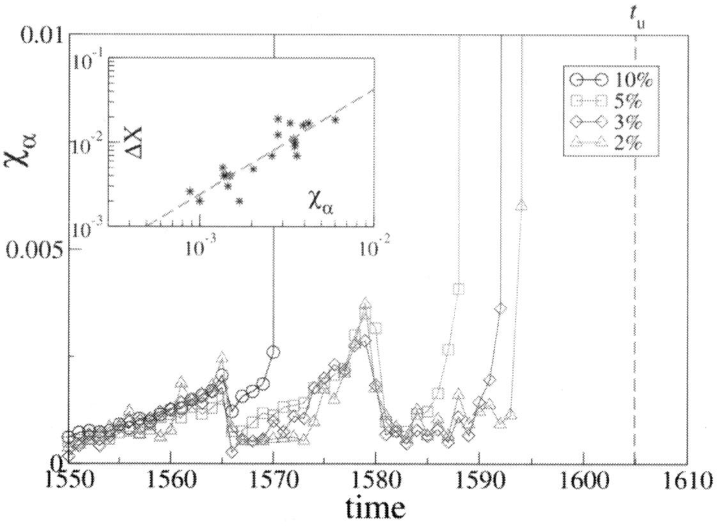

Figure 11. Time evolution of the susceptibility χ_α for different α. (Inset) Size of a microslip ΔX versus the corresponding value χ_α. The straight line is χ_α^b, with $b \simeq 1.2$.

5.3. Response to perturbations: slips versus microslips

The correlation length ξ in equilibrium systems is measured from the response to an external perturbation. The susceptibility, for instance, scales like $\xi^{2-\eta}$ near critical points, where η is the correlation function critical exponent. Here we measure the response of the system to a pressure change at the onset of both slips and microslips. More precisely, at each time t we stop the external drive, setting $V = 0$, and introduce a perturbation in the external pressure P, fixed to $P' = P(1-\alpha)$ for a time interval $\delta t_{perb} = 0.1$. The response χ_α at time t is defined as

$$\chi_\alpha(t) = \frac{1}{\alpha P} \lim_{\tau \to \infty} \left[\frac{1}{N} \sum_i (\mathbf{r}_i^\alpha(t+\tau) - \mathbf{r}_i^0(t+\tau))^2 \right]^{1/2} \quad (8)$$

where $\{\mathbf{r}_i^\alpha\}$ and $\{\mathbf{r}_i^0\}$ are the asymptotic states of the perturbed and of the unperturbed systems. In the unjammed phase, the susceptibility is divergent. In the jammed phase, it measures the linear size of the region of correlated particles that respond to the external perturbation. It therefore provides an estimate of the correlation length. It is a static quantity since the time t only indicates the instant at which the perturbation is applied. χ_α can be also defined without setting $V = 0$, provided that the characteristic response time of the system is much smaller than the timescale over which the applied stress varies. For a wide range of α, the response of the system at times far from t_u is linear in the perturbation, since χ_α does not depend on α (Fig. 11). In particular, χ_α gradually increases in time and drops in correspondence to microslips. The inset shows that the microslip size ΔX depends on χ_α evaluated just before the slip, as $\Delta X \simeq \chi_\alpha^b$. This indicates that the size of a microslip is already encoded in the system state. In fact, considering that χ_α increases in time on approaching a microslip, the measured value of χ_α provides a lower bound for the magnitude of the incoming event.

As the unjamming time is approached, the response is no longer linear. χ_α remains roughly constant until it abruptly increases at a time which depends on α, the sooner the greater α. This increase is consistent with a power-law divergence. Accordingly at each time, namely at each value of the applied shear stress, there is a minimum value of the perturbation intensity for slip triggering. This behavior is in line with the existence of a minimum threshold amplitude in the deformation associated with seismic waves for earthquake remote triggering [34]. A difference between slips and microslips is in how different particles contribute to χ_α in the sum of Eq. 8. Indeed, these contributions are very similar for microslips, and highly heterogeneous for slips. The presence of an heterogeneous response is consistent with previous numerical results found at $\sigma = 0$ [35, 36, 37, 38].

6. Spring-block model

Here we compare the results of the granular model with those of the cellular-automaton version of the spring-block model, originally introduced by Burridge & Knopoff [24]. In particular, we investigate the scaling behaviour of the slip distribution of a spring–block model and its robustness to parameter changes. We focus on the one–dimensional version of the model introduced by Nakanishi [27, 39]. This comparison is instructive since in the granular model the fault is rigid, the particles forming the shearing plate having fixed relative positions, while the Nakanishi model is the simplest representation of an elastic fault under tectonic drive forces.

In the one-dimensional version of this model (Fig. 1, lower panel) L blocks are interconnected by springs with elastic constant k and to a moving plate driven by a spring of constant k_m. Indicating by V the velocity of the moving plate, the force f_i acting on the i-th block is given by

$$f_i = -k_m(x_i - Vt) + k(x_{i+1} + x_{i-1} - 2x_i) \tag{9}$$

where x_i is the position of the i-th block and t is the evolution time. If the force is smaller than the local friction value f_{th}, the block is locked. As soon as f_i overcomes f_{th}, the i-th block is no longer able to sustain the external stress and slips to a new position x'_i. The displacement of the i-th block produces a change of the forces acting on nearest neighbor blocks. If this change leads to $f_{i+1} > f_{th}$ or $f_{i-1} > f_{th}$ the neighbor block becomes unstable, it slips to a new equilibrium position and the event propagates. The evolution of the system is governed by the evolution equation of the forces acting on each block. More precisely, when all $f_i < f_{th}$ the system is locked and forces grow due to the external drive

$$df_i(t)/dt = k_m V. \tag{10}$$

When at the generic site j, f_j becomes larger than f_{th}, the external drive is stopped and the forces f_j and $f_{j\pm1}$ are updated to the new values f'_j and $f'_{j\pm1}$ according to the following rules

$$\begin{aligned} f'_j &= \phi(f_j - f_{th}) \\ f'_{j\pm1} &= f_{j\pm1} + \frac{1}{2}\Delta(f_j - f'_j). \end{aligned} \tag{11}$$

Here

$$\Delta = \frac{2k}{2k + k_m} \tag{12}$$

measures the fraction of elastic energy dissipated. Energy conservation is recovered in the limit $k \gg k_m$. If $f_{j\pm1} > f_{th}$, the update rules Eq.s 11 are iterated and the process goes on until all sites have $f_i < f_{th}$. The total slip S is then given by

$$S = \sum_i \frac{f_i^f - f_i}{k_m}. \tag{13}$$

where f_i^f is the final value of the force acting on the ith block. Stopping the external drive during the propagation process reflects the large separation of time scales between the fracture propagation and the drive load. The function $\phi(x)$ controlling the new value of the force in unstable sites must satisfy the constraint that the new force $f'_j < f_{th}$. In the following we will use the functional form

$$\phi(x) = \frac{(2-\delta f)^2/\alpha}{x + (2-\delta f)/\alpha} - 1 \tag{14}$$

proposed by Nakanishi in Ref. [27] to reproduce velocity weakening, with α and δf two model parameters. The above equations allow a simple interpretation of the role of k_m and k. The elastic constant k enters in the numerical simulations only via the parameter Δ. Conversely the spring constant k_m affects the driving rate (Eq. 10), the total slip S and the parameter Δ. If Δ is fixed, the role of k_m simply reduces to a rescaling of time and S. Therefore if we assume that S scales linearly with the system size L at fixed Δ, slips become independent of k_m/L.

In the following we present results from numerical simulations of the Nakanishi model in order to enlighten similarities and differences with the results obtained for the granular model. In numerical simulations we set $f_{th} = 1$ and $V = 1$. Results are not substantially affected by δf. The parameter α slightly influences the exponent a of the initial power law regime of the slip distribution $p(S)$, and in the following we set $\alpha = 1.75$, which provides $a \simeq 1.65$ in agreement with the experimental GR distribution and results of the MD simulations. We also fix $k = 10^3$ and investigate different values of k_m and L.

In Fig. 12 (upper panel) the size distributions $p(S)$ for $k_m = 10$ and three values of $L = 20, 30, 40$ are plotted. The distribution is very similar to the one observed in Fig. 5: an initial power law decay with an exponent $a \simeq 1.65$, followed by a bump at large sizes S. This bump indicates the existence of a typical slip size S_c. We wish to stress that, as in the granular model, this size cannot be straightforwardly related to the system size L. Indeed, the value of S_c is inversely proportional to k_m. Therefore, by changing k_m and k keeping Δ fixed, S_c can take all values from zero to infinity. However, for Δ and k_m given, the position of the bump depends on L and moves towards larger S as L increases. Differently from the MD model, S_c does not scale with L but rather as $L^{1.5}$ (Fig. 12 lower panel). The quality of the bump collapse is significantly worse than the one observed in Fig. 6 (lower panel).

Next we turn to consider the case of fixed L and different k_m. Since k is fixed, changes in k_m produce two effects: the scaling of S with k_m and the Δ change. In particular for larger k_m, Δ becomes smaller and dissipation increases. Both effects make the probability of large slip size smaller for larger k_m. This is confirmed by Fig. 13 where the largest observed slip decreases by increasing k_m. In Fig. 13 (lower panel) we plot $p(S)$ versus $Sk_m^{1.5}$ and observe that the largest slip size again decreases by increasing k_m, as the consequence of a smaller Δ: As Δ decreases, energy dissipation increases, making less probable the observation of events involving all blocks.

We finally consider the case of different k_m and L but with the fixed ratio $k_m/L = 0.5$. According to the above observations, since k_m varies in a small range, $k_m \in [10, 20]$, Δ is roughly constant and we can expect a scaling with k_m/L. Fig. 14 shows that curves for different L exhibit a similar behavior with the bump located at about the same position. In order to compare the results of the Nakanishi and the granular model, we observe that the scaling with k_m/L_x in the granular model suggests that the elastic coupling among the grains is very stiff, which corresponds for the Nakanishi model to $k \gg k_m$ and $\Delta \simeq 1$, so that small changes in k_m do not modify Δ. This is a reasonable condition since in the granular simulations the typical ratio between the spring constant of the grain-grain interaction, k_n, and the spring constant of the drive, k_m, is $k_n/k_m = 2\ 10^3$. We finally notice that similar slip size distributions have been also observed in the original spring block model with real continuum-time dynamics [40].

7. Conclusions

We have investigated the statistics of slips sizes in a model of fault in which granular materials play the role of fault gouge, as function of model parameters.

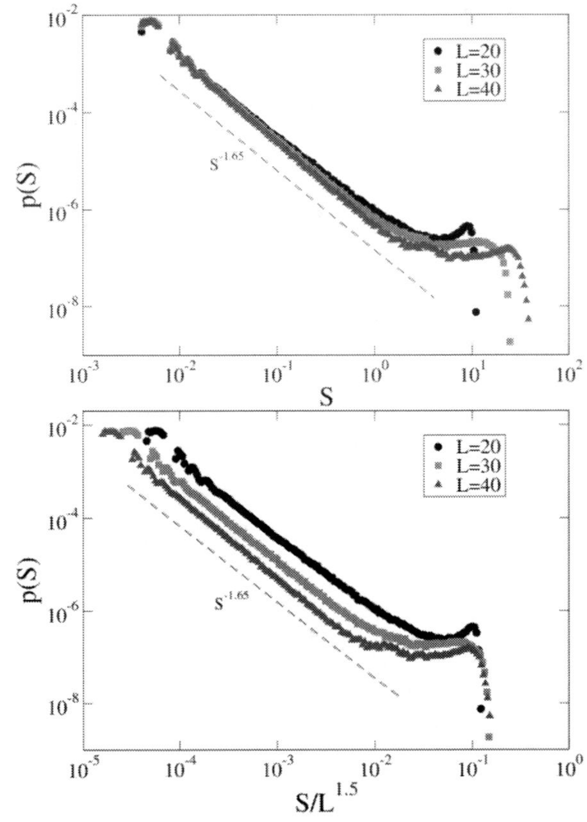

Figure 12. Slip size distribution $p(S)$ versus S for different system size L (upper panel). Same distributions for slip size rescaled by system size as $S/L^{1.5}$ (lower panel).

Our results indicate that the slip distribution is characterized by two regimes, the first one well described by a power law, the second one by a bump at a characteristic slip size. The microslips involve mostly the shearing of the two uppermost layers of the granular medium, which explains why the exponent of the slip size distribution depends on the number of layers h only when $h = 1, 2$. For $h > 2$ the resulting exponent is in very good agreement with the measured value for earthquake catalogs, $a \simeq 1.7$.

Beside being almost independent of the number of layers, the exponent a results to be also independent of other model parameters, and precisely the system size and the elastic constant of the driving spring. This is in agreement with the existence of a universal exponent of the slip size distribution, which is independent of the fault size and the mechanical properties of the rocks the fault walls are made of. Conversely, the bump at large slip sizes depends on L_x and k_m, the relevant variable controlling its location being L_x/k_m. The separation between the power law regime and the bump disappears as L_x/k_m decreases, namely for very stiff faults. This result is particularly interesting since it suggests that large slip behaviour for very large systems can be investigated by tuning the elastic constant of the drive in smaller systems.

The absence of precise structural changes at the unjamming time, and the bursts observed

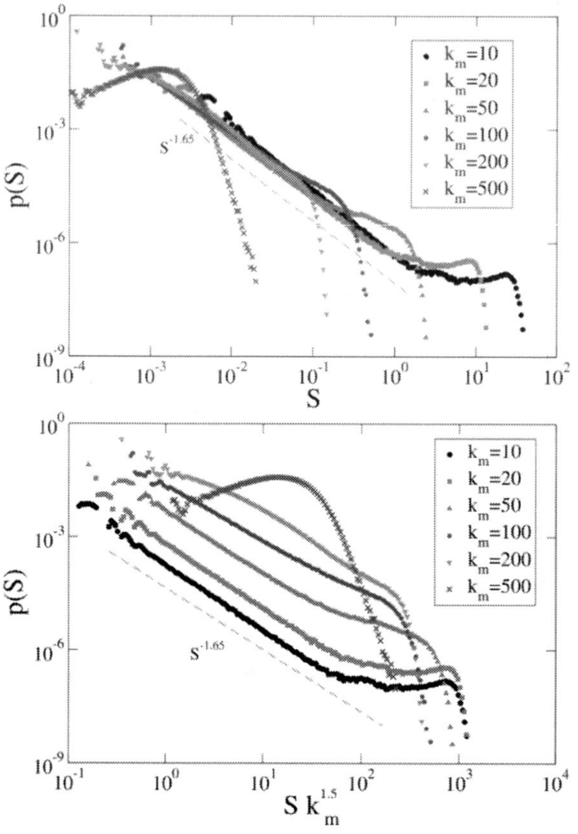

Figure 13. Slip size distribution $p(S)$ versus S for $L = 20$ and different spring elastic constant k_m (upper panel). Same distributions for slip size rescaled by the elastic constant as $Sk_m^{1.5}$ (lower panel).

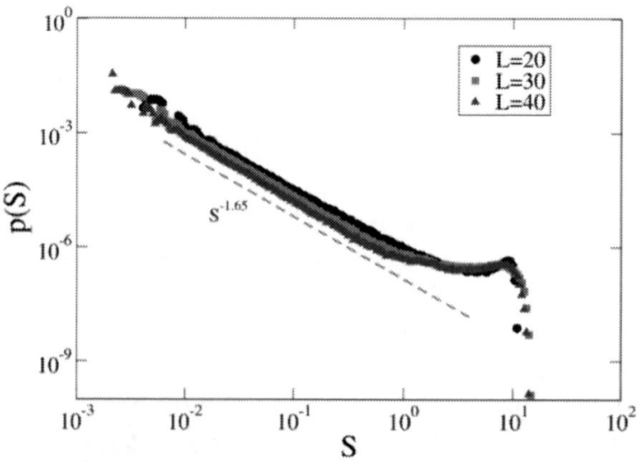

Figure 14. Slip size distribution $p(S)$ versus S for different L and k_m with fixed ratio $k_m/L = 0.5$.

in the prior dynamics, suggest that the increasing external stress progressively modifies the underlying energy landscape. At each time the system is in an equilibrium position which can be seen as local energy minimum of an effective energy landscape which depends on the applied shear stress. Microslips occur when the local energy minimum flattens down as the applied shear stress increases, letting the system fall in a neighbor minimum. If there are no close minima, a slip occurs, and the system jumps to a far away configuration. The deforming energy landscape picture also suggests that soft-modes could be found not only when unjamming is approached at $\sigma = 0$ by decreasing the volume fraction ϕ [35, 38], but also [41] when unjamming is approached increasing σ.

We have also compared the granular model and the Nakanishi model, obtaining very similar results for the probability distribution of slip sizes. This is a surprising finding considering that the two models are very different. Indeed, the first one (granular) models a rigid fault, because of the rigidity of the confining plate, in the sense that the relative positions of the particles of each plate is fixed. Conversely, the Nakanishi model is paradigmatic of an elastic fault. The scaling results suggest that both models behave as a system at the critical point and belong to the same universality class. This behaviour is robust for the granular model, whereas it weakly depends on the choice of the parameter α for the Nakanishi model.

The ability of the granular fault model to reproduce the statistics of slipping events suggests the existence of a deep relation between earthquakes and the jamming and un-jamming transitions of athermal particulate systems (see Ref.s [42, 43] for recent reviews). This opens the possibility of investigating whereas theoretical tools and concepts developed in the study of the jamming transition, such as the presence of soft–modes, of related diverging length scales [44] and hysteric behavior [45], may provide new insights in the earthquakes phenomenology.

[1] Marone C, Raleigh C B and Scholz C H 1990 *J. Geophys. Res.* **95** 7007
[2] Beeler N M, Tullis T E, Blanpied M L and Weeks J D 1996 *J. Geophys. Res.* **101** 8697
[3] Nasuno S, Kudrolli A and Gollub J P 1997 *Phys. Rev. Lett.* **79** 949
[4] Tsai J C, Voth G and Gollub J P 2003 *Phys. Rev. Lett.* **91** 064301
[5] Dalton F and Corcoran D 2001 *Phys. Rev. E* **63** 061312; 2002 *Phys. Rev. E* **65** 031310
[6] Bretz M, Zaretzki R, Field S B, Mitarai N and Nori F 2006 *Europhys. Lett.* **74** 1116
[7] Daniels K and Hayman N W 2008 *J. Geophys. Res.* **113** B11411
[8] Rathbun A P and Marone C 2010 *J. Geophys. Res.* **115** B01204
[9] Johnson P A and Jia X 2005 *Nature* **437** 871
[10] Johnson P A, Savage H, Knuth M, Gomberg J and Marone C 2008 *Nature* **451** 57
[11] Nasuno S, Kudrolli A, Bak A and Gollub J P 1998 *Phys. Rev. E* **58** 2161
[12] Hill D P et al 1993 *Science* **260** 1617
[13] Peña A A, McNamara S, Lind P G and Herrmann H J 2009 *Granular Matter* **11** 243
[14] Mair K and Abe S 2008 *Earth Planet. Sci. Lett.* **274** 72
[15] Abe S and Mair K 2009 *Geophys. Res. Lett.* **36** L23302
[16] Mora P and Place D 1998 *J. Geophys. Res.* **103** 21067
[17] Aharonov E and Sparks D 1999 *Phys. Rev. E* **60** 6890
[18] Aharonov E and Sparks D 2004 *J. Geophys. Res.* **109** B09306
[19] Abe S and Mair K 2005 *Geophys. Res. Lett.* **32** L05305
[20] Pica Ciamarra M, Lippiello E, Godano C and de Arcangelis L 2010 *Phys. Rev. Lett.* **104** 238001
[21] Gutenberg B and Richter C F 1944 *Bull. Seism. Soc. Am.* **34** 185
[22] Kagan Y Y 1991 *Geophys. J. Int.* **106** 123
[23] Godano C and Pingue F 2000 *Geophys. J. Int.* **142** 193
[24] Burridge R and Knopoff L 1967 *Bull. Seismol. Soc. Am.* **57** 341
[25] Carlson J M and Langer J S 1989 *Phys. Rev. Lett.* **62** 2632; 1989 *Phys. Rev. A* **40** 6470
[26] Olami Z, Feder H J S and Christensen K 1992 *Phys. Rev. Lett.* **68** 1244
[27] Nakanishi H 1990 *Phys. Rev. A* **41** 7086
[28] Silbert L E, Ertas D, Grest G S, Halsey T C, Levine D and Plimpton S J 2001 *Phys. Rev. E* **64** 051302
[29] Cundall P A and Strack O D L 1979 *Geotechnique* **29** 47
[30] Kanamori H and Anderson D 1975 *Bull. Seismol. Soc. Am.* **65** 1073
[31] Heussinger C and Barrat J L 2009 *Phys. Rev. Lett.* **102** 218303
[32] Maloney C E and Lemaitre A 2006 *Phys. Rev. E* **74** 016118

[33] Welker P R and Mcnamara S C 2009 *Phys. Rev. E* **79** 061305
[34] Gomberg J and Johnson P A 2005 *Nature* **437** 830
[35] Silbert L E, Liu A J and Nagel S R 2005 *Phys. Rev. Lett.* **95** 098301
[36] Somfai E, van Hecke M, Ellenbroek W G, Shundyak K and van Saarloos W 2007 *Phys. Rev. E* **75** 020301
[37] Ellenbroek W G, Somfai E, van Hecke M and van Saarloos W 2006 *Phys. Rev. Lett.* **97** 258001
[38] Wyart M, Liang H, Kabla A, Mahadevan L 2008 *Phys. Rev. Lett.* **101** 215501
[39] Nakanishi H 1991 *Phys. Rev. A* **43** 6613
[40] Mori T and Kawamura H (2005) *Phys. Rev. Lett.* **94** 058501; (2006) *J. Geophys. Res.* **111** B07302
[41] Pica Ciamarra M and Coniglio A 2009 *Phys. Rev. Lett.* **103** 235701
[42] Liu A J and Nagel S R 2010 *Annu. Rev. of Condens Matter Phys.* **1** 347
[43] van Hecke M 2010 *J. Phys.: Condens. Matter* **22** 033101
[44] Wyart M, Silbert L E, Nagel S R and Witten T A 2005 *Phys. Rev. E* **72** 051306; Silbert L E, Liu A J and Nagel S R 2005 *Phys. Rev. Lett.* **95** 098301; Somfai E, van Hecke M, Ellenbroek W G, Shundyak K and van Saarloos W 2007 *Phys. Rev. E* **75** 020301(R)
[45] Vanel L, Howell D, Clark D, Behringer R P and Clement E 1999 *Phys. Rev. E* **60** R5040; Ciamarra M P, Pastore R, Nicodemi M and Coniglio A e-print arXiv:0912.3140; Grebenkov D S, Ciamarra M P, Nicodemi M and ConiglioA 2008 *Phys. Rev. Lett.* **100** 07801; Forterre Y and Pouliquen O 2008 *Ann. Rev. of Fluid Mech.* **40** 1

Continuum Models and Discrete Systems Symposia (CMDS-12) IOP Publishing
Journal of Physics: Conference Series **319** (2011) 012002 doi:10.1088/1742-6596/319/1/012002

Strong-field macroscopic magneto-transport in a periodic composite medium: Some new results

Yakov M. Strelniker[1] **and David J. Bergman**[2]

[1]Department of Physics, Bar-Ilan University, IL-52900 Ramat-Gan, Israel
[2]Raymond and Beverly Sackler School of Physics and Astronomy, Faculty of Exact Sciences, Tel Aviv University, IL-69978 Tel Aviv, Israel

E-mail: [1]`strelnik@biu.ac.il`, [2]`bergman@post.tau.ac.il`

Abstract. Results are reported of calculations of the macroscopic magnetoresistance of a two-constituent composite electrical conductor where the constituent resistivity tensors are comparable but different, and where the Hall resistivities are much greater than the Ohmic resistivities. Some of the macroscopic Ohmic resistivities fluctuate very strongly with the direction of the externally applied magnetic field $\mathbf{B}$ and keep increasing as $\mathbf{B}^2$ without any saturation.

1. Introduction

That the macroscopic response of a composite conductor with a periodic microstructure can be strongly anisotropic in the presence of a strong magnetic field was first realized in Refs. [1, 2]. In those references a simple-cubic array of insulating inclusions in an otherwise uniform and isotropic conducting host was studied numerically in the case where the Hall resistivity of the host was greater than its Ohmic resistivity. Some time after that some of those results were verified in experiments on a thin semiconducting film with an array of nano-holes [3] and on a bismuth film with a similar array of holes [4].

More recently it was realized that, even if all the constituents are electrical conductors with comparable though different resistivity tensors, the magneto-transport can exhibit dramatic new forms of behavior. These include new critical points in the case of a disordered microstructure [5, 6, 7, 8] and strongly anisotropic magnetoresistance when the Hall-to-Ohmic resistivity ratio is much greater than 1 in all the constituents [9].

This article reports on some new results regarding the macroscopic magneto-transport of a two-constituent composite medium with a periodic microstructure in the regime where the externally applied magnetic field $\mathbf{B}$ is strong enough that the Hall resistivity is much greater than the Ohmic resistivities in both constituents. Both of those constituents are electrical conductors with different but comparable resistivity tensors. Therefore this article reports on an extension of calculations that were described in more detail in Ref. [9].

Among the noteworthy results is the fact that the macroscopic transverse Ohmic resistivities show tremendous fluctuations with changes in the direction of the magnetic field $\mathbf{B}$, and that for some directions the magnitude of those resistivities appears to increase as $\mathbf{B}^2$.

In Section 2 we briefly describe the numerical approach used to compute the macroscopic magnetoresistivity tensor $\hat{\rho}_e$ of a two-constituent periodic composite. In Section 3 we describe some results of those computations. In Section 4 we discuss those results and summarize some of their implications, including some potential applications.

2. Numerical computations for a periodic composite medium

We use a scheme that is described in detail in Ref. [2]. For completeness, we present here a brief description of that scheme, which was also used to calculate all the numerical results reported in Ref. [9].

Ordinarily, the local electric potential $\phi^{(\alpha)}(\mathbf{r})$, produced in the system when the volume-averaged electric field is $\langle\mathbf{E}\rangle = \mathbf{e}_\alpha$, would be found by solving the following standard boundary value problem

$$\begin{aligned} &\nabla\cdot[\hat{\sigma}(\mathbf{r})\cdot\nabla\phi^{(\alpha)}(\mathbf{r})] = 0 \quad \text{in the entire volume } V, \\ &\phi^{(\alpha)}(\mathbf{r}) = r_\alpha \qquad \text{at the external surface } \partial V, \end{aligned}$$

where

$$\begin{aligned} \hat{\sigma}(\mathbf{r}) &= \hat{\sigma}_1\theta_1(\mathbf{r}) + \hat{\sigma}_2\theta_2(\mathbf{r}) \equiv \hat{\sigma}_2 - \theta_1(\mathbf{r})\delta\hat{\sigma}, \\ \delta\hat{\sigma} &\equiv \hat{\sigma}_2 - \hat{\sigma}_1, \\ \theta_i(\mathbf{r}) &= 1 \text{ if } \hat{\sigma}(\mathbf{r}) = \hat{\sigma}_i \text{ and } \theta_i(\mathbf{r}) = 0 \text{ otherwise.} \end{aligned}$$

Instead of doing that, we first use a Green's function of this problem in a uniform conductor $\hat{\sigma}_2$, denoted by $G(\mathbf{r},\mathbf{r}'|\hat{\sigma}_2)$, to transform this partial differential equation with boundary conditions to an integrodifferential equation:

$$\phi^{(\alpha)} = r_\alpha + \hat{\Gamma}\phi^{(\alpha)}, \tag{1}$$

$$\hat{\Gamma}\phi \equiv \int dV'\theta_1(\mathbf{r}')\nabla' G(\mathbf{r},\mathbf{r}'|\hat{\sigma}_2)\cdot\delta\hat{\sigma}\cdot\nabla'\phi^{(\alpha)}(\mathbf{r}'), \tag{2}$$

$$\begin{aligned} G(\mathbf{r},\mathbf{r}'|\hat{\sigma}) &\equiv \int \frac{d^3k}{(2\pi)^3}\frac{e^{i\mathbf{k}\cdot(\mathbf{r}-\mathbf{r}')}}{(\mathbf{k}\cdot\hat{\sigma}\cdot\mathbf{k})} = \frac{1}{4\pi(\sigma_{xx}\sigma_{yy}\sigma_{zz})^{1/2}} \\ &\times\left(\frac{(x-x')^2}{\sigma_{xx}} + \frac{(y-y')^2}{\sigma_{yy}} + \frac{(z-z')^2}{\sigma_{zz}}\right)^{-1/2}. \end{aligned} \tag{3}$$

If the microstructure is spatially periodic, then $\psi^{(\alpha)}(\mathbf{r}) \equiv \phi^{(\alpha)}(\mathbf{r}) - r_\alpha$ is a similarly periodic function of $\mathbf{r}$ and we expand it in an appropriate Fourier series,

$$\psi^{(\alpha)}(\mathbf{r}) = \sum_{\mathbf{g}\neq 0}\psi^{(\alpha)}_{\mathbf{g}}e^{i\mathbf{g}\cdot\mathbf{r}}, \tag{4}$$

where the sum is over all the nonzero vectors $\mathbf{g}$ of the appropriate reciprocal lattice. Equation (1) is thereby transformed into an infinite set of linear algebraic equations for the coefficients $a^{(\alpha)}_{\mathbf{g}} \equiv i(\mathbf{g}\cdot\hat{\sigma}_2\cdot\mathbf{g})^{1/2}\psi^{(\alpha)}_{\mathbf{g}}$, where $\mathbf{g}\neq 0$,

$$a^{(\alpha)}_{\mathbf{g}} = r^{(\alpha)}_{\mathbf{g}} + \sum_{\mathbf{g}'\neq 0}\Gamma_{\mathbf{g}\mathbf{g}'}a^{(\alpha)}_{\mathbf{g}'} \quad \text{for} \quad \mathbf{g}\neq 0, \tag{5}$$

$$r^{(\alpha)}_{\mathbf{g}} \equiv \frac{g_\beta\delta\sigma_{\beta\alpha}}{(\mathbf{g}\cdot\hat{\sigma}_2\cdot\mathbf{g})^{1/2}}\theta_{\mathbf{g}}, \tag{6}$$

$$\theta_{\mathbf{g}} = \frac{1}{V_a}\int_{V_a} d^3r\,\theta_1(\mathbf{r})e^{i\mathbf{g}\cdot\mathbf{r}}, \tag{7}$$

$$\Gamma_{\mathbf{g}\mathbf{g}'} \equiv \frac{(\mathbf{g}\cdot\delta\hat{\sigma}\cdot\mathbf{g}')}{(\mathbf{g}\cdot\hat{\sigma}_2\cdot\mathbf{g})^{1/2}(\mathbf{g}'\cdot\hat{\sigma}_2\cdot\mathbf{g}')^{1/2}}\theta_{\mathbf{g}-\mathbf{g}'}, \tag{8}$$

where V_a is the volume of one unit cell. After a truncated subset of Eqs. (5) are solved for $a_{\mathbf{g}}^{(\alpha)}$, the fields $\phi^{(\alpha)}$, $\mathbf{E}^{(\alpha)}$ and the elements of the macroscopic conductivity tensor $\hat{\sigma}_e$ can be calculated as follows:

$$\phi^{(\alpha)} = r_\alpha + \psi^{(\alpha)} = r_\alpha + \sum_{\mathbf{g}\neq 0} \frac{a_{\mathbf{g}}^{(\alpha)}}{i(\mathbf{g}\cdot\hat{\sigma}_2\cdot\mathbf{g})^{1/2}} e^{i\mathbf{g}\cdot\mathbf{r}}, \tag{9}$$

$$E_\beta^{(\alpha)} = \nabla_\beta \phi^{(\alpha)} = \delta_{\alpha\beta} + \sum_{\mathbf{g}\neq 0} \frac{g_\beta a_{\mathbf{g}}^{(\alpha)}}{(\mathbf{g}\cdot\hat{\sigma}_2\cdot\mathbf{g})^{1/2}} e^{i\mathbf{g}\cdot\mathbf{r}}, \tag{10}$$

$$\sigma_{\alpha\beta}^{(e)} = \langle\sigma_{\alpha\beta}\rangle - \sum_{\mathbf{g}\neq 0} \frac{\delta\sigma_{\alpha\beta} g_\beta}{(\mathbf{g}\cdot\hat{\sigma}_2\cdot\mathbf{g})^{1/2}} \theta_{-\mathbf{g}} a_{\mathbf{g}}^{(\beta)}. \tag{11}$$

Note that the coefficient multiplying $a_{\mathbf{g}}^{(\beta)}$ in the last sum looks similar to $r_{\mathbf{g}}^{(\alpha)*}$, but actually differs from it, because $\delta\sigma_{\alpha\beta} \neq \delta\sigma_{\beta\alpha}$.

Because the characteristic function $\theta_1(\mathbf{r})$ is discontinuous at the $\hat{\sigma}_1$-$\hat{\sigma}_2$ interface, the local electric field also is discontinuous there. Consequently the Fourier coefficients of both functions decrease very slowly for large $\mathbf{g}$. For this reason a very large number of the coefficients $a_{\mathbf{g}}^{(\alpha)}$ need to be used in Eqs. (9)–(11) in order to get accurate results. This means that a large number of the coupled linear algebraic Eqs. (5) must be kept and solved, which requires huge computing time and memory resources.

An alternative procedure, which is less demanding in terms of those resources, is to solve those equations by iterative application of the $\Gamma_{\mathbf{g}\mathbf{g}'}$ matrix, i.e., by computing the nth term $a_{\mathbf{g}n}^{(\alpha)}$ in the infinite series

$$a_{\mathbf{g}}^{(\alpha)} = \sum_{n=0}^{\infty} a_{\mathbf{g}n}^{(\alpha)} \tag{12}$$

iteratively by

$$a_{\mathbf{g}0}^{(\alpha)} = r_{\mathbf{g}}^{(\alpha)}, \quad a_{\mathbf{g}n}^{(\alpha)} = \sum_{\mathbf{g}'\neq 0} \Gamma_{\mathbf{g}\mathbf{g}'} a_{\mathbf{g}'n-1}^{(\alpha)}. \tag{13}$$

The expression [Eq. (12)] obtained by this procedure is a series in powers of the $\Gamma_{\mathbf{g}\mathbf{g}'}$ matrix elements. Those elements are sometimes very large, consequently the series is often strongly divergent. In those cases advanced methods must be employed in order to get useful results from that series. Because of this, the computations reported in this paper are all based upon numerical solution of a truncated subset of Eqs. (5), in which more than 9000 of the $a_{\mathbf{g}}^{(\alpha)}$ coefficients were calculated. That is, the components of $\mathbf{g}$ in each direction went up to $\pm 2\pi n/a$, where a is the unit cell edge and the integer n was between 10 and 15. Because this limited number of Fourier components is not sufficient to represent the discontinuous functions $\theta_1(\mathbf{r})$ and $\mathbf{E}(\mathbf{r})$ accurately, therefore the detailed numerical results for $\mathbf{E}(\mathbf{r})$, obtained by using Eq. (10), often exhibit spurious spatial oscillations—see Ref. [9]. Nevertheless, when the computed Fourier coefficients are used to calculate the macroscopic conductivity tensor $\hat{\sigma}^{(e)}$, using Eq. (11), the results are quite accurate. This was demonstrated in Ref. [9] by comparing the numerically computed results for $\hat{\sigma}^{(e)}$ with the results of an exact asymptotic theory for large values of the Hall-to-Ohmic resistivity ratio.

3. Results of computations for the macroscopic magnetoresistivity tensor $\hat{\rho}_e \equiv 1/\hat{\sigma}_e$

We consider a simple-cubic array of spherical inclusions with radius R given by $R/a = 0.4$, where $a \times a \times a$ is the unit cell size, and characterized by the resistivity tensor $\hat{\rho}_1$. These inclusions are embedded in an otherwise uniform host characterized by the resistivity tensor $\hat{\rho}_2$. Both constituents are assumed to be isotropic media, thus their resistivity tensors can be written in the form

$$\hat{\rho}_i = \rho_0 \begin{pmatrix} \alpha_i & -\beta_i & 0 \\ \beta_i & \alpha_i & 0 \\ 0 & 0 & \lambda_i \end{pmatrix}, \quad i = 1, 2. \tag{14}$$

These two tensors are different but comparable, and their form assumes that the z-axis lies along the external magnetic field $\mathbf{B}$. Clearly, $\rho_0\alpha_i$ are the transverse Ohmic resistivities, $\rho_0\lambda_i$ are the longitudinal Ohmic resistivities, $\rho_0\beta_i = R_{Hi}|\mathbf{B}|$ are the Hall resistivities, where R_{Hi} are the Hall coefficients. We assume that ρ_0 characterizes the common magnitude of the four relevant Ohmic resistivities, i.e., α_1, α_2, λ_1, λ_2 are all dimensionless parameters of order $\mathcal{O}(1)$. We also assume that the externally applied uniform magnetic field is strong enough so that both β_1 and β_2 are much greater than 1, i.e., that the Hall-to-Ohmic resistivity ratio β_i/α_i is much greater than 1. From Eq. (14) we immediately get the following form for the conductivity tensors of the two constituents:

$$\hat{\sigma}_i \equiv \frac{1}{\hat{\rho}_i} = \frac{1}{\rho_0} \begin{pmatrix} \frac{\alpha_i}{\alpha_i^2+\beta_i^2} & \frac{\beta_i}{\alpha_i^2+\beta_i^2} & 0 \\ \frac{-\beta_i}{\alpha_i^2+\beta_i^2} & \frac{\alpha_i}{\alpha_i^2+\beta_i^2} & 0 \\ 0 & 0 & \frac{1}{\lambda_i} \end{pmatrix}. \tag{15}$$

This is used to compute the elements of the macroscopic conductivity tensor $\hat{\sigma}_e$, and subsequently those of the macroscopic resistivity tensor $\hat{\rho}_e \equiv 1/\hat{\sigma}_e$.

In Fig. 2 we show results for the diagonal elements of $\hat{\rho}_e$ for the case where the y axis is fixed along the (010) lattice axis of the cubic array, while the magnetic field $\mathbf{B}$ is continuously rotated in the perpendicular plane, along with the axes x' and z', by the angle θ—see Fig. 1(a). Results are shown for two different values of the β_i, while keeping fixed the ratio β_1/β_2 and also all of the other constituent parameters. In lines 1 and 3 the elements of $\hat{\rho}_e$ are taken with respect to the fixed coordinate system, namely $x \parallel (100)$, $y \parallel (010)$, $z \parallel (001)$, even though the direction of $\mathbf{B} \parallel z'$ keeps changing with changing θ, since θ is the angle between the rotating z' axis and the fixed $z \parallel (001)$ axis. In lines 2 and 4 the elements of $\hat{\rho}_e$ are taken with respect to the rotating coordinate system, where the z' and x' axes are inclined by θ with respect to the fixed $z \parallel (001)$ and $x \parallel (100)$ axes respectively.

In Fig. 3 we show similar plots for another kind of rotation of the direction of $\mathbf{B}$: The y' axis is fixed along the (011) lattice axis, while the other two fixed axes are $x \parallel (100)$, $z' \parallel (0\bar{1}1)$. The magnetic field $\mathbf{B}$ is then rotated in the plane perpendicular to the y' axis, which remains fixed along the (011) lattice axis, and θ is the angle between $\mathbf{B}$ and the fixed $z' \parallel (0\bar{1}1)$ axis—see Fig. 1(b),(c). In lines 1 and 3 the elements of $\hat{\rho}_e$ are taken with respect to the fixed lattice coordinates $x \parallel (100)$, $y \parallel (010)$, $z \parallel (001)$ and plotted vs. θ. In lines 2 and 4 the elements of $\hat{\rho}_e$ are taken with respect to the rotating coordinates x'', y', z'', i.e., although $y' \parallel (011)$ is fixed, the x'' and z'' axes rotate around y' so that z'' remains oriented along the rotating $\mathbf{B}$. The results in all parts of Fig. 3 are plotted vs. the angle θ between $\mathbf{B} \parallel z''$ and the fixed $z' \parallel (0\bar{1}1)$ axis.

4. Discussion of the results

When $y \parallel (010)$ is kept fixed, $\rho_{yy}^{(e)}$ exhibits strong oscillations—see Fig. 2(b),(e),(h),(m). Comparing Fig. 2(b),(e), (which are identical) and Fig. 2(h),(m) (which are also identical) shows that, for most values of θ, $\rho_{yy}^{(e)}$ appears to increase as β_i^2 when the ratio β_1/β_2, as well as all the other constituent parameters, are kept fixed. Because the Hall resistivities of electrical conductors are usually simply proportional to the magnitude of the magnetic field $\mathbf{B}$, therefore this means that usually $\rho_{yy}^{(e)} \propto |\mathbf{B}|^2$. Deviations from this behavior occur when $\mathbf{B}$ lies along a high symmetry lattice axis, e.g., a (001)-like axis or a (011)-like axis. At those orientations $\rho_{yy}^{(e)}$

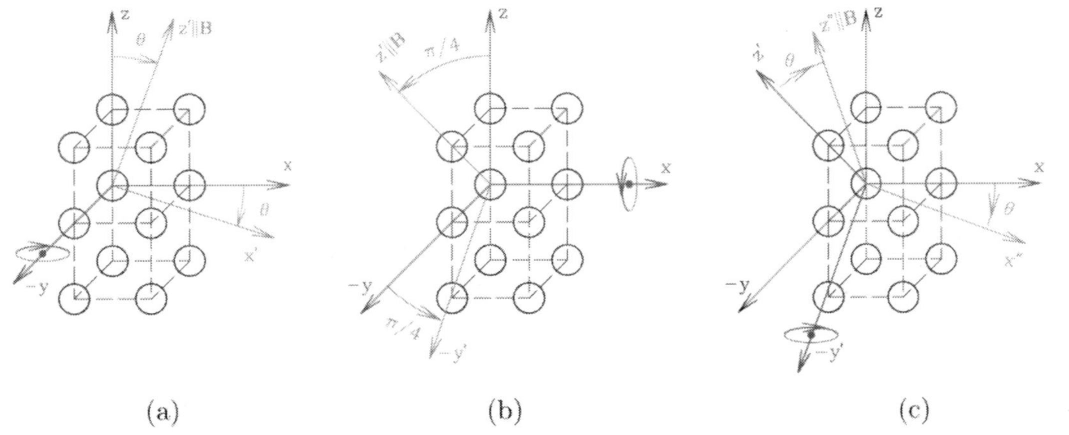

Figure 1. (Color online) The geometry described in (a) is used to get the results shown in Fig. 2, while the geometry described in (b) and (c) is used to get the results shown in Fig. 3. (a) The axis $y \parallel (010)$ is kept fixed while the axes x' and z' subtend a changing angle θ with the fixed $x \parallel (100)$ and $z \parallel (001)$ axes, respectively, and remain perpendicular to y. (b) New fixed axes $y' \parallel (011)$, $z' \parallel (0\bar{1}1)$ are obtained by a $\pi/4$ rotation about the x axis. (c) The $y' \parallel (011)$ axis is kept fixed while the axes x'' and z'' subtend a changing angle θ with the fixed axes $x \parallel (100)$ and $z' \parallel (0\bar{1}1)$, respectively, and remain perpendicular to y'.

also exhibits a sharp minimum as function of θ. By contrast, the longitudinal component $\rho^{(e)}_{zz}$ in Fig. 2(f),(n), i.e., the Ohmic component of $\hat{\rho}_e$ along $\mathbf{B}$, is nearly independent of θ and of the two β_i if the ratio β_1/β_2 is kept fixed. The other transverse Ohmic resistivity component $\rho^{(e)}_{xx}$—see Fig. 2(d),(l)— does oscillate as function of θ and changes with the β_i, but does not increase significantly when the β_i are increased by a factor of 10.

Turning to the other kind of rotations, shown in Fig. 1(b),(c), we note that now $y' \parallel (011)$ and $z' \parallel (0\bar{1}1)$ are kept fixed. In this case too $\rho^{(e)}_{yy}$ exhibits oscillations—see Fig. 3(b),(e),(h),(m). It also increases with the β_i, but not as rapidly as β_i^2. Again, the longitudinal Ohmic component $\rho^{(e)}_{zz}$ in Fig. 3(f),(n) is constant as function of θ and of the two β_i if the ratio β_1/β_2 is kept fixed. The most interesting behavior is exhibited by the other transverse Ohmic resistivity component $\rho^{(e)}_{xx}$ in Fig. 3(d),(l): For most values of θ it is quite small and does not increase significantly when the β_i are increased by a factor of 10. However, when $\mathbf{B}$ lies along some special directions $\rho^{(e)}_{xx}$ exhibits a sharp maximum where its value is very large and appears to increase as β_i^2—see the cases $\theta = 0$ and $\theta = \pi/2$, i.e., $\mathbf{B}$ lies along $(0\bar{1}1)$ and along (001), respectively, and the cases $\theta = \pi/4$ and $\theta = 3\pi/4$. In the last two cases the direction of $\mathbf{B}$ does not correspond to any simple or high symmetry lattice axis. As we mentioned before, this behavior means that, at those peaks, we usually will expect to find $\rho^{(e)}_{xx} \propto |\mathbf{B}|^2$.

The strongly oscillating behavior described in the preceding two paragraphs is not yet well understood. It may possibly be exploitable in making a magnetic field sensor for a strong magnetic field $\mathbf{B}$ that is sensitive to the magnitude as well as to the direction of $\mathbf{B}$. This is particularly true of the behavior exhibited by $\rho^{(e)}_{xx}$ in Fig. 3(d),(l), where the peaks at $\theta = 0$ and $\theta = \pi/2$ are extremely high, narrow, and isolated, and their height appears to increase as $|\mathbf{B}|^2$.

The results and predictions reported here all arise from numerical computations of the local electric field and current distributions in a periodic composite conductor in the presence of a strong magnetic field. Those results and predictions need to be tested in experiments. We hope this challenge will be taken up by experimentalists who have the necessary equipment and

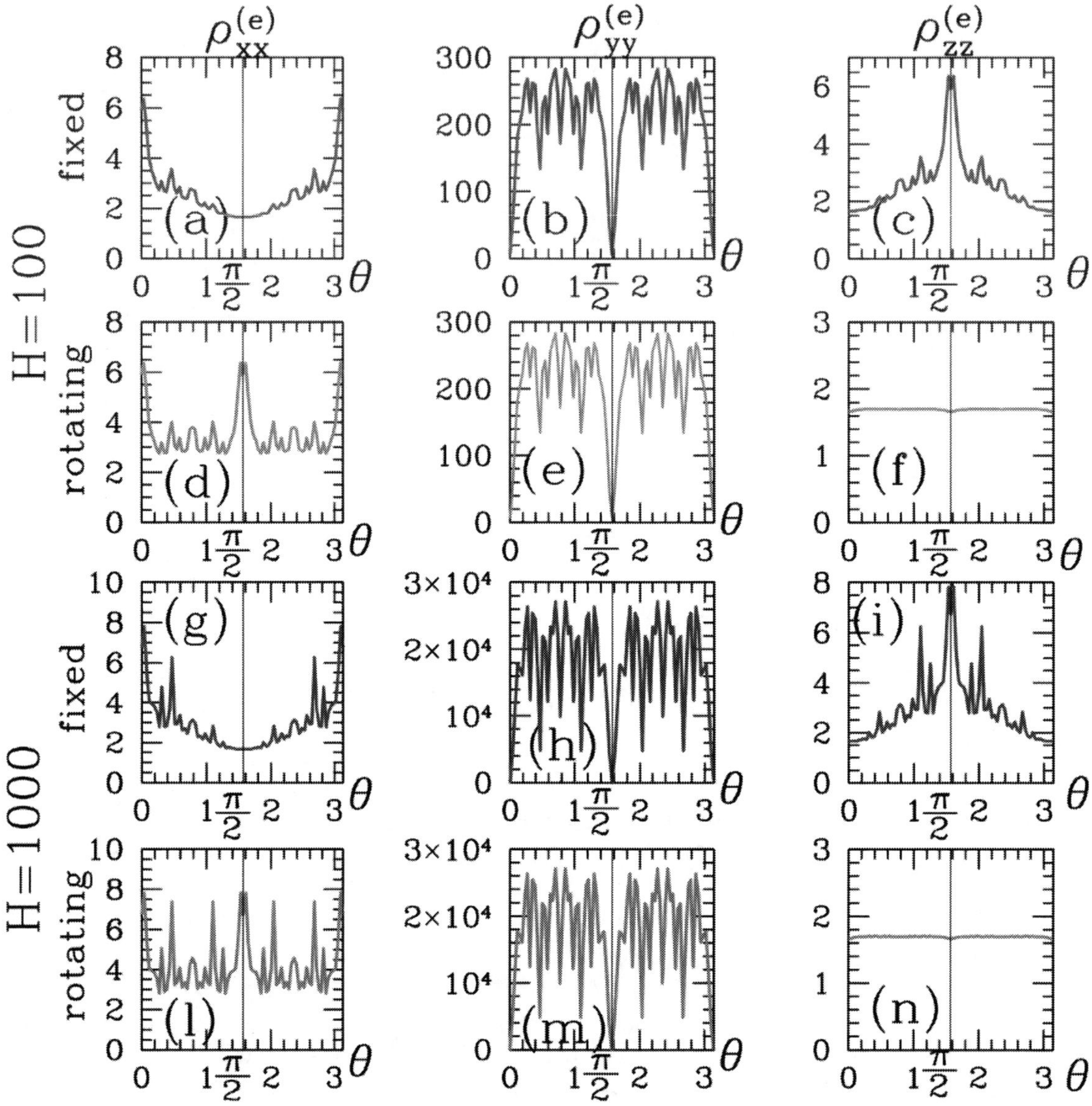

Figure 2. (Color online) All the results are for a simple-cubic array of spherical inclusions with radii R given by $R = 0.4a$, where a is the edge length of a unit cell. The inclusions are characterized by the resistivity tensor $\hat{\rho}_1$ and are embedded in an otherwise uniform host characterized by the resistivity tensor $\hat{\rho}_2$. Lines 1 and 2 show results when $\rho_0 = 1$, $\alpha_1 = 1$, $\alpha_2 = 2$, $\lambda_1 = 1$, $\lambda_2 = 2$, $\beta_1 = 100$, $\beta_2 = 200$. Lines 3 and 4 show results for the same physical parameters with the exception of $\beta_1 = 1000$, $\beta_2 = 2000$, i.e., both Hall resistivities are larger by a factor 10. (a)–(c) $\rho^{(e)}_{xx}$, $\rho^{(e)}_{yy}$, and $\rho^{(e)}_{zz}$ plotted vs. the rotation angle θ of Fig. 1(a) but where these components are taken along the *fixed coordinates* x, y, z of Fig. 1(a),(b). (d)–(f) Similar to panels (a)–(c), but now the components of $\hat{\rho}_e$ are taken along the *rotating coordinates* x, y', z' of Fig. 1(b). (g)–(n) The same as panels (a)–(f) but with values of the two β_i's that are larger by a factor 10. It is useful to note that panels (e), (f), (d) correspond to α_e, λ_e, $\tilde{\alpha}_e$ which are shown as polar plots in the top line of panels in Fig. 13 of Ref. [9]. Similarly, panels (m), (n), (l) correspond in a similar fashion to α_e, λ_e, $\tilde{\alpha}_e$ which are shown in the top line of panels in Fig. 14 of that reference.

Journal of Physics: Conference Series **319** (2011) 012002 doi:10.1088/1742-6596/319/1/012002

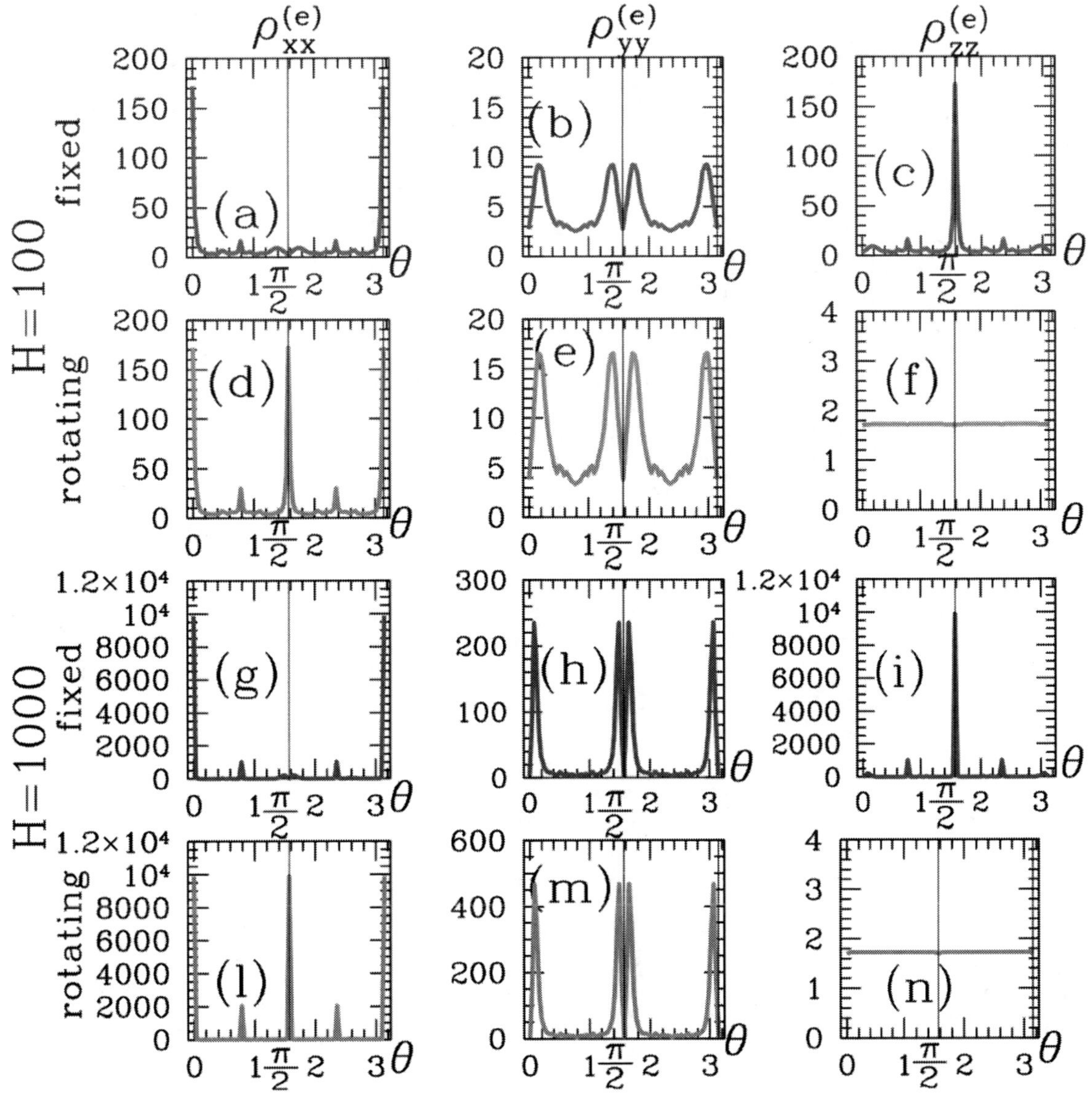

Figure 3. (Color online) Similar to Fig. 2 but now the fixed coordinate systems are $x \parallel (001)$, $y \parallel (010)$, $z \parallel (001)$ and $x \parallel (001)$, $y' \parallel (011)$, $z' \parallel (0\bar{1}1)$, while x'' and z'' subtend a changing angle θ with x and z', respectively, and remain perpendicular to y'—see Fig. 1(b) and (c). Note that the θ axis in all the panels stretches from -0.1 up to $\pi + 0.1$ in order to clarify the violent behavior of some of the plots near $\theta = 0$ and $\theta = \pi$. We note that the orientations of **B** reached in these plots were not considered in any previous reference, including Ref. [9].

expertise.

Acknowledgments
Partial support for this research was provided by grants from the U.S.-Israel Binational Science Foundation and the Israel Science Foundation and the Russia-Israel research program of the Science Ministry of the State of Israel, and by KAMEA Program (for YMS) of the Ministry of Absorption of the State of Israel.

References

[1] Bergman D J and Strelniker Y M 1994 *Phys. Rev. B* **49** 16256
[2] Strelniker Y M and Bergman D J 1994 *Phys. Rev. B* **50** 14001
[3] Tornow M, Weiss, D Klitzing K. v., Eberl K, Bergman D J, and Strelniker Y M 1996 *Phys. Rev. Lett.* **77** 147
[4] Strijkers G J, Yang F Y, Reich D H, Chien C L, Searson P C, Strelniker Y M, and Bergman D J 2001 *IEEE Trans. Magn.* **37** 2067
[5] Guttal V and Stroud D 2005 *Phys. Rev. B* **71** 201304
[6] Guttal V and Stroud D 2006 *Phys. Rev. B* **73** 085202
[7] Magier R and Bergman D J 2006 *Phys. Rev. B* **74**, 094423
[8] Magier R and Bergman D J 2008 *Phys. Rev. B* **77** 144406
[9] Bergman D J and Strelniker Y M 2010 *Phys. Rev. B* **82** 174422

Journal of Physics: Conference Series **319** (2011) 012003 doi:10.1088/1742-6596/319/1/012003

Hydraulic Jump

J. K. Bhattacharjee
S. N. Bose National Centre for Basic Sciences, JD Block, Sec-III, Salt Lake, Kolkata - 700098, India
E-mail: jkb@bose.res.in

Arnab K. Ray
Jaypee University of Engineering and Technology, Guna - 473226, Madhya Pradesh, India

Abstract. We consider the issue of the mechanism of the hydraulic jump. Combining viscosity with the existence of a horizon in the inviscid limit which allows only one way passage of information gives a reasonable picture. Some recent experiments which seem to support such a point of vieware discussed.

1. Introduction

Hydraulic jump, discussed in almost all text books [1, 2, 3, 4], is a reasonable ubiquitous phenomena occurring in kitchen sinks, spillway of dams, in tidal bores and so on. The basic phenomena consists of a fast flowing fluid on a substrate which after traversing a distance jumps by a certain amount and thereafter flows out as a reduced speed. The flow could be circular as shown in Fig. 1 (the kitchen sink) where it originates from a point (where the jet from the tap hits the sink) and flows radially out or it could be channel like originating along a line like element (Fig. 2). The basic question is that given the constant volumetric flow rate (to be denoted by Q) and the parameters of the fluid, is it possible to predict the position (radius) of the jump and the amount of jump [5, 6, 7, 8]. This has turned out to be an impossibly high requirement and there is no clear cut answer. In the early nineties, Bohr, Dimon and Putkaradze [9, 10] offered an extremely useful way of analyzing the jump. In effect, this treated th jump as a *shock* and recent experiments [11, 12] have shown the relevance of the shock quite clearly in the high viscosity, low surface tension regime.

A different twist was brought to this problem in the last decade by Schützhold and Unruh [13] who showed that for an irrotational flow in a channel, a small perturbation around the steady flow satisfies a Klein-Gordon equation with a Schwarzschild metric. Subsequently, the jump position became with the horizon of the Schwarzschild metric [14, 15] leading to an interesting physical picture where the shock and the horizon are combined with help of viscosity to lead to the formation of the jump [16, 17]. An interesting outcome of the inclusion of viscosity is an instability [17] where an incoming perturbation from outside the jump increases in amplitude as the horizon is approached.

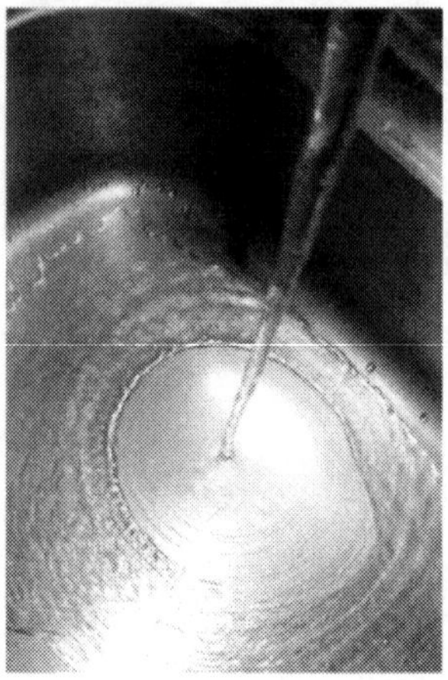

Figure 1. Circular hydraulic jump in a kitchen sink.

Figure 2. An experimental set up to study one dimensional hydraulic jump.

2. Channel Flow

We consider a one dimensional channel flow to begin with where a fast moving fluid layer(velocity v_1, height h_1) suddenly jumps to height, h_2 and flows with a lower velocity, v_2. We ask the question whether the laws of inviscid hydrodynamics would allow such a motion. Referring to Fig. 3 and using as a control volume, the region bounded by the dashed lines, we can write the mass conservation as (incompressible fluid),

$$v_1 h_1 = v_2 h_2 = \frac{Q}{L}, \tag{1}$$

where L is the channel width and Q the volumetric flow rate. Newton's law applied to the control volume would require the momentum flowing out per unit time, $L\rho v_2^2 h_2 - L\rho v_1^2 h_1$, to be equal to the force due to the pressure difference i.e. $\frac{1}{2}L\rho g(h_1^2 - h_2^2)$ and hence,

$$v_2^2 h_2 - v_1^2 h_1 = \frac{1}{2} g(h_1^2 - h_2^2). \tag{2}$$

With $h_2/h_1 = X$ and $F = v_1^2/gh_1$, the Fronde number, Eqs. (1) and (2) yield,

$$\begin{aligned} 2F &= X(1+x), \\ or \quad 2X &= -1 + \sqrt{1+8F}. \end{aligned} \tag{3}$$

It is possible to have $X \geq 1$ for $F \geq 1$. This implies that for the jump to occur the fluid velocity

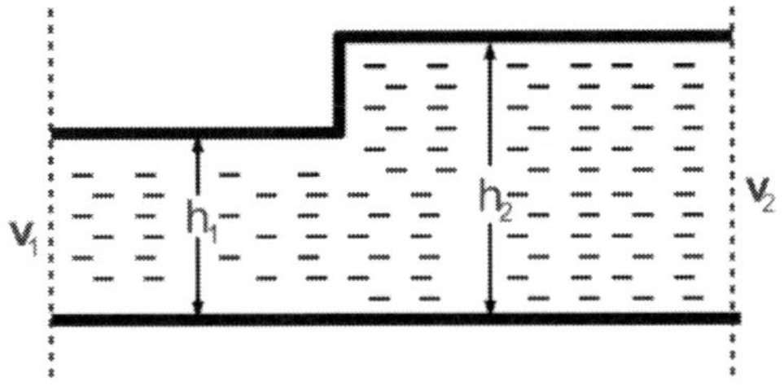

Figure 3. The volume element...

has to be exceed $\sqrt{gh}$, which is the speed of capillary waves in a thin fluid layer. This fact will become important as we go along. However, our approach so far cannot locate the position or calculate the magnitude of the jump. For that, we need to look at a more detailed picture.
The steady state differential equations for incompressible flow in a channel of width L are (assuming the width L is large enough to ignore motion in the y-direction),

$$\frac{\partial u}{\partial x} + \frac{\partial w}{\partial z} = 0, \tag{4}$$

$$u\frac{\partial u}{\partial x} + w\frac{\partial u}{\partial z} = -\frac{\partial P}{\partial x} + \nu\nabla^2 u. \tag{5}$$

The constant volumetric flow rate is,

$$Q = Lhu, \tag{6}$$

where $h(x)$ is the height of the fluid layer. The pressure P is clearly $\Pi + (h(x) - z)\rho g$, where Π is the atmospheric pressure and the local coordinates is (x, z). In $\nabla^2 u$, the primary variation is a along z-direction and hence

$$u\frac{\partial u}{\partial x} + w\frac{\partial u}{\partial z} = -g\frac{\partial h}{\partial x} + \nu\frac{\partial^2 u}{\partial z^2} \tag{7}$$

In eqs. (4), (6) and (7), we have three equations for the three unknowns u, w and h. The boundary conditions are the so called *no-slip* conditions at $z = 0$, which means $u(x, 0) = w(x, 0) = 0$ for all x and stress free at the surface i.e. $w(x, h(x)) = 0$ and $\partial u/\partial z = 0$
It is difficult to solve these equations with the given boundary conditions. Numerically a decent scheme for solving these equations have been devised by Dasgupta and Govindarajan [18] but even their scheme does not yield an answer for the amount of the jump. To get a hold on both the position and magnitude of the jump, we resort to an approximation introduced by Bohr et al.
The technique of reference [9], is to introduce variables which are averaged over the z-direction, i.e. we use $\langle f(x, z)\rangle = \tilde{f}(x)$, where

$$\langle f(x, z)\rangle = \frac{1}{h(x)}\int_0^h (x) f(x, z)dz. \tag{8}$$

Consequently, we average Eq. (7) over z and find

$$\left\langle u\frac{\partial u}{\partial x}\right\rangle + \left\langle w\frac{\partial u}{\partial z}\right\rangle = -g\frac{dh}{dx} + \nu\left\langle\frac{\partial^2 u}{\partial z^2}\right\rangle. \tag{9}$$

An integration by parts shows $\langle w\frac{\partial u}{\partial z}\rangle = \langle u\frac{\partial u}{\partial x}\rangle$ on using the continuity equation. Further, $\left\langle\frac{\partial^2 u}{\partial z^2}\right\rangle = -\frac{\partial u}{\partial z}\frac{1}{h}\Big|_{z=0} \simeq -\frac{\langle u\rangle}{h^2}$, on using the stress free boundary condition. Thus,

$$2\left\langle u\frac{\partial u}{\partial x}\right\rangle = -g\frac{dh}{dx} - \nu\frac{\langle u\rangle}{h^2}. \tag{10}$$

We use a factorization approach to write,

$$\left\langle u\frac{\partial u}{\partial x}\right\rangle = \alpha\langle u\rangle\left\langle\frac{du}{dx}\right\rangle, \tag{11}$$

where α is a number of $\mathcal{O}(1)$. Noting that an averaging of Eq. (6) leads to $Q = \langle u\rangle hL$, we find

$$\left(\alpha\frac{Q^2}{L^2} - gh^3\right)\frac{dh}{dx} = \frac{\nu Q}{L}. \tag{12}$$

The above equation can easily be integrated to obtain $h(x)$. It is obvious that a blind use of the solution, which can be written as

$$\alpha\frac{Q^2}{L^2}h - g\frac{h^4}{4} = \frac{\nu Q}{L}x + C \tag{13}$$

would lead to a double valued solution. This is a clear cut indication that there are two branches, which need to be connected by a shock. The anticipated form is shown in Fig. 4. Before proceeding, we would like to write Eq. (13) in a dimensionless form. This can be done in the following fashion. First, we write Eq. (13) as

$$\begin{aligned} \alpha h - \frac{g}{4}\frac{L^2}{Q^2}h^4 &= \frac{\nu L}{Q}\cdot x + \frac{CL^2}{Q^2} \\ \text{or}\quad \alpha h - \frac{1}{4}\left(\frac{h}{h_0}\right)^3\cdot h &= \frac{\nu L}{Q}\cdot x + \frac{CL^2}{Q^2} \end{aligned} \tag{14}$$

where $h_0 = (Q^2/L^2 g)^{1/3}$ has the dimension of length and provides the scaling for the height h. The dimensionless height $H = h/h_0$ and hence we have

$$\begin{aligned} \alpha H - \frac{1}{4}H^4 &= \frac{\nu g^{1/3}}{(Q/L)^{5/3}} + \bar{C} \\ &= X + \bar{C} \end{aligned} \tag{15}$$

where $X = x/x_0$, $x_0 = (Q/L)^{5/3}\nu^{-1}g^{-1/3}$ and $\bar{C}$ is a dimensionless constant of integration. We need two branches and hence two boundary conditions. For the flow to the left of the shock (located at X_c) in Fig. 4, we can use $H = 0$ at $X = 0$, so that

$$\alpha H - \frac{1}{4}H^4 = X \qquad (X < X_c). \tag{16}$$

For $X > X_c$, we need to fix the flow at a distance. If $H \to 0$ at $X = D$, then

$$\frac{1}{4}H^4 - \alpha H = D - X \qquad (X > X_c). \tag{17}$$

We now plot H vs X from Eq. (17). This is the curve marked 'a' in Fig. 5. Across the shock (inviscid approximation good), mass and momentum balance imply Eqs. (1) and (2). For a

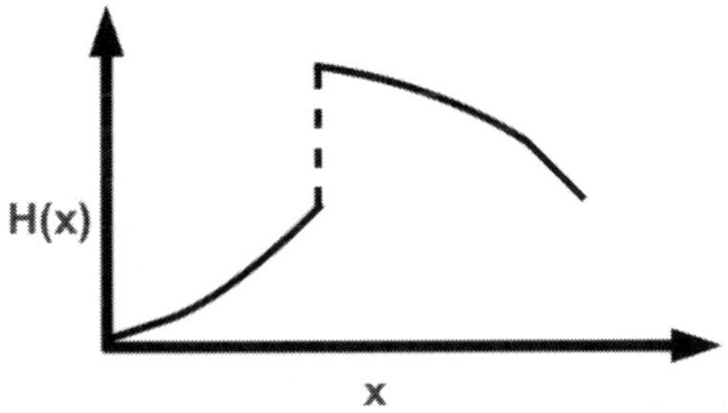

Figure 4. Schematic diagram of shock.

given X, H $(X > X_c)$ and thus h_2 is found from Eq. (17) and the corresponding v_2 from Eq. (1) and then Eq. (2) yields,

$$\frac{2v_2^2}{gh_2} = \frac{h_1}{h_2}\left(1 + \frac{h_1}{h_2}\right), \tag{18}$$

from which h_1/h_2 and hence $h_1(x)$ is obtained. This $h_1(x)$ is shown as the curve marked 'b' in Fig. 5. We now plot the solution $h_1(x)$ which we find from Eq.(16) and this is the curve marked 'c' in Fig. 5. The point P, where the two curves 'b' and 'c' meet is the point where the solution to the governing equation of motion (Eq. (12)) gives the correct amount of discontinuity and hence P is the location of the shock. The $h(x)$ that is physically relevant follows curve 'c' upto point P, then has a discontinuous jump and finally follows the curve marked 'a'. It is clear that the radius of the jump (i.e. the position corresponding to X_c) scales as $(Q/L)^{5/3}\nu^{-1}g^{-1/3}$ while the magnitude of the jump scales as $(Q/L)^{2/3}g^{-1/3}$. It is the magnitude of jump that is not addressed usually. The procedure of locating the shock gives the magnitude of the jump. The scaling property of the radius of the jump for the two dimensional flow (Fig. 1) was found by Bohr et al to be $r_c \propto Q^{5/8}\nu^{-3/8}g^{-1/8}$. Thus we have,

$$\begin{aligned} \text{One dimensional flow :} \quad & r_c \propto \left(\frac{Q}{L}\right)^{5/3}\nu^{-1}g^{-1/3} \\ \text{Two dimensional flow :} \quad & r_c \propto Q^{5/8}\nu^{-3/8}g^{-1/8} \end{aligned} \tag{19}$$

In the next section, we will take, in some sense, a more physically motivated look at the jump

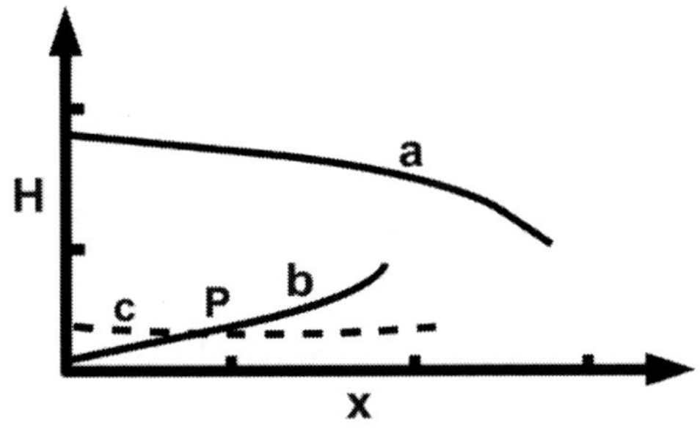

Figure 5. Schematic digram...

and this will help get both the above relations from the same picture.

3. Circular Jump

For the circular jump (Fig. 1), the approximation carried out by Bohr et al had led to the steady state equations

$$rvh(r) = \text{Const.} = \frac{Q}{2\pi} \tag{20}$$
$$v\frac{\partial v}{\partial r} = -g\frac{\partial h}{\partial r} - \nu\frac{v}{h^2} \tag{21}$$

The procedure for analyzing the above as carried through by Bohr et al is actually very similar to locating the shock as described for the one dimensional flow given in Sec. 2. We assume that $v_0(r)$ and $h_0(r)$ are the steady state solutions that describe the hydraulic jump. We consider time dependent perturbation around this solution. To do so we need the dynamic equations and these are,

$$\frac{\partial h}{\partial t} + \frac{1}{r}\frac{\partial}{\partial r}(rvh(r)) = 0, \tag{22}$$
$$\frac{\partial v}{\partial t} + v\frac{\partial v}{\partial r} = -g\frac{\partial h}{\partial r} - \nu\frac{v}{h^2}. \tag{23}$$

We can use these equations to discuss perturbations around the static flow $h = h_0(r)$, $v = v_0(r)$. This is best done in a linear stability analysis in terms of the variable,

$$f = rvh, \tag{24}$$

which is a constant f_0 in the static limit. If we denote the variations around f_0, h_0, v_0 by f', h', v' and restrict ourselves to terms which are linear in f', h' and v', then

$$f' = r(v_0 h' + h_0 v') \tag{25}$$

From Eq. (22),

$$\frac{\partial h'}{\partial t} = -\frac{1}{r}\frac{\partial f'}{\partial r} \tag{26}$$

and using this with Eq.(25), we are led to

$$\frac{\partial v'}{\partial t} = \frac{1}{rh_0}\left(\frac{\partial f'}{\partial t}\right) + \frac{v_0}{r_0 h_0}\left(\frac{\partial f'}{\partial r}\right) \tag{27}$$

The linearization of Eq.(23) yields

$$\frac{\partial v'}{\partial t} + v_0\frac{\partial v'}{\partial r} + v'\frac{\partial v_0}{\partial r} = -g\frac{\partial h'}{\partial r} - \nu\frac{v'}{h_0^2} + 2\nu v_0\frac{h'}{h_0^3} \tag{28}$$

taking the partial derivatives with respect to time in Eqs. (27) and (28) and equating the two expressions of $\frac{\partial^2 v'}{\partial t^2}$ so obtained leads after some manipulation to

$$\frac{\partial}{\partial t}\left[v_0\frac{\partial f'}{\partial t}\right] + \frac{\partial}{\partial t}\left[v_0^2\frac{\partial f'}{\partial r}\right] + \frac{\partial}{\partial r}\left[v_0^2\frac{\partial f'}{\partial t}\right] + \frac{\partial}{\partial r}\left[v_0(v_0^2 - gh_0)\frac{\partial f'}{\partial r}\right] = -\nu\frac{v_0}{h_0^2}\left(\frac{\partial f'}{\partial t} + 3v_0\frac{\partial f'}{\partial r}\right) \tag{29}$$

This result can be cast in a compact form

$$\partial_\alpha\left(F^{\alpha\beta}\partial_\beta f'\right) = -\nu\frac{v_0}{h_0^2}\left(\frac{\partial f'}{\partial t} + 3v_0\frac{\partial f'}{\partial r}\right) \tag{30}$$

where $F^{\alpha\beta}$ is the 2×2 symmetric matrix,

$$F^{\alpha\beta} = \begin{pmatrix} 1 & v_0 \\ v_0 & v_0^2 - gh_0 \end{pmatrix} \tag{31}$$

Let us focus on the inviscid limit, when the right hand side of Eq. (30) is zero and the perturbation f' satisfies the wave equation

$$\Box^2 f' = \frac{1}{\sqrt{-g}}\partial_\alpha\left(\sqrt{-g}g^{\alpha\beta}\partial_\beta f'\right) = 0 \tag{32}$$

where we make the identification $F^{\alpha\beta} = \sqrt{-g}g^{\alpha\beta}$ and $g = Det F^{\alpha\beta}$, so that near critical condition $v_0^2 = gh_0$, we can identify

$$g^{\alpha\beta} = \begin{pmatrix} 1 & v_0 \\ v_0 & v_0^2 - gh_0 \end{pmatrix}. \tag{33}$$

This is a remarkable result. The perturbation around the static jump solution in the inviscid limit satisfies a Klein-Gordon equation for a massless particle in the Schwarzschild background if we identify the velocity of light with $\sqrt{gh_0}$. It should be noted that our derivation here following references [16] and [17] is different in character from that of the pioneering work of Schützhold and Unruh. It is specific to the jump problem and consequently able to throw light on the mechanism of the jump. An identical calculation holds for the one dimensional case as shown in [16].
The important thing to note is that the perturbation to the jump profile satisfies a wave equation in a background metric which has a horizon defined by $v_0^2 = gh_0$. Now the wave propagating in a shallow water of height h is the usual capillary wave with the speed $\sqrt{gh}$. The horizon condition above implies that the flow speed of the fluid equals the wave speed. Let us combine the picture with what we saw in the last section. There is a fast flowing fluid on one side of the jump (the interior or the supercritical region) and then a slower flow on the other (exterior or subcritical region) side. The jump was modelled by fitting a shock. We can identify the shock with the horizon. In the interior of the jump the velocity exceeds the wave speed $\sqrt{gh}$ and hence the wave cannot propagate in that region. The waves carry information and hence there can be no information carried from outside the jump to the inside. This ties up with the picture of the horizon interpreted as a *white hole* - the information can only come out, it cannot go in. This is crucial to the jump formation. Everywhere interior to the jump $v_0 > \sqrt{gh_0}$, in spite of the fact that the fluid is slowing down due to viscosity (an ingredient not considered in the horizon calculation). The fast moving layers at small values of $r(< r_c)$ do not know that the fluid in front has slowed down (no information transmission) and when they reach $r = r_c$, they "jump" over the slowed down fluid causing a jump at $r = r_c$. There are two essential points in the jump:

(i) In the inviscid limit, the perturbation to the static flow live in a space with a Schwarzschild like metric with c replaced by $\sqrt{gh}$ - the velocity of capillary waves. The metric leads to the existence of a horizon where $v = \sqrt{gh_0}$. This is the barrier for information transmission. No information crosses to the left of the barrier where $v > \sqrt{gh}$.

(ii) The jump is caused by the viscosity which slows down the flow from its origin to the horizon but due to (i) above, this information does not reach the fast moving flow. When at the "horizon" ($v = \sqrt{gh}$) there is sudden realization and thus there is an accumulation that leads to the jump.

Clearly, the shock of the last section is the horizon of this and a clear experimental realization of this can be seen from the work of Jannes et al [12]. The formation of the Mach cone shown

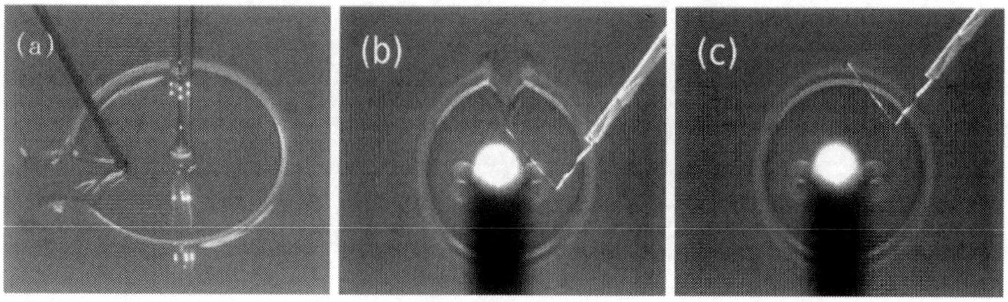

Figure 6. Ty

in Fig. 6, taken from the reference [12] clearly shows that the jump occurs at $v = \sqrt{gh}$ and $v > \sqrt{gh}$ in the interior of the jump.

We now need to obtain a qualitative position of the jump that will be valid for any dimension. For this we need to know the viscous time scale. To do so, we write a possible solution for $f'(r,t)$ as $e^{-\lambda t}\phi(r)$ and use Eq.(29) to obtain

$$\lambda^2\phi - \lambda\left[2\frac{\partial}{\partial r}(v_0\phi) + \frac{\nu\phi}{h_0^2}\right] + \frac{1}{v_0}\left[v_0(v_0^2 - gh_0)\phi'\right] + 3\nu\frac{v_0\phi'}{h_0^2} = 0. \tag{34}$$

Taking a slowly varying ϕ, so that ϕ' can be dropped in comparison and integrating over a region at the boundaries of which ϕ vanishes (localized $\phi(r)$), the last two terms of Eq.34) are dropped and then (we use $v_0\phi$ to multiply before integrating)

$$\begin{aligned}\lambda^2\int v_0\phi^2 dr - \lambda\nu\int\frac{v_0\phi^2}{h_0^2}dr &= \lambda\int\frac{\partial}{\partial r}(\phi v_0)^2 dr = 0,\\ \text{giving}\quad \lambda &\simeq \frac{\nu}{h_0^2},\end{aligned} \tag{35}$$

if h_0 is slowly varying.

Thus we obtain the time scale in which viscosity slows down the velocity. This is consistent with what one obtains from Eq.(23) if only the viscous term is kept. That gives $\dot{v} \simeq -\nu v/h^2$, showing $v \approx v_0 e^{-\nu t/h^2}$ yielding Eq.(35). The viscous time scale is always $\tau = h_0^2/\nu$. Now, we need the time scale for the flow to reach the jump point. In the inviscid limit, the only velocity scale is $\sqrt{gh}$ and hence the time to the jump point is $\tau' \simeq r_c/\sqrt{gh}$. The jump in the above scenario occurs when $\tau = \tau'$, i.e.

$$\frac{r_c}{\sqrt{gh}} \simeq \frac{h_0^2}{\nu} \quad \text{or} \quad r_c \simeq g^{1/2}h_0^{5/2}\nu^{-1}. \tag{36}$$

in a base flow in D-dimensions, the volumetric flow rate is

$$Q = r^{D-1}v - 0h_0. \tag{37}$$

Evaluating every quantity at $r = r_c$,

$$Q = r_c^{D-1}\sqrt{gh_0}h_0 = r_c^{D-1}g^{1/2}h_0^{3/2}. \tag{38}$$

Substituting for h_0 from Eq.(38) into Eq.(38) we finally get,

$$Q = r_c^{D-1}g^{1/2}(r_c\nu g^{-1/2}) \simeq r_c^{D-2/5}g^{1/5}\nu^{3/5}. \tag{39}$$

For $D = 1$, we get

$$r_c \simeq Q^{5/3} g^{-1/3} \nu^{-1}, \tag{40}$$

while for $D = 2$, we have

$$r_c \simeq Q^{5/8} g^{-1/8} \nu^{-3/8}, \tag{41}$$

exactly in agreement with Eq.(19).

4. Viscosity Induced Instability

We return to Eq.(30), with a view to looking at travelling wave solutions which should certainly be the interesting ones. Writing

$$f' = Ae^{iS(r,t)} = Ae^{i(\omega t - \psi(r))}. \tag{42}$$

Straightforward calculations yield

$$\omega^2 v_0 - 2\omega v_0^2 \psi' + v_0(v_0^2 - c^2)\psi'^2 + i\omega \partial_r v_0^2 - i\partial_r \left[v_0(v_0^2 - c^2)\psi'\right] = \frac{\nu v_0}{h_0^2}(i\omega + 3iv_0\psi') \tag{43}$$

In the above $c^2 = gh(r)$ and $\psi' = \partial_r \psi$. If we assume that $\psi(r)$ is primarily real i.e. $\psi(r) = \psi_R(r) + i\psi_I(r)$ with $\psi_I \ll \psi_R$, then

$$\psi_R = \omega \int \frac{dr}{v_0 \pm c} \tag{44}$$

and

$$2\left[\omega v_0^2 - v_0(v_0^2 - c^2)\psi_R'\right]\psi_I' = -\frac{\nu v_0}{h_0^2}\left(\omega + 3v_0\psi_R'\right) \tag{45}$$

The real part is singular when $v_0 = c$ i.e. at the jump radius and since this is the dominant contribution, we will focus on $\psi_R' = \frac{\omega}{v_0 - c}$. This leads to

$$\psi_I' = -\frac{\nu}{2ch_0^2}\left[1 + \frac{3v_0}{v_0 - c}\right]. \tag{46}$$

For $v_0 < c$, psi_R' is negative and hence we are discussing a wave that is travelling towards

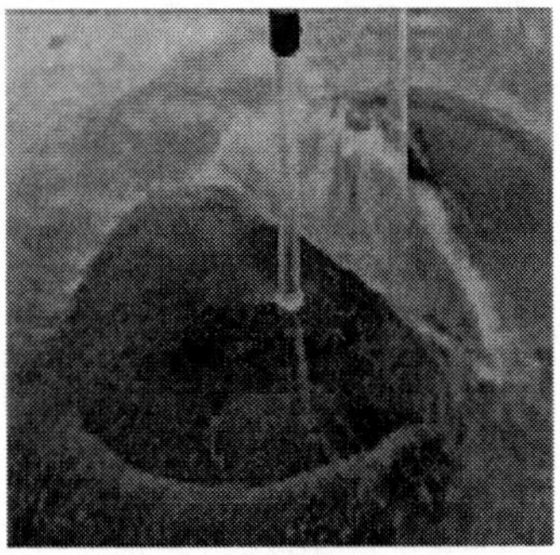

Figure 7. xxxxx

the jump radius. The growing part of ψ_I' (see Eq.(46)) is positive at this point and hence the travelling wave gets magnified by the factor $e^{\frac{3v_0\nu}{2ch_0^2}\cdot\frac{1}{c-v_0}}$ which blows up for $c = v_0$. Hence the amplitude of the perturbation grows as one approaches the jump. There is a large amplitude of the perturbation at the jump (this could be some kind of an analogy of the Miles instability) - certainly an instability triggered by viscosity. This dramatic effect of viscosity has been captured by Kate et al [19]. A hydraulic jump set up as shown in Fig. 1. Now one needs to perturb this flow from outside the jump. To do that a second tap is arranged setting up a second jump whose distance from the first is greater than the radius of the first. The flowing water from the second tap is a perturbation on the first, while the flow from the first is a boundary for the second. At the junction the viscosity makes the fluid pile up as exhibited in Eq.(46) and so clearly demonstrated in Fig. 7.

5. Conclusion

We thus arrive at a picture where the inviscid limit gives rise to a horizon across which the region external to the jump (slow moving) cannot communicate with the internal region (fast moving). Adding viscosity to this requires fast moving fluid to know that fluid to know that the fluid in front has slowed down. Since the communication cannot be transmitted, the knowledgs is not received till one is at the horizon and hence the jump as the fast moving fluid "climbs over" over the slowed down flow.

References

[1] Landau L D and Lifshitz E M 1987, *Fluid Dynamics*, (Oxford: Butterworth-Heinemann).
[2] Walshaw C and Jobson D A 1979, *Mechanics of Fluids*, (London: Longman).
[3] Granger S 1987, *Engineering Fluid Mechanics*, (New York: Dover).
[4] Faber T E 1995, *Fluid Dynamics for Physicists*, (Cambridge: Cambridge Univ. Press).
[5] Tani I 1949, *J. Phys Soc. Japan* **4** 212.
[6] Watson E J 1966, *J. Fluid Mech.* **20** 481.
[7] Olsson R G and Turkdogan E T 1966, *Nature* **211** 813.
[8] Bush J W M and Aristoff J M 2003, *J. Fluid Mech.* **489** 229.
[9] Bohr T, Dimon P and Putkaradze V 1993, *J. Fluid Mech.* **254** 635.
[10] Bohr T, Putkaradze V and Watanbe S 1997, *Phys. Rev. Lett.* **79** 1038.
[11] Hansen S H, Horlück S, Zanner D, Dimon P, Ellegard C and Creagh S C 1997, *Phys. Rev. E* **55** 7048.
[12] Jannes G, Piquet R, Maissa P, Mathis C and Rousseaux G 2010, *Phys. Rev. E* **83** 056312.
[13] Schützhold R and Unruh W 2002, *Phys. Rev. D* **66** 044019.
[14] Volovik G E 2005, *JETP Lett.* **82** 624.
[15] Volovik G E 2006, *J. Low temp. Phys.* **145** 337.
[16] Singha S B, Bhattacharjee J K and Ray A K 2005, *Eur. Phys. J B* **48** 417
[17] Ray A K and Bhattacharjee J K 2007, *Phys. Lett. A* **371** 241.
[18] Dasgupta R and Govindarajan R 2010, *Phys. of Fluids* **22** 112108.
[19] Kate R P, Das P K and Chakraborty S 2007, *J Fluid Mech.* **590** 355.

Journal of Physics: Conference Series **319** (2011) 012004 doi:10.1088/1742-6596/319/1/012004

A fractal model of earthquake occurrence: Theory, simulations and comparisons with the aftershock data

Pathikrit Bhattacharya

Department of Earth Sciences, Indian Institute of Technology, Roorkee, Uttarakhand, *247 667*, India

Department of Earth Sciences, University of Western Ontario, London, Ontario, *N6A 5B7*, Canada. (At present)

Department of Geosciences, Princeton University, Guyot Hall, Princeton, NJ, *08544*, USA. (From September 2011)

E-mail: `pathikri@princeton.edu`

Bikas K. Chakrabarti

Theoretical Condensed Matter Physics Division and Centre for Applied Mathematics and Computational Science, Saha Institute of Nuclear Physics, Kolkata, West Bengal, *700064*, India.

E-mail: `bikask.chakrabarti@saha.ac.in`

Kamal

Department of Earth Sciences, Indian Institute of Technology, Roorkee, Uttarakhand, *247 667*, India

E-mail: `kamalfes@iitr.ernet.in`

Abstract. Our understanding of earthquakes is based on the theory of plate tectonics. Earthquake dynamics is the study of the interactions of plates (solid disjoint parts of the lithosphere) which produce seismic activity. Over the last about fifty years many models have come up which try to simulate seismic activity by mimicking plate plate interactions. The validity of a given model is subject to the compliance of the synthetic seismic activity it produces to the well known empirical laws which describe the statistical features of observed seismic activity. Here we present a review of one such, purely geometric, model of earthquake dynamics, namely The Two Fractal Overlap Model. The model tries to emulate the stick-slip dynamics of lithospheric plates with fractal surfaces by evaluating the time-evolution of overlap lengths of two identical Cantor sets sliding over each other. As we show later in the text, some statistical aspects of natural seismicity are naturally captured by this simple model. More importantly, however, this model also reveals a new statistical feature of aftershock sequences which we have verified to be present in nature as well. We show that, both in the model as well as in nature, the cumulative integral of aftershock magnitudes over time is a remarkable straight line with a characteristic slope. This slope is closely related to the fractal geometry of the fault surface that produces most of thee aftershocks. We also go on to discuss the implications that this feature may have in possible predictions of aftershock magnitudes or times of occurrence.

1. Introduction

Many models of natural seismicity have been proposed over the last half-decade and more. These cover a very broad range of approaches ranging from purely physical to purely statistical. The relative successes of these very diverse models is testament to the great complexity involved in the production of natural seismicity and points to the fact that the correct model is probably neither extreme and is some elusive combination of these approaches. However, an approach that has been less pursued to some extent is the investigation of the role of the fractal surface geometry of a fracture surface topography (or self affine for most natural surfaces) in producing the known statistical properties of natural seismicity. The model we propose in this article incorporates this fractal topography of the fracture surface in a simplistic way.

1.1. Earthquake statistics

The overall frequency distribution of earthquakes is given by the Gutenberg-Richter (GR) Law [1] which states

$$\log N(m) = a - bm, \tag{1.1}$$

where $N(m)$ is the frequency of earthquakes with magnitude greater than or equal to m occurring in a specified area. The parameter b, the so called b-value, has some regional variation (the value of the exponent b-value has been seen to change from one geographical region to another) but the value of b estimated from global catalogs or regional catalogs spanning many fault systems is deemed to be universal and close to unity. The constant a is a measure of the regional level of seismicity at the given lower magnitude threshold for the catalog. In particular, for aftershock sequences, the a-value points to the total aftershock productivity of the mainshock to which the sequence corresponds to. Owing to the log-linear relationship between seismic energy released and the magnitude of the earthquake, there is another form in which the Gutenberg-Richter law is stated:

$$N(\varepsilon) \sim \varepsilon^{-\alpha}, \tag{1.2}$$

where $N(\varepsilon)$ is defined in analogy to the previous form but for events which release energy greater than or equal to ε. This is due to the fact that usually magnitude is defined as logarithm of the trace amplitude on a seismogram and hence bears a log-linear relationship with energy. The temporal distribution of aftershocks of magnitude m greater than or equal to some threshold value M_c is given empirically by another well known power law, namely the Omori Law [2], saying

$$\frac{dN(t)}{dt} = \frac{1}{t^p}, m \geq M_c. \tag{1.3}$$

Here $dN(t)/dt$ gives the rate of occurrence of aftershocks at time t after the occurrence of the mainshock. The value of the exponent p is generally larger than but close to unity for tectonically active regions though a large range of variation in the p-value has been observed [3].

1.2. Modeling earthquake dynamics

The principal objective in constructing models of earthquake dynamics is to reproduce the above two empirical (statistical) laws by simulating the dynamics of a fault or of a system of interconnected faults. Different types of models have been proposed to capture this dynamics which focus on different aspects of fault dynamics. One class mimics the dynamics by slowly driving an assembly of locally connected spring-blocks over a rough surface. This essentially captures the stick slip scenario involved in generation of earthquakes. The first successful model of this kind was proposed by Burridge and Knopoff [4]. This model and all its variants [5, 6] have been reliably shown (numerically) to recreate the GR Law but the Omori Law has not been clearly demonstrated from this class of models. The underlying principle for this class of models has been found to be Self Organized Criticality [7]. There is another traditional class of models based on the mechanical properties of deformable materials that break under a critical stress. Fiber bundle models are typical of this class.

The main class of models that we will discuss here are a relatively new type. This class of models deals with the fractal geometry of fault surfaces. We shall discuss in the next chapter some of the available observations indicating that fault surfaces are fractals and how faults are distributed in a fault zone with a fractal size-distribution. These are two very well established facts. Naturally, a few of the geometrical models of earthquakes capture the fractal effects of one fault surface sliding over the other by considering two fractals sliding over each other and by taking into account the stresses developed and released due to such overlaps. Fig. 1.1 shows a cartoon depicting this scenario. This is the basic motivation behind fractal overlap models. There have been attempts at using random fractional Brownian profiles as the fractals involved (in the so called Self-affine Asperity Model) in [8, 9]. The model yields the GR law readily and relates the b-value to the geometry of the fault. A more generalized version of the model discussed in [9] also recreates the Omori law but with a universal exponent. But in nature the exponent value varies considerably. Also, the exponent is very different in value from the exponent observed for real earthquakes (for this Self-affine Asperity Model [9] the value of the exponent is 0.37, while in nature we observe values close to unity for seismically active zones as mentioned before). Our focus though, will be on yet another geometric model which has been reasonably successful in capturing most of the observed statistical features of earthquake processes reproducing values of the parameters of these empirical laws much in agreement to what we see in nature. We call this the 'Two fractal overlap model'. The simplest scenario of a fractal sliding over its complementary set involves a Cantor set sliding over its complement. But the scenario considered here is even more simplified. We consider the overlap statistics of a Cantor set sliding over its replica. Although the model does not, to start with, consider a real fault profile the main strength of the model lies in the fact that it is completely analytically tractable and gives all the well established statistics that real earthquakes demonstrate. We will, through the length of this discussion, show these results and compare them with real earthquake data. The reader will readily recognize that these results require a knowledge of no more than high school mathematics to derive and in simplicity lies the true strength of this model.

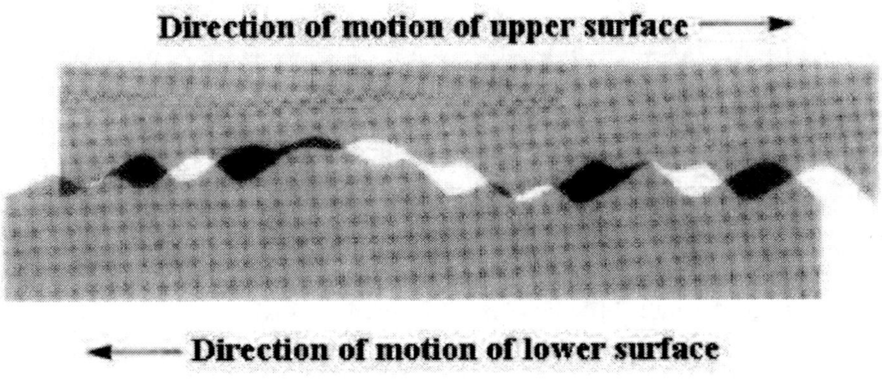

Figure 1.1: A cartoon showing overlap of two fractal surfaces. The sticking is due to interlocking of the asperities. Stress energy is accumulated and released at every slip (Adapted from [10]).

1.3. Fractal geometry of fault surfaces

Before we undertake a study and modeling of seismic activity, it is of interest to review what we imply by fractures, joints and faults. Any crack or fissure on the surface of a rock is a fracture. If the two blocks separated by the fracture are laterally displaced creating a plane across which the rock beds are discontinuous then, in strict terms of structural geology, the locus of the discontinuity in the various rock beds is the fault. Figure 1.2 shows a fault exposure in the Dixie Valley in the United States.

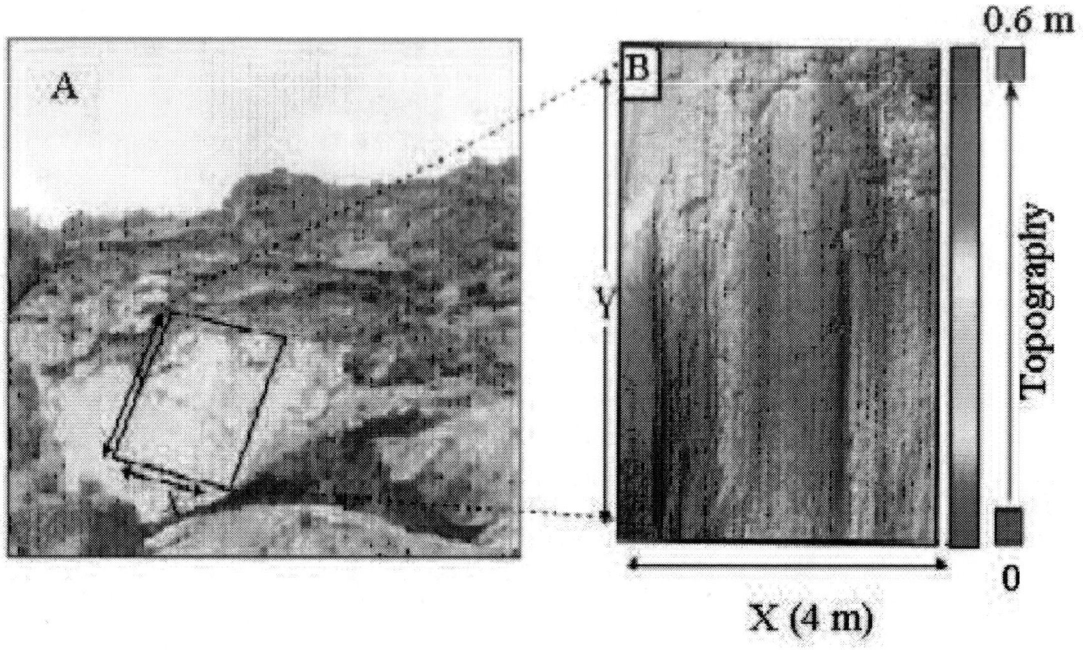

Figure 1.2: A) Chapter of a partly eroded slip surface at the Mirrors locality on the Dixie Valley fault. B) LiDAR fault surface topography as a color-scale map rotated so that the X-Y plane is the best-fit plane to the surface (Adapted from [11]).

If there has been no lateral offset across the fracture then the structure is generally referred to a joint. Faults and joints often do not come singly but in a complex system of interconnected structures. Such a system of interconnected faults is called a fault zone. In other words, it is basically a highly fractured system of fault networks all of which have been formed by the same tectonic process. It has for long been suggested that fractured rock surfaces are fractals. The fractal geometry implies a balance between two competing processes: strain weakening and strain hardening. This balance is critically tuned to produce neither positive nor negative feedback mechanisms during deformation. In such a case, the long-term deformation is accommodated statistically, at all time intervals, by structures that have no preferred size scale, i.e., structures following a scale free (due to the lack of feedback) frequency-size distribution. Fractal geometry has been reported to characterize brittle deformation structures in the crust over several bands of length scales, from regional fault networks through main traces of individual faults to the internal structure of fault zones.

In fact, fault surfaces are fractals. It was shown by [13] that the surfaces of joints are fractal. They studied the surface topography of naturally occurring joints by analyzing the power spectra of the profiles. They studied fresh joints (a fresh surface in structural geological context implies an unweathered surface) in both sedimentary and crystalline rocks, a frictional wear surface due to glacial activity and a bedding plane surface. The power spectrum of all these surfaces showed a 'red noise' spectrum over the entire spatial frequency bandwidth employed in the study with the amplitude falling off 2 to 3 orders of magnitude per decade increase in spatial frequency. This was explained using a fractal model of the topography. The dimension D was

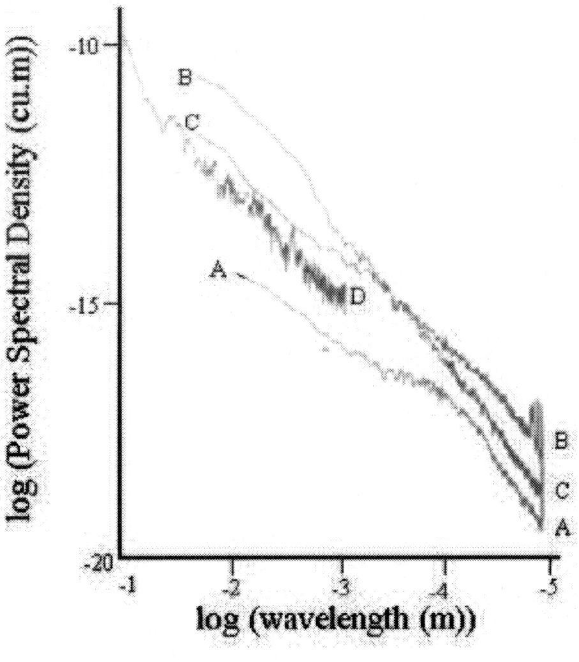

Figure 1.3: Power spectra for the fault surfaces studied in the Dixie Valley by [12]. A, B, C are from 10 – 20 mm long lab profiles. D is part of a spectrum from a 1 m long field profile. A – smoothest, unweathered hand sample of surface. B – sugary weathered surface. C – surface that apparently a is composite of sub-parallel surfaces (Adapted from [12]).

found to vary with spatial frequency. [12] did a similar analysis on the surface of faults in the western United States and found fault surfaces to be fractal over eleven orders of magnitude in wavelength. They found that the amplitude of the spectrum increased roughly in proportion to the wavelength under consideration. The power spectra for the fault surfaces in Dixie Valley (western United States) are shown in Figure 1.3 as reported in [12]. Such studies have been strengthened by modern techniques of imaging like the LiDAR profile shown in Figure 1.2. The topography of fault surfaces is now generally considered as fractal. So it is very reasonable to consider the movement of fault surfaces on and relative to one another as two fractals sliding over one another. This forms the basis of our motivation behind studying the overlap statistics of a Cantor set sliding over its replica.

2. Two Fractal Overlap model

2.1. The model

As discussed already in Section 1, earthquakes are physically caused by the slip movements of adjacent fault planes along the contact of hanging wall and footwall asperities and the release of the stress energy accumulated due to friction during the period of sticking. But as (i) faults surfaces are fractals (ii) friction is purely a surface phenomenon and (iii) the motion is, in general, in a given direction, the process that causes release of the stored elastic energy can be analyzed effectively in one dimension. Therefore a fractal embedded in one dimension can provide us a suitable geometry to investigate the overall process. The sliding of one fractal over another thus would mimic a stick-slip scenario where the slip occurring after a stick would effectively be the physical process through which the stored energy would be released. The one dimensionality of the problem previously discussed means that we have to consider a fractal embedded in 1-D and the natural choice is a Cantor set. This is especially valid due to the fact that the projection of any fractal surface in a 1-D space is clearly a Cantor set (albeit a random one in most cases we encounter in nature). But for the sake of analytical tractability of the process we adopt, in this case, the middle third removal algorithm to generate it (fractal dimension is $\frac{\log 2}{\log 3}$). The dynamical model involves one such Cantor set moving with uniform relative velocity over its replica and one looks for the time variations of the measure of the overlapping sets common between the two at any instant of time. The model was initially given by [14]. They tackled the problem of determining the overlap statistics using a renormalization group method. We, however, present a modified analysis following [15]. This model has been extended to Sierpinski gaskets and carpets and random Cantor sets as well [16]. The model considered here however, as we said earlier, employs two regular Cantor sets of the same generation and dimension sliding over each other with uniform velocity as shown in Figure 2.1. For the nth generation, the step size is $1/3^n$ and the time taken to cover each step is taken as unity. Stress energy is accumulated at each overlap of the non-empty intervals of the upper (moving) Cantor set with the non-empty intervals of the lower (stationery) Cantor set. The extent of such overlaps (the number of such overlapping non empty intervals) is represented by the 'overlap magnitude'. This measure may represent the stress (or stress energy) accumulated due to friction within the surfaces which gets released through slips. The energy released at each such 'slip' is proportional to the overlap magnitude during the 'stick' period. We therefore need to evaluate the overlap time series. At any finite generation, the time series is exactly solvable in this model.

2.2. *Analysis of the time series*

As mentioned already, we present here a modified version of the analysis of the Chakrabarti Stinchcombe model (or Two Fractal Overlap model) following [15]. We employ periodic boundary conditions to formulate the time series. The overlap magnitude is evaluated in terms of the number of pairs of non-empty intervals overlapping at a time. Therefore the overlap magnitude $Y_n(t)$ can only assume values in a geometric progression given by $Y_n(t) = 2^{n-k}, k = 0, 1, ..., n$. Clearly $Y_n(0) = 2^n$ and, due to the periodic boundary conditions, taking unit time to be the time required to take one step of size $1/3^n$ we obtain

$$Y_n(t) = Y_n(3^n - t), 0 \leq t \leq 3^n \tag{2.1}$$

owing to the symmetric structure of the finite generation Cantor set.

A detailed analysis of the time series reveals a straightforward recursive structure. If we simulate the overlap time series for the nth generation, after 3^{n-1} time steps we have the overlap time series for the $(n-1)^{\text{th}}$ generation. Again after 3^{n-2} time steps beginning from the 3^{n-1} time steps previously taken we have the overlap time series for the $(n-2)^{\text{th}}$ generation and recursively so on. In other words the entire time series for the 1st generation ($n = 1$) is contained in the time series for the 2nd generation ($n = 2$) starting from the time step $t = 3$ of the 2nd generation time series) and ending at the time step $t = 6$ (of the 2nd generation time series). Again the entire 2nd generation time series is contained in the 3rd generation time series starting from the time step $t = 9$ (of the 3rd generation time series) and ending at the time step $t = 18$ (of the 3rd generation time series). Also the entire 1st ($n = 1$) generation time series is contained in the 3rd generation ($n = 3$) time series starting from the time step $t = 12$ (of the 3rd generation time series) and ending at the time step $t = 15$ (of the 3rd generation time series). This nested recursive structure is present throughout the time series of any n^{th} generation. Generalizing, we may state that the entire time series of the $(n-1)^{\text{th}}$ generation is contained in the time

Journal of Physics: Conference Series **319** (2011) 012004 doi:10.1088/1742-6596/319/1/012004

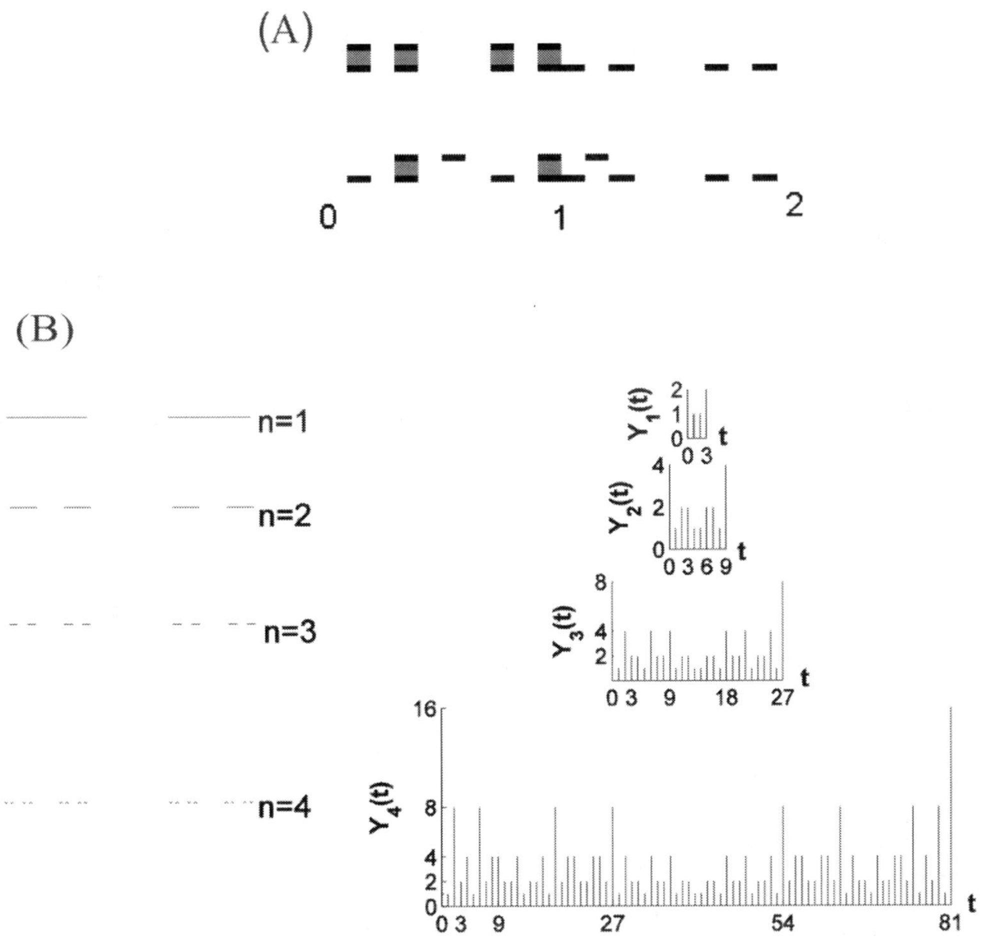

Figure 2.1: A) The recursive structure of the time series for the first four generations in the fractal-fractal overlap model. On the left the respective Cantor set generations are shown. Noticeable is the fact that the time series of all preceding generations are embedded within the time series at a given generation. B) A realization of the model for the second generation at t=0 and at t=2. The overlapping segments are shaded in grey. The lower Cantor set is repeated between 1 and 2 to employ the periodic boundary condition. The upper Cantor set slides over the lower (cf. [17]).

series of the nth generation starting from the time step $t = 3^{n-1}$ (of the nth generation time series) and ending at the time step $t = 2 \times 3^{n-1}$ (of the nth generation time series). Again, the entire time series of the $(n-2)^{\mathrm{th}}$ generation is contained in the time series of the nth generation starting from the time step $t = 4 \times 3^{n-2}$ (of the nth generation time series) and ending at the time step $t = 5 \times 3^{n-2}$ (of the nth generation time series). Again, the entire time series of the $(n-3)^{\mathrm{th}}$ generation is contained in the time series of the nth generation starting from the time step $t = 13 \times 3^{n-3}$ and ending at the time step $t = 14 \times 3^{n-3}$ and so on. This can be understood very clearly from the illustrations in Figure 2.1.

There is however a finer recursive structure in the time series that leads to the analytical evolution of the number density distribution. At any given generation n, a pair of nearest line segments form a doublet and there are 2^{n-1} such doublets in the Cantor set. Within a given doublet, each segment is two time steps away from the other segment. This means that an overlap of 2^{n-1} occurs when one of the sets is moved two time steps relative to the other. Similarly, an overlap of magnitude 2^{n-1} also occurs if one considers a quartet and a relative shift of 2×3 time steps between the two Cantor sets. Again we can consider an octet and a relative shift of 2×3^2 time steps to obtain an overlap of magnitude 2^{n-1}. In general if we consider pairs of blocks of 2^{r_1} nearest segments ($r_1 \leq n-1$), an overlap magnitude of 2^{n-1} occurs for a relative time shift of 2×3^{r_1} time steps:

$$Y_n(t = 2 \times 3^{r_1}) = 2^{n-1}; r_1 = 0, \ldots\ldots, n-1. \tag{2.2}$$

The complementary sequence is obtained using (2.1). We can create such rules for each of the possible overlap magnitude values $Y_n(t) = 2^{n-k}$. Rules like these give us the frequency distribution of overlap magnitudes. For example, from (2.2) we can see that as r_1 can have n possible values. Also, for each of these times at which an overlap of magnitude 2^{n-1} occurs we have another time step in the complementary sequence (due to (2.1)) at which again an overlap of magnitude 2^{n-1} occurs. Therefore the frequency of occurrence $Fr(Y_n)$ of an overlap magnitude $Y_n = 2^{n-1}$ is $2n$, that is $Fr(Y_n = 2^{n-1}) = 2n$. The complete distribution can be obtained by studying the aforementioned recursive structure carefully (the mathematical details can be found in [17]) and using simple combinatorics. The probability distribution of overlap magnitudes for the model comes out to be a binomial distribution:

$$Pr(2^{n-k}) = \begin{pmatrix} n \\ n-k \end{pmatrix} \left(\frac{1}{3}\right)^{n-k} \left(\frac{2}{3}\right)^{k} \tag{2.3}$$

where $Pr(Y_n) = \frac{Fr(Y_n)}{3^n}$, that is $Pr(Y_n)$ gives the probability of occurrence of an overlap of magnitude Y_n in a total of 3^n time steps. Now, remembering that the overlap magnitude 2^{n-k} is proportional to energy we can put $\log_2 Y_n = n - k = m$ where m is the magnitude analog for the model. It must however be kept in mind that while analyzing the model n is a constant as we are considering the model at a specific generation number and m changes as k changes. Then the frequency distribution for the model in terms of magnitude becomes

$$Pr(m) = \begin{pmatrix} n \\ m \end{pmatrix} \left(\frac{1}{3}\right)^{m} \left(\frac{2}{3}\right)^{n-m}. \tag{2.4}$$

2.3. The Gutenberg Richter law

In the limit of large n the Cantor set becomes a true mathematical fractal and we have the standard normal approximation of (2.4) which gives:

$$F(m) = \frac{3}{2\sqrt{n\pi}} \exp\left(-\frac{9}{4}\frac{(m-n/3)^2}{n}\right), \tag{2.5}$$

where $F(m)$ is the probability distribution of magnitudes $m = \log_2(Y_n)$. The objective is to show that the $\int_m^\infty F(m)dm$ gives the Gutenberg Richter (GR) law. We shall attempt to evaluate the following integral using (2.5):

$$F_{cum}(m) = \int_m^\infty F(m')dm' = \int_m^\infty \frac{3}{2\sqrt{n\pi}} \exp\left(-\frac{9}{4}\frac{(m'-n/3)^2}{n}\right) dm'. \tag{2.6}$$

This is not a standard Gaussian integral and we evaluate this integral by parts. Making the substitution $p = \frac{3(m'-n/3)}{\sqrt{2n}}$ in (2.6) we get:

$$F_{cum}(m) = \frac{3}{4\sqrt{n}} \int_{\frac{3(m-n/3)}{\sqrt{2n}}}^{\infty} \frac{2}{\sqrt{\pi}} \exp\left(\frac{-p^2}{2}\right) dp. \tag{2.7}$$

Invoking the generalized definition for the complementary error function, erfc(x), we obtain,

$$F_{cum}(m) = \frac{3}{4\sqrt{n}} \text{erfc}\left(\frac{3(m-n/3)}{2\sqrt{n}}\right). \tag{2.8}$$

For large values of the argument x, erfc(x) has the asymptotic expansion (see 7.1.23 in [18])

$$\text{erfc}(x) = \frac{\exp(-x^2)}{\sqrt{\pi}x}\left(1 + \sum_{m=1}^{\infty}(-1)^m \frac{1.3...(2m-1)}{(2x^2)^m}\right). \tag{2.9}$$

For very large x, (2.9) gives

$$\text{erfc}(x) = \frac{\exp(-x^2)}{\sqrt{\pi}x}. \tag{2.10}$$

Therefore in the limit, $m \to \infty$ we have

$$F_{cum}(m) = \frac{3}{4\sqrt{n}} \frac{\exp\left(-\frac{9}{4}\frac{(m-n/3)^2}{n}\right)}{\sqrt{\pi}\left(\frac{3(m-n/3)}{2\sqrt{n}}\right)}. \tag{2.11}$$

Equation (2.11) can be further simplified as

$$F_{cum}(m) = \frac{1}{2\sqrt{\pi}} \exp(-n/4) \exp\left(\frac{-9m^2}{4n} + \frac{3m}{2}\right)(m-n/3)^{-1}. \tag{2.12}$$

In the model, the largest value of m possible is n, therefore in the limit of large m we can assert $m^2/n \approx m$. Using the above alongwith the fact that the total number of events for a Cantor set of generation n is 3^n and $\delta m = 1$ in the model, it is easy to see that

$$\log N(m) = A - \frac{3}{4}m - \log(m - n/3), \tag{2.13}$$

where $N(m)$ is the number of earthquakes greater than a given magnitude m for the model. The term A denotes all the constants and the purely generation (n) dependent terms taken together.

Now we shall show that equation (2.13) depicts all the features of the cumulative count distribution of the model observed in simulations and seen in Figure 2.2. The third term on the right hand side shows that this form of the distribution holds only for $m \geq n/3$. For larger values of m, the linear term dominates. For smaller values of m the cumulative count becomes smaller than what is expected purely on the account of the linear term due the increasing effect of the logarithmic term. But the turning is smeared out gradually over a range of magnitudes due to the logarithmic term (as seen in the figure). The model also predicts a small bump in the cumulative statistics when the argument of the logarithmic term becomes smaller than one (due to negative values of the logarithmic term adding to the linear term). A little bit of deliberation

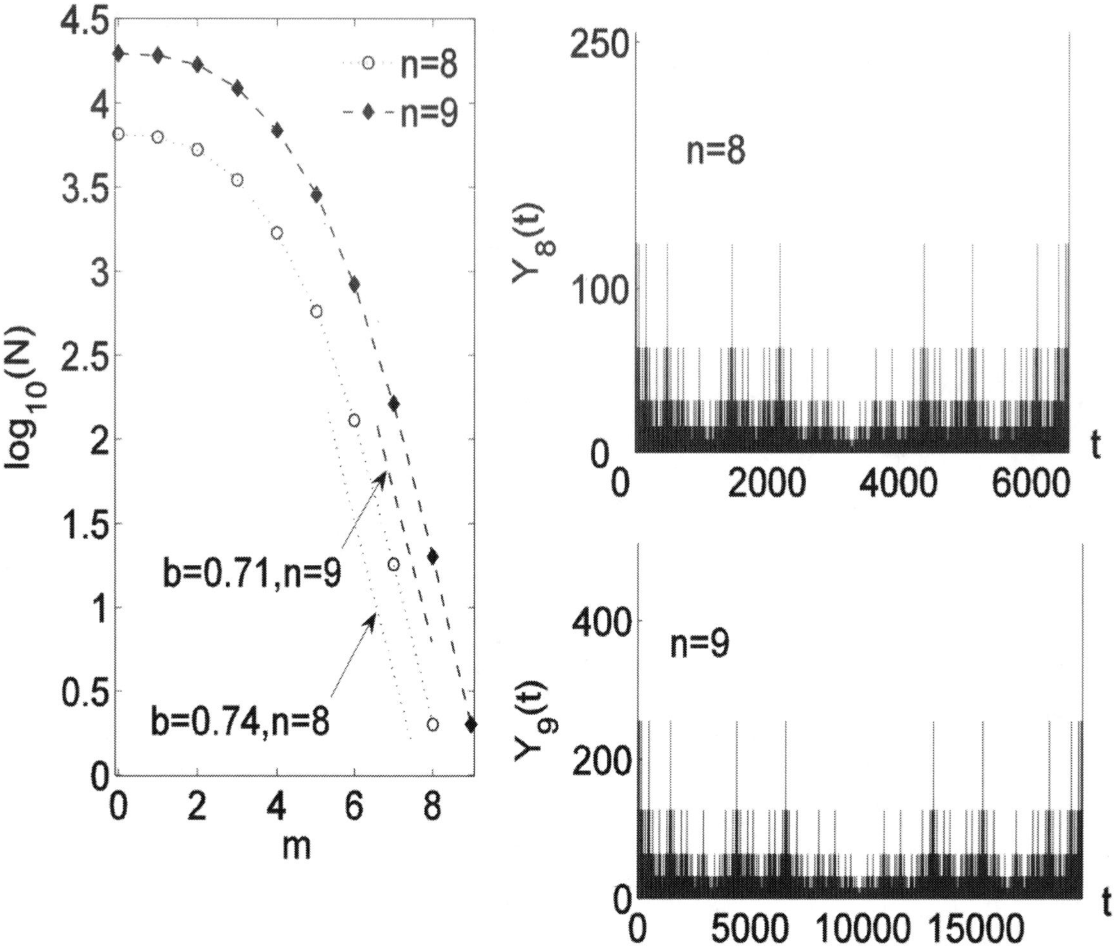

Figure 2.2: The Frequency-Magnitude (FM) plots ($\log N(m)$ vs. magnitude m, N being the number of earthquakes with magnitude greater than or equal to magnitude m) for the model for generations 8 and 9. The overlap time series are also shown on the right for the respective generations. The low-magnitude roll-off is evident for both the generations. The lines are drawn as visual aid to understand the linear trend. The b-values were obtained by fitting a linear polynomial to the data. The values of the exponents are the slopes of the indicated straight lines. The b-values thus obtained are also indicated for each of the generations (cf. [17]).

on this issue tells us that the requirement of integral magnitudes in the model does not let this condition persist for more than one magnitude value for a given value of n and even then the effect is negligible as seen from simulations. This gives the synthetic seismic sequence a smooth roll off towards the lower end of the magnitude scale. This is the precise form of the GR law as observed in earthquake seismology.

Now the so-called b-value from our theoretical distribution is $3/4$ and not unity as generally reported for natural seismicity. The value $3/4$ arises out of the fact that we have constructed our Cantor set by the middle third removal procedure. In fact for a Cantor set with dimension $\frac{\log(q-1)}{\log(q)}$ the exponent would be $\frac{q}{q+1}$. This of course would be effectively unity for higher dimensional Cantor sets. Now this means that there will be region to region variation in the b-value and this is important as most other theoretical models [4, 5, 7, 19] give universal values for the exponent. In practice the b-value has shown some variability from unity. The b-value generally varies from 0.5 to 1.5 depending on the tectonic setting, tectonic stress and the magnitude ranges but normally comes close to 1 for seismically active regions [20–23]. The GR law holds good for aftershock sequences also which is really what our model describes [24, 25]. In our model however the range is from slightly smaller than 0.75 to 1 (the lower bound on the exponent is smaller than 0.75 in practice as m^2/n is slightly smaller than m in reality).

Further, the constant A in (2.13) is dependent on the generation number n and the value of n determines, for a given similarity dimension, the seismicity in our model i.e. the number of earthquakes increases with increasing n. Mathematically, A is equivalent to the constant a in the GR law. It is notable that in the GR law too a characterizes seismic activity. So A is a reasonable proxy for the a value in GR law. Figure 2.2 shows the GR law plot from the model for generations 8 and 9. The values obtained for the exponent are also indicated in the plot. The values obtained by fitting ($b = 0.74$ for $n = 8$ and $b = 0.71$ for $n = 9$) support our analysis presented above. The low magnitude roll-offs are also quite conspicuous for both $n = 8$ and 9. Comparison with the frequency-magnitude plot for Sumatra shown in later in Figure 3.1 clearly brings home the similarities between our theoretical distribution and the form observed in nature.

2.4. The Omori Law

Previously a theoretical study derived the Omori formula from a preliminary statistical model where aftershocks are produced by a random walk on a pre-existing fracture system [26]. The derived result shows a direct connection between p and the fractal dimension of the pre-existing fracture system. This study showed that the fractal properties of aftershocks are determined by the fractal geometry of the pre-existing fracture system. The Omori law comes out naturally from our fractal overlap statistics as well.

Physically, our model corresponds to an aftershock sequence for a mainshock of magnitude n. So it is of inherent interest to check for the Omori Law in our model by studying the temporal distribution of these synthetic aftershocks. The time series of overlap magnitudes in our model has built-in power law behavior. The entire magnitude-time sequence is a nested structure of geometric progressions as pointed out earlier. This makes it difficult to enumerate an exact value of the exponent p in general. But there is, however, departure from this in two limiting cases. Omori Law in practice gives specific value of p for a given magnitude threshold. We observe that for any generation, when the threshold is the minimum overlap magnitude 1 in our model, the p value is 0. This is because by the virtue of the assumption of uniform velocity there is an aftershock at every time step. A very interesting fact is however unearthed on putting the magnitude threshold at the second highest possible value $n - 1$ (that means we are considering aftershocks only of magnitude $n - 1$ and higher). Now the times of occurrences of aftershocks of magnitude $n - 1$ are at each value of $t = 2 \times 3^{r_1}$ where r_1 varies from 0 to $n - 1$. Therefore when the lower magnitude threshold is $n - 1$, we have, not considering the constant prefactor 2, consecutive aftershocks occurring at times which follow this geometric progression (2.3) with common ratio 3 (that is if at any t there is an aftershock of magnitude $n - 1$ then at $3t$ the next aftershock will occur and at 3^2t the one after that and so on). This gives the general rule $N(3t) = N(t) + 1$ leading to:

$$N(t) = \log_3 t \qquad (2.14)$$

where $N(t)$ is the cumulative number of aftershocks (of magnitude $m \geq n - 1$ for a mainshock of magnitude $m = n$). To observe this, one has to remember that the symmetry in the time series is an artificially introduced artefact of the periodic boundary conditions and in the above

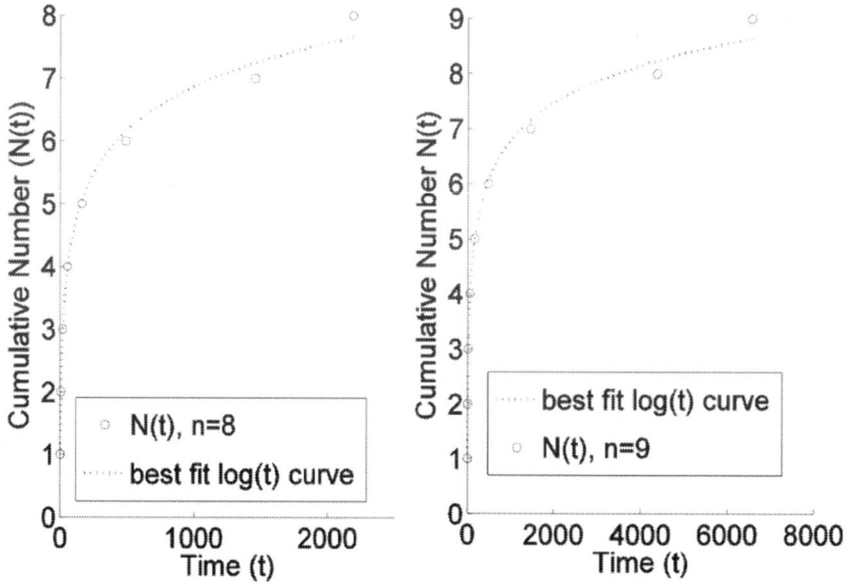

Figure 2.3: Omori Law from the model for generations 8 and 9 respectively. Dashed lines show best logarithmic fits. Plots are for $N(t)$ vs. t, $N(t)$ being cumulative number of aftershocks at time t where t is the time since the mainshock. Time parameter for the model being as defined in the text i.e. unit time for a step of size 3^{-n} (cf. [17]).

analysis we only consider the first half of the times series (e.g. see Figure 2.1). Integration of the Omori relation gives, $N(t) = 1/t^p$.

From (2.14) this gives us $p \sim 1$ which is the traditional Omori exponent value. The model therefore gives a range of p-values from 0 to 1 which systematically increases within the range with increasing threshold. Figure 2.3 shows the plots (from the model) for cumulative number of aftershocks $N(t)$ of magnitude greater than equal to $n-1$ versus t for $n = 8$ and $n = 9$.

The fact that the Omori exponent p is not universally unity is a very well documented fact and some workers have reported variability in p from 0.5 to 2.5 [27]. But for seismically active zones p is generally larger than but close to unity. This variability is present in our model too. The variability in p in our model apparently stems from implementing different magnitude thresholds.

But, there is a more subtle feature of this analogy with the real world. The magnitude threshold for Omori Law calculations is always put above the completeness magnitude M_c. Completeness magnitude is that magnitude below which the frequency-magnitude statistic rolls-off from the GR like power law. It is generally believed that the number of earthquakes is not exhaustively recorded below this magnitude and this is the reason for the roll-off [28, 29]. In other words the complete record of earthquakes below this magnitude is not available in the sense that the frequency level below this magnitude is less than what really should be according to the GR law. For real earthquakes Omori Law exponents are calculated only in the power law region of the magnitude scale. In our model such a roll off occurs naturally. The roll-off occurs at approximately below a magnitude $n/3$ as discussed in Section 2.3. A meaningful comparison with the Omori statistics for real data sets can be done only for the power law region and that means our threshold can be no smaller than $n/3$. This implies that the p-value can never be observed to be zero. And for a higher generation or a higher dimension fractal, at the same magnitude cut-off, the p exponent will be higher than a lower dimension or lower generation fractal. Values of p closer to unity will be seen as we take up Cantor sets of progressively higher dimensions and/or generations at the magnitude cut-off $n/3$. The higher the generation and/or dimension of the fractal we consider the higher will be the mainshock magnitude and more number of aftershocks will be observed in the model. Thus the seismic activity will increase. At the same time the exponent p will yield values closer to unity even at magnitude cut-offs lower than $m = n - 1$. Thus for seismically active zones p-values will be closer to unity. One might be tempted to extend this analogy to real seismicity and conclude that the p-value is close to 1 as observed in seismically active regions. This would be far-fetched as the p-value is generally observed to be larger than and nearly equal to unity for seismically active regions. Our model approaches 1 from the left hand side on the real number line and hence the analogy is not very direct in terms of the values of the exponent. But the analogy of the form of the temporal decay is direct and unmistakeable and shows that the Omori Law behavior might also be linked to the fractal geometry of the fault.

2.5. Temporal distribution of magnitudes of an aftershock sequence

There is another very important observation that comes out from the model. If we evaluate the time cumulant of magnitude, i.e. $\int_0^t m(t')dt'$ where t is the time since the mainshock and $m(t)$ is the magnitude at t, it comes out to be a remarkable straight line. In other words:

$$Q(t) = \int_0^t m(t')dt' = St \tag{2.15}$$

where S is the slope of the straight line. This temporal distribution of the $Q(t)$ statistic is very significant. The slope S is a function of both the generation number as well as the dimensionality of the Cantor set. It is however quite difficult to enumerate the slope exactly due to the presence of the nested geometric progressions in the time series as stated earlier but an approximate estimate of the slope is given by

$$S_n^q = \left(\frac{q-1}{q}\right)\frac{n}{2} \tag{2.16}$$

for the model where the Cantor set has been formed by removing the middle block from q blocks and the generation number is n. Now the important fact coming out of (2.16) is the dependence of S on both the dimension and the generation number of the model. The model predicts that the slope S for real aftershock sequences would be fault dependent as we expect the generation number and/or the dimension of the fractals involved to vary from fault to fault. Thus in a sense, the slope S is a kind of a 'fractal fingerprint' of the fault zone. The slope is a very characteristic local feature of the aftershock sequence and hence of particular interest as a diagnostic feature of aftershock sequences. In effect this provides us a new approach in analyzing the temporal behavior of aftershock sequences from which we can, at least from the model, clearly extract information about the fault geometry. The precise meaning of the slope comes out to be something a bit different when we put the value of q into (2.16) for the middle third removed Cantor set we have used here. Here $q = 3$ giving the slope to be equal to approximately $n/3$.

A look back at (2.5) immediately brings home the meaning of the slope. It is simply the average magnitude of the Gaussian distribution of the aftershock magnitudes generated by the model. The significance of this will become even more clear when we look for this feature of aftershock statistics in real earthquakes. Let us suspend this discussion for then and look at Figure 2.4 for such $Q(t)$ vs t plots for the model for $n = 7, 8, 9$ respectively. From the figure one can clearly see the increase in slope with successive increases in generation number n. The increase in generation number is something that we might expect in an active seismic zone. This can take place due to re-rupturing of an existing rupture zone. Such re-rupturing has been

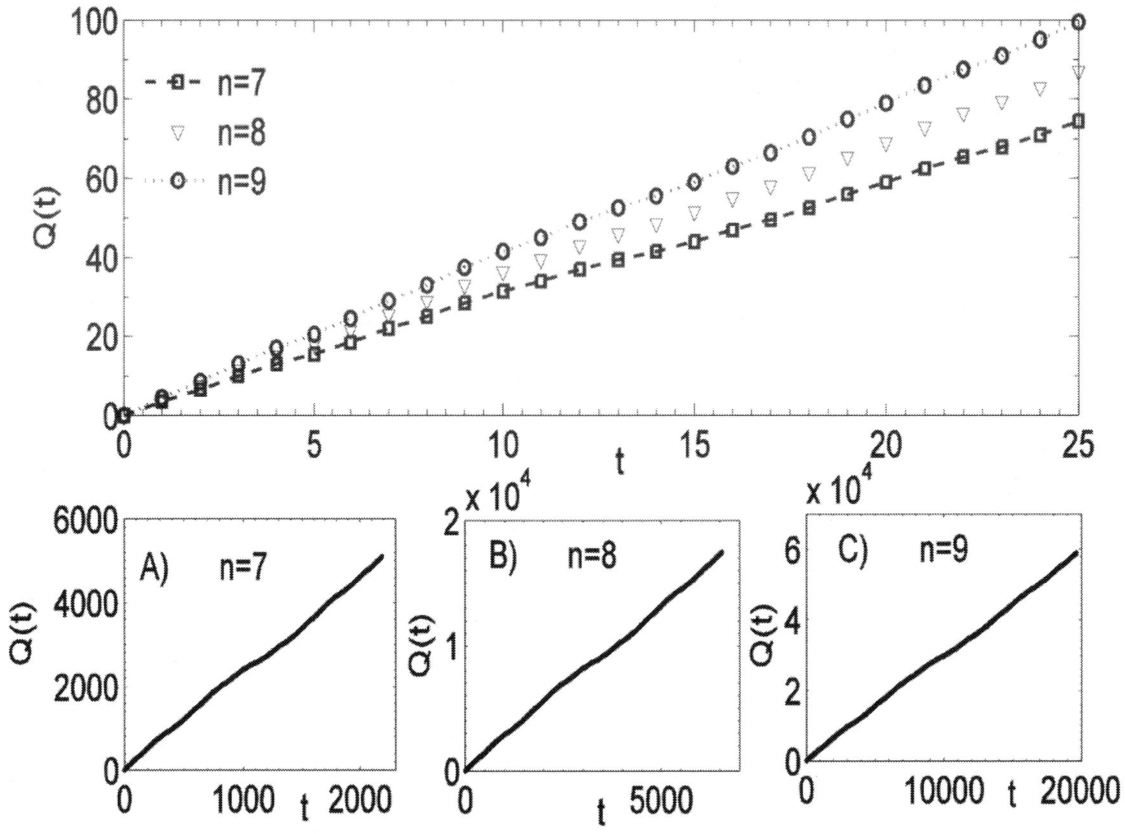

Figure 2.4: The $Q(t)$ vs. t statistic for the model for generations 7, 8 and 9. At the top plots for all the three generations (for the first 25 time steps) are shown together to show the increase in slope with increase in generation number n. At the bottom plots A), B) and C) show the entire $Q(t)$ time series for generations 7, 8 and 9 respectively (cf. [17]).

reported very often and happens when an earthquake occurs at or near the hypocenter of a previous large earthquake (hypocenter is the assumed point from which seismic waves emanate) years afterwards. We discuss such an event and the resultant $Q(t)$ vs. t plot in Section 3.3.

3. Comparison with observations

3.1. The Gutenberg Richter law

In Figure 3.1 we have considered the frequency-magnitude distribution for two real aftershock sequences to compare our theoretical formulation (see Section 2.3) with real earthquake data. The data sets considered were 1) The 2004 Sumatra earthquake aftershock sequence (26/12/2004, $M_w = 9.0$, Epicenter latitude: 3.30°, Epicenter longitude: 95.98°, source catalog: NEIC (PDE) catalog (`http://neic.usgs.gov/neis/epic/`) and 2) The 1995 Kobe earthquake aftershock sequence (17/01/1995, $M_{JMA} = 7.2$, Epicenter latitude: 34.6°, Epicenter longitude: 135.0°, source catalog: JUNEC catalog (`http://www.eic.eri.u-tokyo.ac.jp/CATALOG/junec/monthly.html`). Aftershocks of a major event were considered to be events within a given region, geographically defined as boxes or polygons constrained by suitable latitudes and longitudes, and the magnitudes were recorded over a length of time (of the order of a year or more) over which the region has not yet relaxed to its background seismicity (tentatively within the first 1000 days). Now one point needs to be made clear with respect to the Sumatra dataset. The dataset was inhomogeneous in the sense that it reported earthquake magnitudes in different magnitude units. So we had to convert all the magnitudes reported to one uniform magnitude scale using inter-magnitude conversion relationships. We chose the uniform magnitude scale for our work to be the moment magnitude M_w as defined in [30]. For the Sumatra event we used the conversion relationships used in [31]. These relationships were specifically designed for the aftershock sequence of the Sumatra event extracted from the PDE catalog and hence serve our purpose. The fact that the conversion relationships were designed for nearly the same dataset as we have used here is important as such conversion models are in general regressional models and hence their use in our work is validated by the fact that here we use them on the same population for which they were originally designed. But errors in magnitude reporting as well as those induced due to magnitude conversions can severely affect the estimation of the GR law exponent. These errors have been discussed in a bit more detail in Section 3.3.

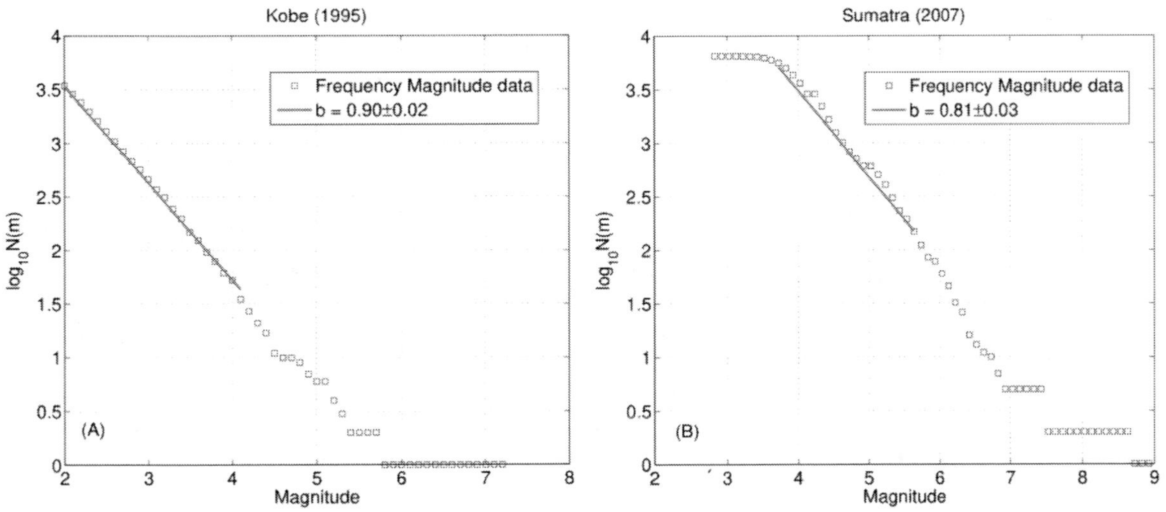

Figure 3.1: GR or Frequency-Magnitude distributions for the aftershock sequences described in the text A) the 1995 Kobe earthquake and B) the 2004 Sumatra earthquake. We clearly see the lower magnitude roll-off from the power law distribution in Sumatra. $N(m)$ represents number of earthquakes with magnitude greater than or equal to m, m represents magnitude.

As we remarked earlier, the frequency-magnitude plot for the Sumatra aftershock sequence clearly shows the roll off from GR statistics at the low magnitude end. This is similar visually to the roll off observed in our model (see Section 2.3 and Figure 2.2). But the Kobe sequence does not show any such clear roll-off. According to conventional wisdom in statistical seismology, this means that the Kobe sequence is probably complete upto $m = 2.0$. According to the model however, this means that the instrumentation deployed by the network only records events larger than the physical roll-off threshold. As the JUNEC catalog is well known for its accuracy, one might assume that this means that the threshold itself is low or the fault is of small generation number. The roll-off phenomenon will be observed again when we present the GR plots for the various aftershock sequences used in Section 3.3.

3.2. The Omori law

Our model shows that the Omori exponent p (see equation (1.3)) increases with increase in the lower magnitude threshold. We tried to check for this trend of increase in p with increase in lower magnitude threshold for three real aftershock data sets. The aftershock sequences chosen were 1) 1989 Loma Prieta earthquake aftershock sequence (18/10/1989, M_w= 7.1, Epicenter latitude: 37.0°, Epicenter longitude: -121.88°, source catalog: same dataset as [32]); 2) 1999 Chamoli earthquake aftershock sequence (29/03/1999, M = 6.6, Epicenter latitude: 30.51°, Epicenter longitude: 79.40°, source catalog: Wadia Institute of Himalayan Geology catalog (same dataset was used in [33]) and 3) 2004 Sumatra earthquake aftershock sequence described before. The results are given as $\log n(t)$ vs. $\log t$ plots in Figure 3.2 where the cut-off thresholds are denoted as M_c and the p values are indicated. Here $n(t)$ denotes number of aftershocks per unit time and t denotes time since the mainshock in days. As is evident from Figure 3.2, the increase in p with increase in M_c is clearly seen in Chamoli and in Sumatra. However in Loma Prieta, which is a very well characterized data set, the same trend is not seen. The reason for widely different values of M_c for the three data sets is that the completeness level (as explained earlier meaningful analysis can only be done above the completeness magnitude) for the three catalogs are very much different mainly due to the nature of the seismic networks implemented.

Most statistical seismologists now agree that the Omori law is too simplistic a model to describe the true nature of the temporal decay of aftershock sequences. It is now standard to describe the decay of aftershock rates by the more complicated modified Omori law [3]

$$\frac{dN}{dt} = \frac{1}{\tau[1+t/c]^p}, m \geq M_c \tag{3.1}$$

where t is time elapsed since the mainshock, M_c is a lower magnitude cutoff above which earthquakes are taken into account, τ and c are characteristic times, and p is an exponent specifying how fast the sequence is decaying in time. The mechanism behind the time dependence in Omori law has been debated. Elastic effects work on short time scales and cannot explain the gradual decay of aftershock sequences over very long times. Therefore, the value p is thought to reflect the mechanical conditions of Earth's crust. There have been reports of systematic regional variations of the p-value in Japan which could be attributed to regional variation of surface heat-flow values [27, 34]. The central argument behind this hypothesis is that aftershock activity decays faster as the stress relaxes faster in regions of higher crustal temperature. Many other mechanisms have been proposed for this temporal decay e.g. subcritical crack growth [35], visco-elastic relaxation [36], post-seismic creep due to stress corrosion in the regions of stress concentration after the mainshock [37], static fatigue [38], pore fluid flow [39], post-seismic slip [40] and earthquake nucleation under rate- and state-variable friction [41]. In general, it is assumed that the parameters c and τ are constants and are specific to a given aftershock sequence. The significance of the parameter c has been somewhat debated. It may reflect the poor detection level during the early post-seismic period when coda waves from the mainshock make it difficult to identify aftershocks or large magnitude event swarms mask smaller events [38, 42]. It may have more physical connotations as well; e.g. due to finite duration of earthquake nucleation time [35, 41] or, for post-mainshock seismicity driven by afterslip, due to the response of aseismically creeping zones to the co-seismic stress change [43]. It recently has been suggested that τ and c may be considered functions of the lower magnitude cutoff M_c, and thus may be written as $\tau(M_c)$ and $c(M_c)$ [44]. In view of this discussion, it must be recognized that our model does not capture the modified Omori Law. Hence the analogy of the behavior of the p-value in the model with that in nature is very approximate at best in the physical sense and is only valid if one accepts the Omori Law decay of aftershocks rather than the modified Omori Law decay as the mathematical description of the temporal decay of aftershock rate.

Journal of Physics: Conference Series **319** (2011) 012004 doi:10.1088/1742-6596/319/1/012004

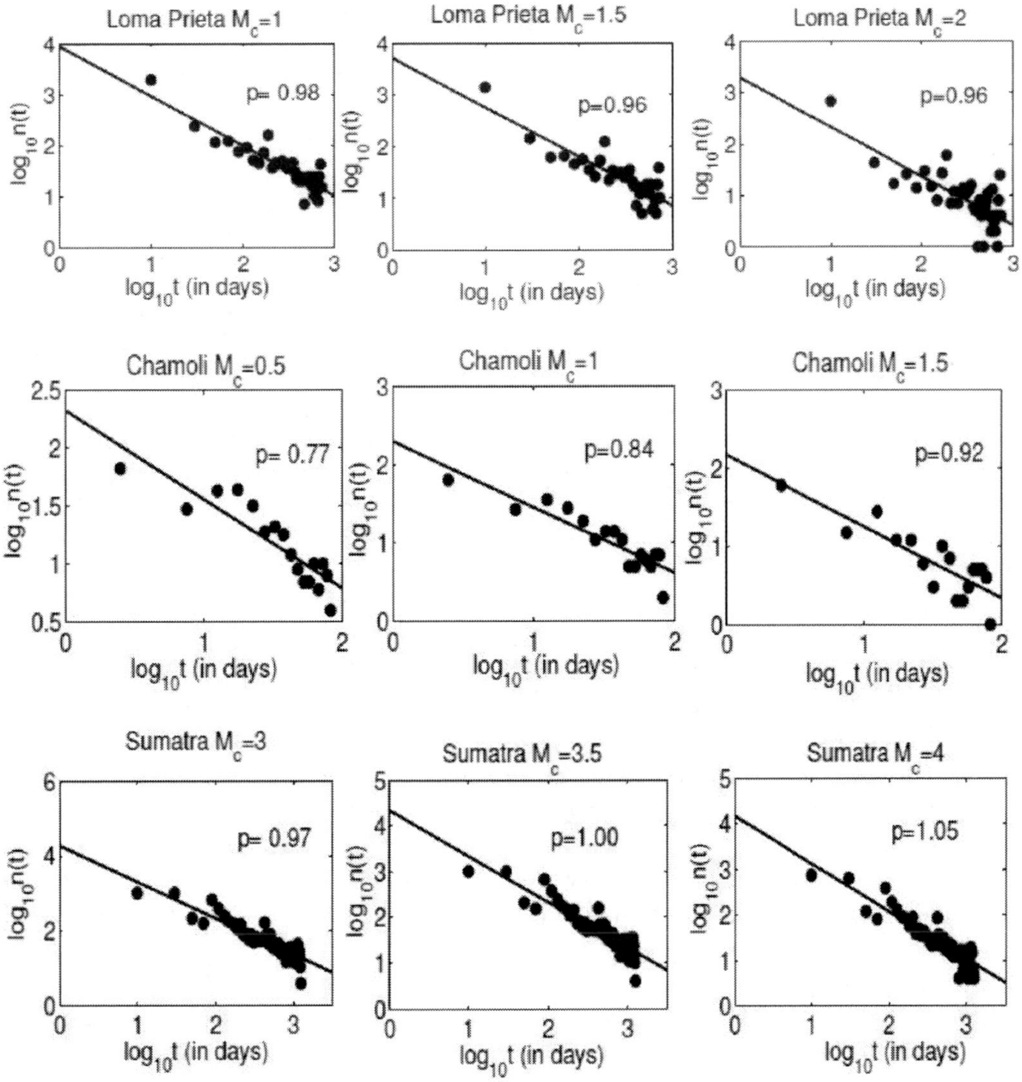

Figure 3.2: The plots for number of aftershocks per unit time $n(t)$ vs. time since the mainshock in days, t, for the Loma Prieta, Chamoli and Sumatra data sets. The cut-off magnitudes M_c are indicated in each plot title. The corresponding p values are shown within the plots. The solid lines give the linear fits to the data with slope p (cf. [17]).

3.3. The temporal distribution of aftershock magnitudes

To establish the observations obtained for the $Q(t)$ statistic. We first collected the aftershock magnitude-time sequences $m(t)$ of eleven major earthquakes from different geographical regions of the world. The earthquakes were selected carefully from all over the globe to ensure that no regional bias was introduced due to the choice of a specific catalog or a specific geological setting. We also intentionally selected some multiple events in the same geological region on a) different fault zones b) the same fault zone at a different time. We then evaluated a cumulative integral $Q(t)$ of the aftershock magnitudes over time. Numerically, we evaluated the integral $Q(t) = \int_0^t m(t')dt'$, the time cumulant of magnitude where t denotes the time since the main shock. The integration is done over all event magnitudes above a threshold level, in this case chosen to be the completeness magnitude M_c. The datasets we used for our analyses are as follows (some of the datasets used here have already been cited in Sections 3.1 and 3.2 and used for calculations and we restate them again for the sake of completeness):

a) The 1989 Loma Prieta earthquake (18/10/1989, M_w = 7.1, 37.0°, - 121.88°). The data set used was the same as the one used in Section 3.2.

b) The 1995 Kobe earthquake (17/01/1995, M_{JMA} = 7.2, 34.6°, 135.0°). The aftershock region was chosen on the basis of the work of [45] (latitudes 34°-36°, longitudes 133.5°-137°). The data were taken from the JUNEC catalog (`http://www.eic.eri.u-tokyo.ac.jp/CATALOG/junec/monthly.html`) for the period 17/01/1995-31/12/1995.

c) The 2004 Sumatra earthquake (26/12/2004, M_w= 9.0, 3.30°, 95.98°). The box was chosen in accordance to the earthquake summary poster prepared by the USGS (available at `http://earthquake.usgs.gov/eqcenter/eqarchives/poster`). The box chosen was latitudes 0°-20°, longitudes 90°-100°. The data were taken from the NEIC (PDE) catalog (`http://neic.usgs.gov/neis/epic/`) for the period 26/12/2004-28/05/2008.

d) The 2005 Muzaffarabad (Kashmir, North India) earthquake (08/10/2005, M_s= 7.7, 34.52°, 73.58°). The box chosen is defined by latitudes 33.5°-35.5° and longitudes 72.2°-74.2°. The data were once again from the NEIC (PDE) catalog (`http://neic.usgs.gov/neis/epic/`) for the period 08/10/2005-28/02/2008.

e) The 1999 Chamoli earthquake (29/03/1999, M_s = 6.6, 30.51°, 79.40°), the aftershocks were obtained from a highly localized network employed by the Wadia Institute of Himalayan Geology (same dataset as used in [33]).

f) The 2003 Bam earthquake (26/12/2003, M_s = 6.8, 29.00°, 58.31°).The box was chosen to be latitudes 27.5°-30.5° and longitudes 57.5°-59.5°, the time interval being from 26/12/2003-26/12/2005 on the IIEES listing (`http://www.iiees.ac.ir/iiees/EQsearch/(4tswcof3bokb2r2z0gca4245)/EventQuery.aspx`) reported only in local magnitude M_L.

g) The 2005 Zarand earthquake (22/02/2005, M_s = 6.5, 30.80°, 56.76°), the aftershock sequence was chosen over latitude extent 29.5°-32.5° and longitude extent 55.5°-59.5° for the time period 22/02/2005-22/02/2007. The catalog used was again IIEES (`http://www.iiees.ac.ir/iiees/EQsearch/(4tswcof3bokb2r2z0gca4245)/EventQuery.aspx`).

h) The 2002 Denali fault earthquake in central Alaska (03/11/2002, M_S=8.5, 63.52°, -147.44°). The listing was taken from the NEIC (PDE) listing (`http://neic.usgs.gov/neis/epic/`). The box was: latitude 65.0°-60.0° and longitude -141.0°-(-151.0°) and the time window was 23/10/2002-02/05/2008. This sequence is referred to as Alaska 1 henceforth in the text.

i) The 2003 Rat Islands, Aleutian Islands earthquake in Alaska (17/11/2003, M_w = 7.8, 51.15°, 178.65°). The source was again the NEIC (PDE) catalog (`http://neic.usgs.gov/neis/epic/`). The latitude longitude box was defined by latitudes 54°-50° and longitudes 174°-(-174°) and the time series was taken between 17/11/2003-02/05/2008. This dataset is henceforth referred to as Alaska 2.

j) The 2002 Taiwan earthquake (31/03/2002, M_w = 7.1, 24.13°, 121.19°) for which the listing was taken from BATS (Broadband Array in Taiwan for Seismology) CMT catalog (`http://tecws.earth.sinica.edu.tw/BATS/cmtbyform.php`). The relevant box was latitudes 20°-23° and longitudes 119°-122° for the time period 31/03/2002-31/03/2005. This sequence is called Taiwan 1 in the text.

k) Another 2006 Taiwan earthquake on the same plate boundary, viz. the Eurasian plate and the Philippines plates, (26/12/2006, $M_w = 6.7$, 21.89°, 120.56°). The latitude longitude box was defined as 20°-23° and 119°-122° and the time window was 26/12/2006-23/05/2008. The catalog was once again the BATS CMT catalog (`http://tecws.earth.sinica.edu.tw/BATS/cmtbyform.php`).

(All dates are expressed as dd/mm/yyyy, positive latitude implies northern hemisphere, positive longitude implies eastern hemisphere, negative latitude implies southern hemisphere and negative longitude implies western hemisphere.) The spatial distribution of aftershocks for each of these sequences is shown in Figure 3.3. $Q(t)$ for each of the above aftershock sequences was estimated using numerical integration.

The important limitation of our analysis, while evaluating the aforementioned integrals, is the fact that more often than not most catalogs which give the most exhaustive list of aftershocks report the various events in different magnitude scales. This warrants the need for using conversion relationships to convert the various magnitude scales to a uniform scale. This, wherever we have inhomogeneous catalogs, we have chosen to be M_w, the moment magnitude as defined by [30] analogous to the previous sections. To this end we have used well defined and previously employed conversion relationships. The datasets extracted from the NEIC (PDE) catalog are all inhomogeneous with respect to the magnitude scales used to report the various events.

The PDE listing was used in cases of the Sumatra, Muzaffarabad and the Alaska events (see Table 3.1). For the Sumatra event we used the conversion relationships used in Section 3.1. These relationships were specifically designed for the aftershock sequence of the Sumatra event extracted from the PDE catalog and hence serve our purpose. For the Muzaffarabad event we used conversion relations given in [46] which were again designed specifically for the region and is based on the PDE listing. The fact that the conversion relationships were designed for nearly the same datasets as we have used here is important as such conversion models are in general regressional models and hence their use in our work is validated by the fact that here we use them on the same Alaska 1 and Alaska 2 we could not obtain valid conversion relationships. Our strategy for these two sequences is described later in the text. Once we had homogeneous magnitude-time listings, we calculated the b-values and completeness magnitudes (M_c) for each of the recorded aftershock sequence. Our estimate of M_c is based on the assumption that the GR law [1] explains a large percentage of the frequency-magnitude distribution above a given completeness magnitude. We assumed the percentage to be an ad hoc 90% and followed the methodology in [28].

It has been recently noted [29] that the assignment of a single completeness magnitude to an aftershock sequence is oversimplified and the M_c for such sequences have temporal variability. Within the first hours to days of an aftershock sequence, M_c tends to decrease systematically. This is caused by improvements of the station network and the fact that the frequent larger aftershocks eclipse smaller events. In our case however, as we track an aftershock sequence over a large period of time exceeding at least a year, this initial variability has been disregarded and a unique M_c has been ascribed to each of the homogenized sequences. Such an assignment is problematic for the Alaska 1 sequence due to the lack of homogeneity (and hence increased possibility of bias in determination of the b-value) discussed previously. So we progressed following a different methodology. For Alaska 1, the most numerous listing of events was in M_L. So we chose this subset of the data (1848 events out of a total 2031 events) and calculated the b-value and M_c on this set. This was done keeping in mind that the lower magnitudes are more regularly reported in M_L in the PDE catalog.

Journal of Physics: Conference Series **319** (2011) 012004 doi:10.1088/1742-6596/319/1/012004

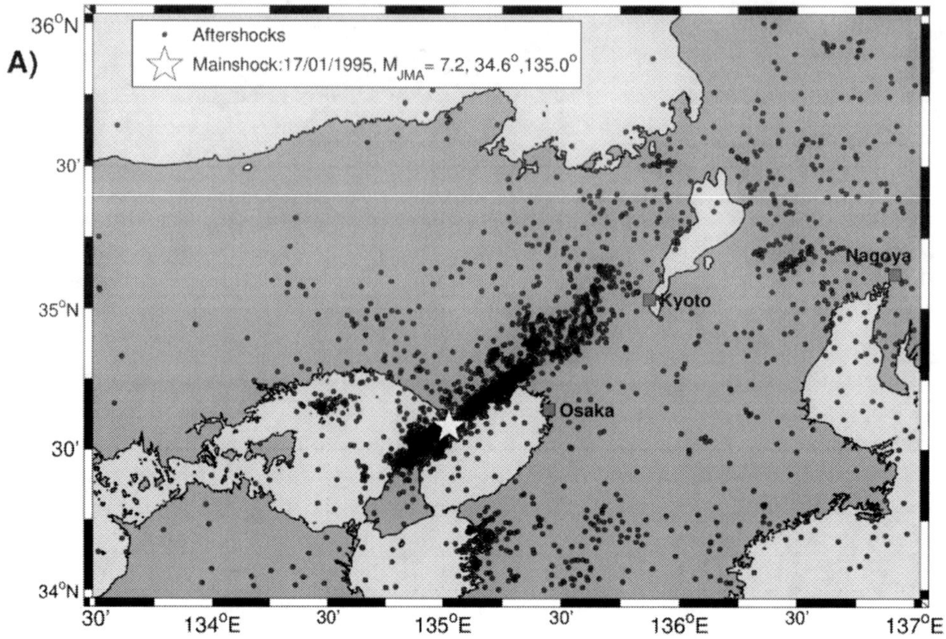

(a)

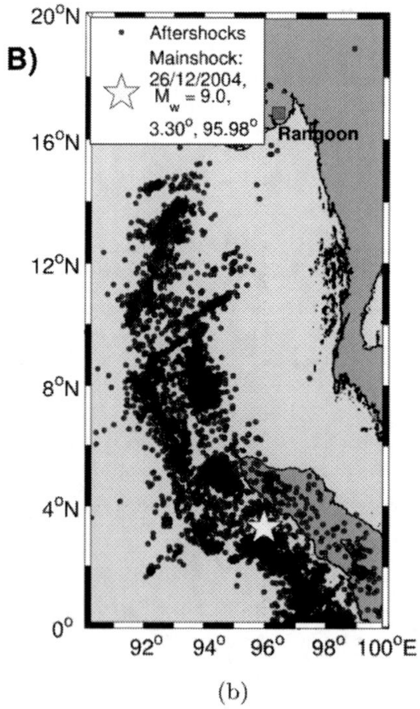

(b)

Figure 3.3: Maps showing spatial distribution of aftershocks chosen for A) Kobe, B) Sumatra datasets.

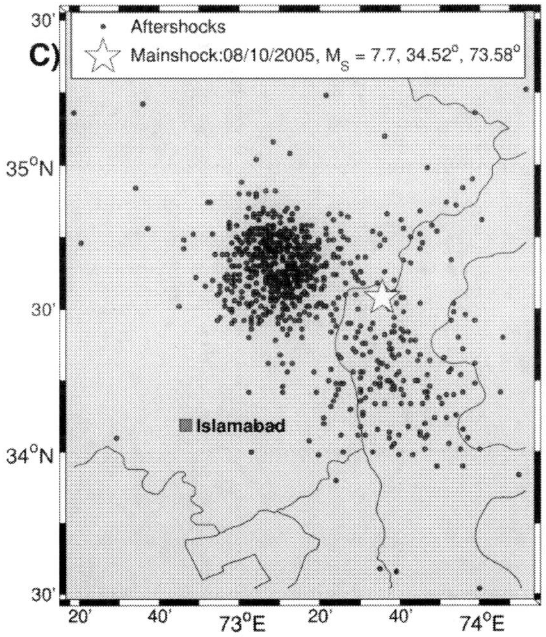

(c)

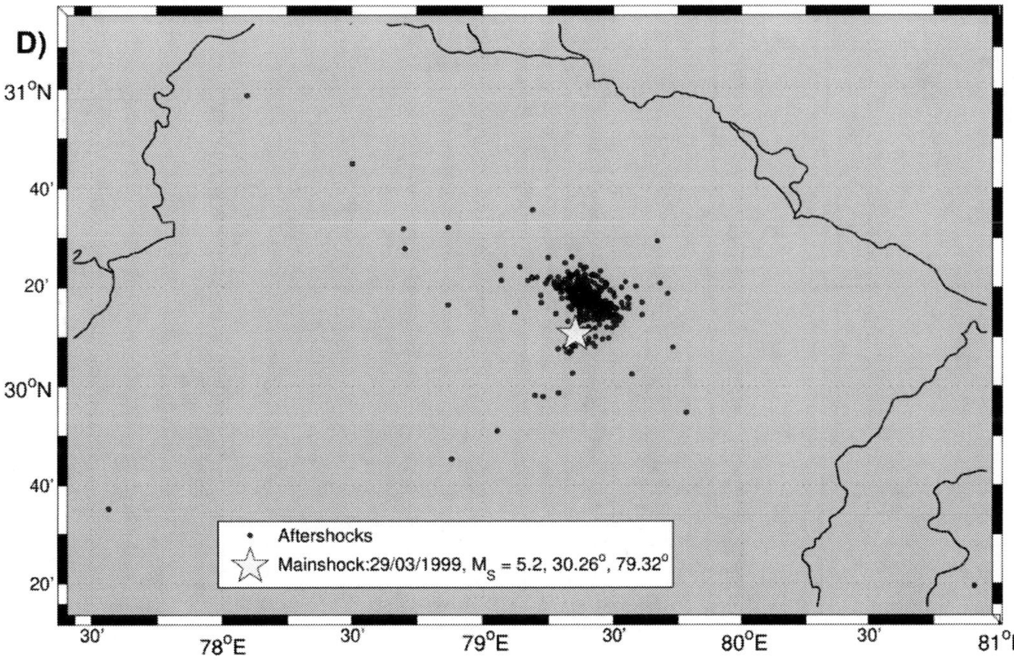

(d)

Figure 3.3: Maps showing spatial distribution of aftershocks chosen for C) Muzaffarabad, D) Chamoli datasets.

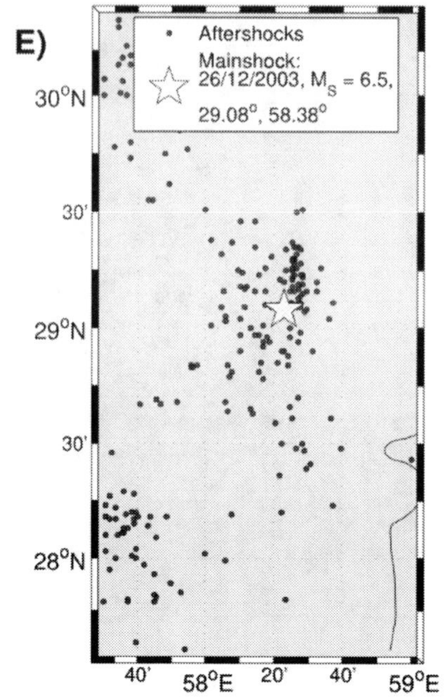

(e)

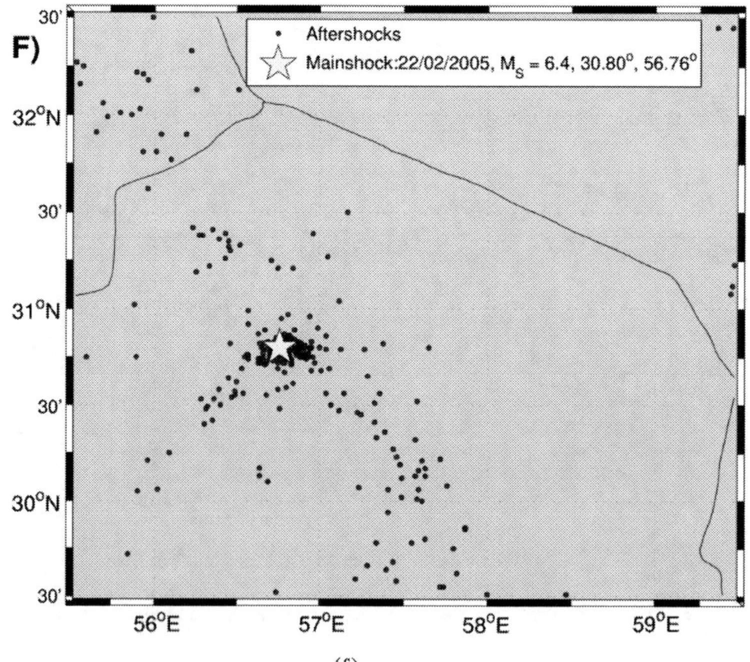

(f)

Figure 3.3: Maps showing spatial distribution of aftershocks chosen for E) Bam, F) Zarand datasets.

G)

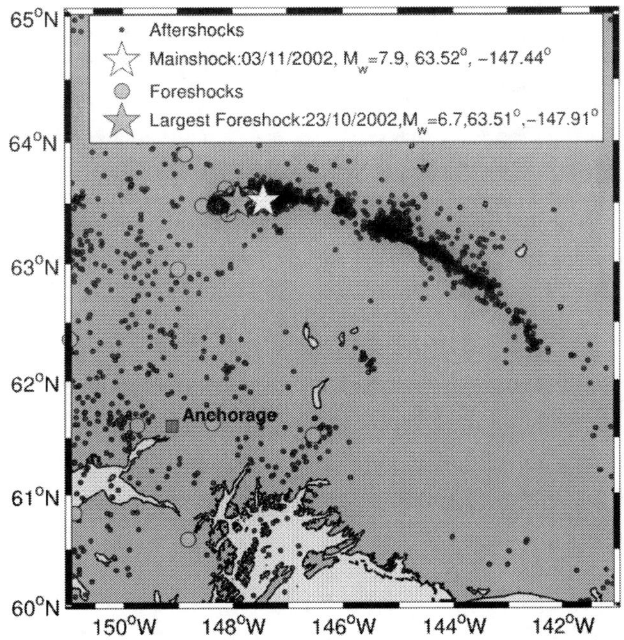

(g)

H)

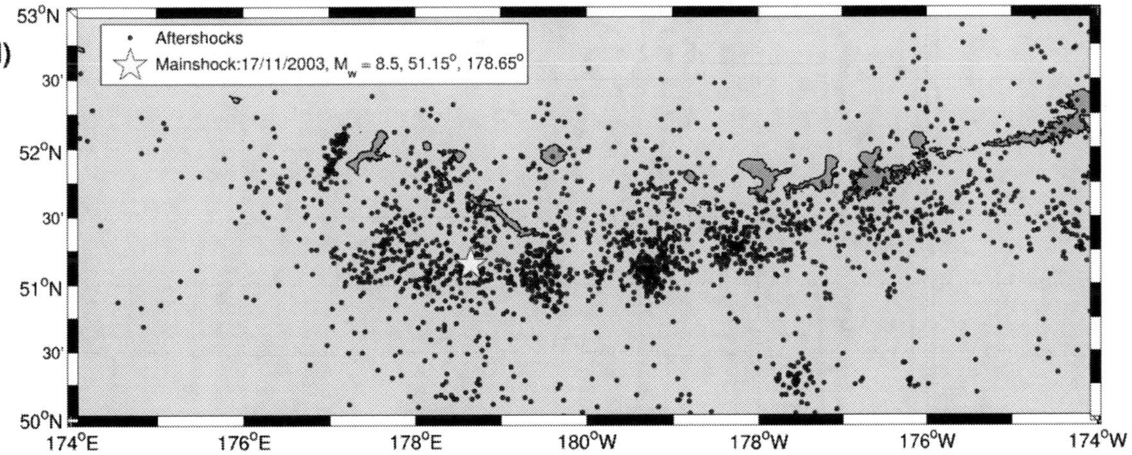

(h)

Figure 3.3: Maps showing spatial distribution of aftershocks chosen for G) Alaska 1, H) Alaska 2 datasets.

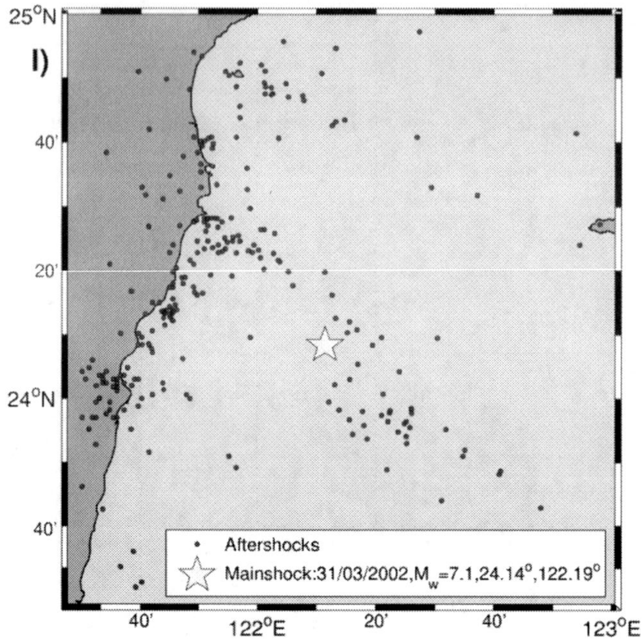

(i)

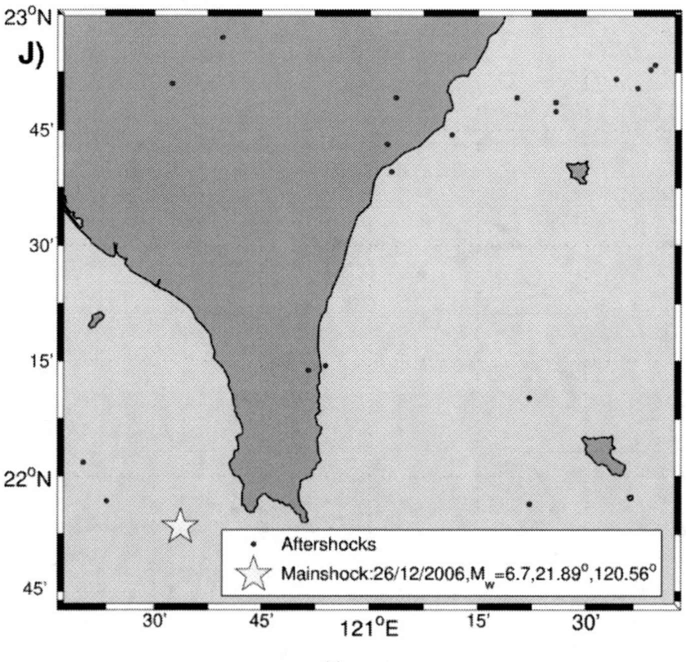

(j)

Figure 3.3: Maps showing spatial distribution of aftershocks chosen for I) Taiwan 1 and J) Taiwan 2 datasets.

Event Name (Event Tag)	M_c	b	S_1	S_2	RMS error	R square
Loma Prieta (L)	0.50	0.67	1.15	-	2.22	0.99
Kobe (K)	2.00	0.90	2.41	-	1.11	0.99
Sumatra (S)	3.63	0.81	4.16	-	5.18	0.99
Muzaffarabad (M)	3.33	0.79	4.05	-	7.38	0.99
Chamoli (C)	-0.20	0.32	1.39	-	2.02	0.99
Bam (B)	2.70	0.90	3.48	-	45.28	0.99
Zarand (Z)	2.80	0.99	3.47	-	10.11	0.99
Alaska 1 (Al 1)	3.00	1.08	-	3.47	12.78	0.99
Alaska 2 (Al 2)	2.60	0.46	-	3.47/4.08	9.41/11.37	0.99/0.99
Taiwan 1 (T 1)	3.42	0.77	4.30	-	39.55	0.99
Taiwan 2 (T 2)	3.63	0.75	4.32	-	17.73	0.99

Table 3.1. Event names are used to refer to respective sequences in the text. The event tags correspond to those in the plot in Figure 3.4 and Figure 3.5. S_1 corresponds to the slope of the linear fit with the homogeneous listings while S_2corresponds to the linear fit with the inhomogeneous datasets. In Alaska 2, the slope changes midway (see Figure 3.6(C)) and the two slopes, rms errors and R square values depict the values obtained while fitting for the earlier part before the slash and for the later part after the slash.

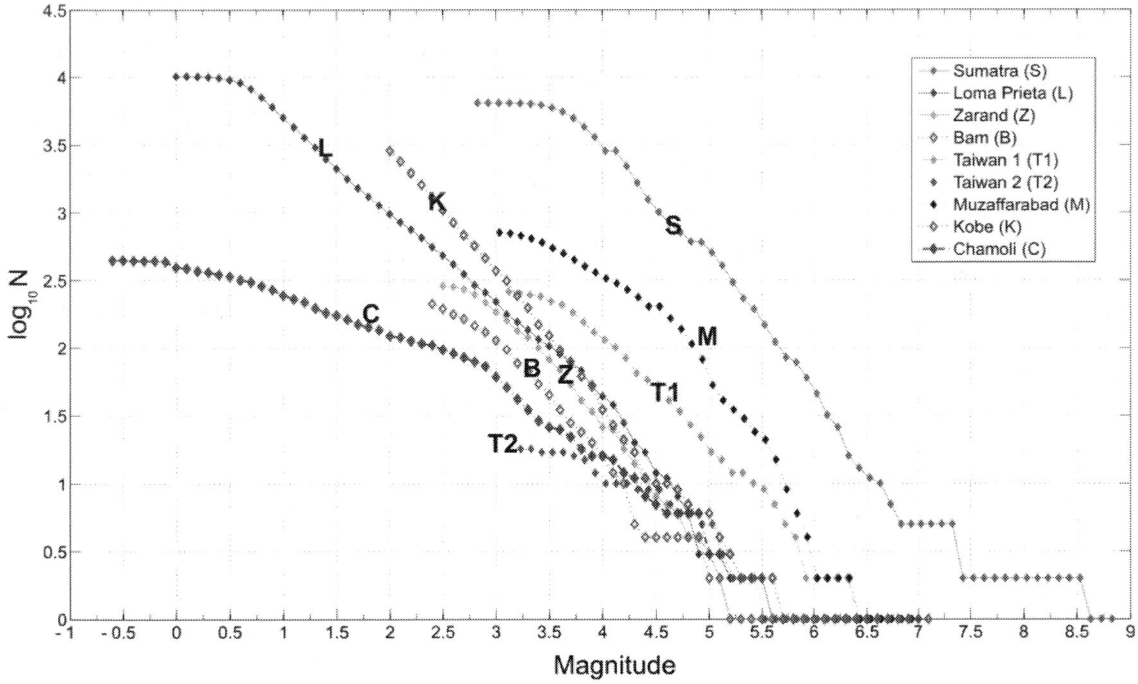

Figure 3.4: The Frequency-Magnitude distributions for the various aftershock sequences enlisted in Table 3.1 (except for the Alaska 1 and Alaska 2 sequences).

However, for the Alaska 2 listing no such clear maximal homogeneous subset was available and we calculated M_c and b based on the entire inhomogeneous dataset. This is followed by calculation of $Q(t)$ where the cumulative integral is carried out only over events that have magnitude $m \geq M_c$. Therefore in our case the minimum magnitude or lower cut off is M_c unless otherwise specified. Our analyses indicate a clear linear relationship $Q(t) = St$ where S denotes the slope. Wherever we have a homogeneous listing, we have attempted to fit a straight line to the observed $Q(t)$ curve and we call the slope thus obtained as S_1. Wherever we have a inhomogeneous listing the slope obtained by fitting is renamed as S_2. This is the nomenclature followed in Table 3.1 in listing the slopes for each sequence. The results of our analysis mentioned in Table 3.1 and the plots in Figure 3.1 point clearly to the linear variations mentioned above. The straight line (fit) retains this slope for years. Also the slope changes significantly over different fault zones. This indicates that the slope, S, is characteristic of the fault zone. This was further checked by integrating from anywhere in the time series (i.e. shifting our $t = 0$ to any randomly chosen aftershock) after the mainshock. The slope was found to be the same and the linearity of $Q(t)$ was not affected by shifting the origin of integration. A wide variety of events can lead to systematic errors in the reported magnitudes (events as varied as a change in instrumental calibration to addition or removal of seismograph stations) and such systematic errors can be very large going up to as much as 0.5 magnitude units [47]. Such errors would set the eventual error bound for the slope as the errors due to fitting are much smaller as stated already. Additionally, the conversion relationships themselves induce some errors in the magnitudes. This can also lead to systematic errors in the slope estimate. With the available catalogs, the errors in slope estimation would be thus about 6-10% [47, 48].

3.3.1. Analyzing the properties of the slope As we said earlier, it is our contention that these slopes are characteristic of the fault zone. Firstly, there is significant global variation in the values of the slopes. Each aftershock sequence seems to have a different slope for the $Q(t)$ statistic. Significantly, aftershock sequences caused by the same fault zone or on the same geologic and tectonic setting yield similar slopes. These two facts are clear from the results in Table 3.1. To illustrate the second point further, we draw attention to the two Taiwan sequences, Taiwan 1 and 2. Both of these took place on the Eurasian and Philippines plate boundary (from the Earthquake Summary Poster for the Taiwan event (26/12/2006, $M_W = 6.7$) available at (`http://earthquake.usgs.gov/eqcenter/eqarchives/poster/2006/20061226.php`). It is only natural that the two corresponding slopes would be nearly identical in view of our proposed error bounds due to the geological similarities and precisely similar tectonism. In Iran though, on the contrary, the Bam and Zarand earthquakes took place on two different faults belonging to a highly developed fault system. The Bam event occurred on the Bam fault whereas the Zarand event took place in close proximity of a previous event on the Gowk system (1981 July 28, Sirch earthquake $M_W = 7.1$) at a distance of about 60 km from the northern extremity of the rupture zone of the Sirch event [49]. But still the slopes were found to be the same (within proposed error bounds). This might be because the Gowk system and the Bam fault are part of a highly developed fault system. Further, the slope does not change with unusually large aftershocks in the sequence e.g. the Sumatra sequence had a few very large aftershocks including one great earthquake on March 28, 2005 ($M_W = 8.7$) which occurred about 150 km SE of the earlier giant earthquake epicenter ($M_W = 9.3$) of December 26, 2004. This further reveals the characteristic nature of the slope.

More importantly it is critical to note that the integration process is akin to an averaging process on the magnitudes and the slope, as we will later see, is approximately an average magnitude. So the reader may immediately think that the slope is more a signature of the catalog and the completeness level of the listing rather than the fault itself. But then if we have examples where reactivation or re-rupturing or similar events on a preexisting fault system

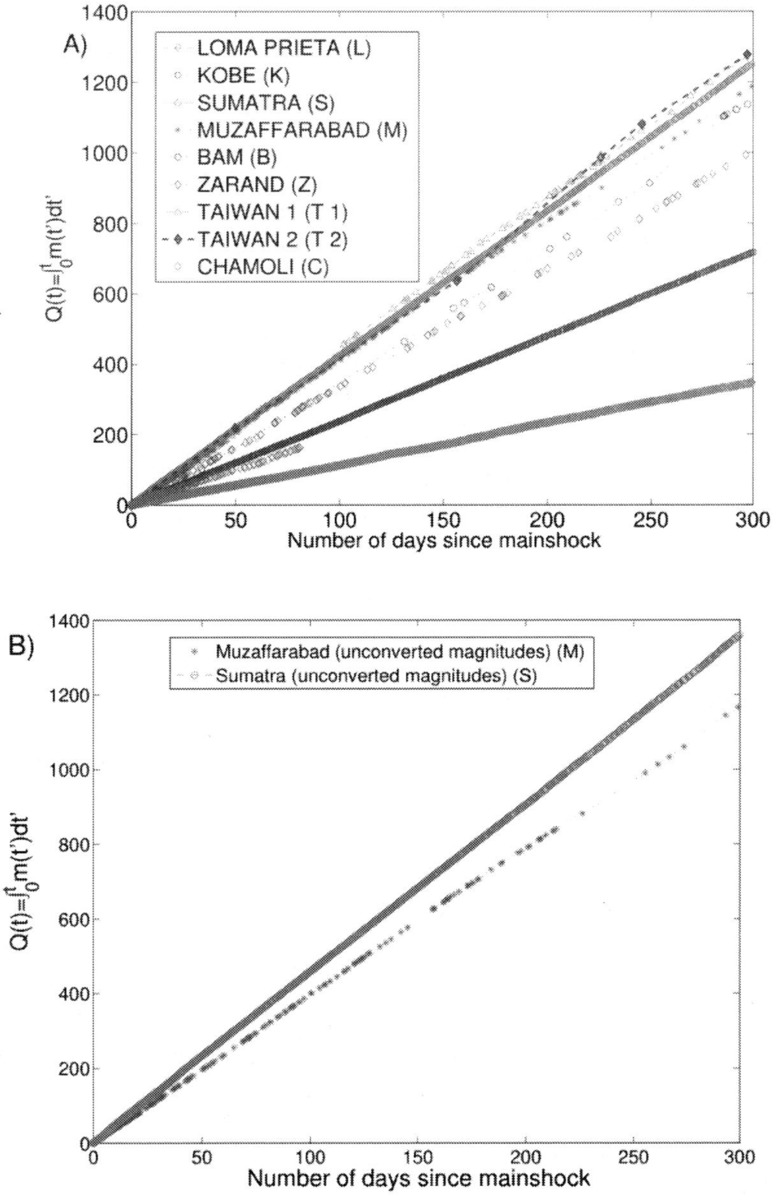

Figure 3.5: (A) Plots of time cumulant of magnitude $Q(t)$ vs. t (in number of days since the mainshock) for the datasets described in the text and in Table 1 for the first 300 days. The tags for the events used in the plot are the same as in Table 3.1. (B) Plots of $Q(t)$ vs. t for the Sumatra and Muzaffarabad sequences before conversion of magnitudes according to [31] and [46] respectively. Again, only the first 300 days are plotted.

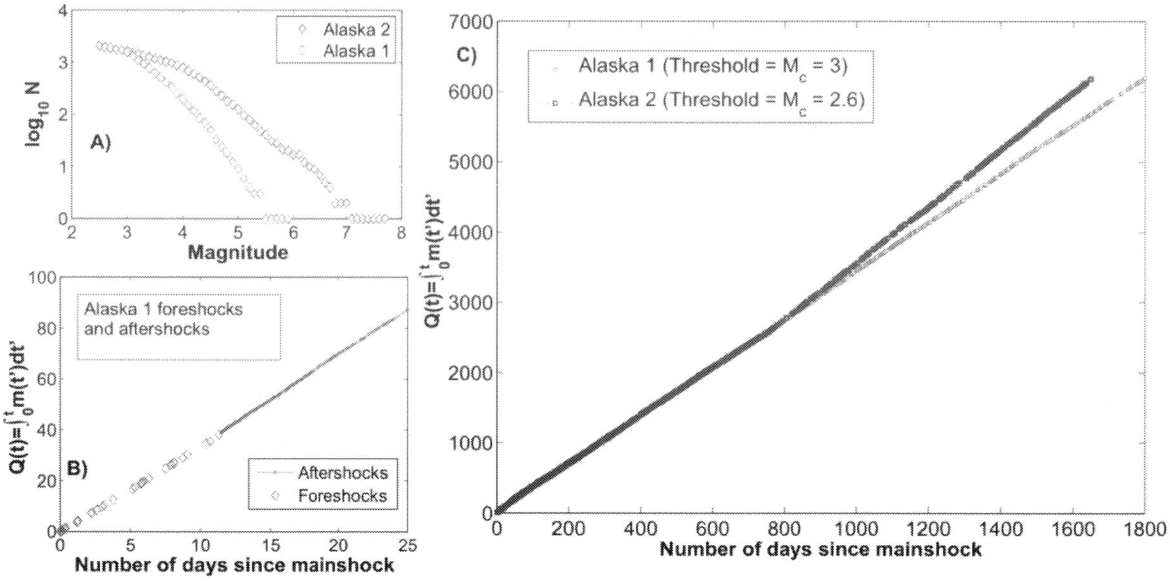

Figure 3.6: A) The frequency magnitude distribution (FMD) plots for Alaska 1 and Alaska 2. As noted in the text, the FMD for Alaska 1 is computed on the M_L listing. For Alaska 2 the FMD was calculated on the entire inhomogeneous listing. B) $Q(t)$ for the first 25 days for Alaska 1 showing that foreshocks and aftershocks exhibit same slope. C) The entire sequences for Alaska 1 and 2. Note slope change in Alaska 2.

changes the slope of $Q(t)$ over time, that would establish the characteristic nature of the slope. The reason for this is two fold. Firstly, as we will show later, we have considerable reason to believe that the slope of the $Q(t)$ statistic depends significantly on the geometry of the faults and the slope change is an expected result of the change in geometry of the fault surface due to the re-rupturing. Secondly, the inherent changes in completeness level in aftershock sequences are limited to within days of the occurrence of the mainshock. If we observe a change in slope a long time after the mainshock, when the M_c has stabilized to a steady level, then this change in slope cannot be ascribed to a change in completeness level. This slope change is then due to the changes in asperity distribution and stress patterns brought about by the re-rupturing. The reader will quickly recognize that this statement supports the analytical expression for the slope obtained in (2.16). Re-rupturing will clearly modify the slope as it would modify the nature of the fractal involved, either the dimension or q or the generation or n. If the slope changes due to these factors then of course it is characteristic of the fault involved. This part will become very clear later when we discuss the statistical interpretation of the slope as average magnitude. At present let us look at an example of such an occurrence which strengthens our claims. To this end we shall use the results for the two sequences obtained in Alaska. Alaska 1 was an event on the inland Denali fault and the $Q(t)$ statistic gives a slope $S_2 = 3.47$. One important aspect came out during the analysis of the Alaska 1 dataset. The first shock considered here was not the Denali fault mainshock but a previous shock in the same region. This was done because this event is a very well established foreshock of the Denali fault event and is followed by a series of foreshocks till the occurrence of the mainshock. The cumulant $Q(t)$ for the foreshocks retains the same linearity as the cumulant $Q(t)$ for the aftershocks (see Figure 3.6 (B)). This is expected though if one believes in the characteristic nature of the slope for a given fault system.

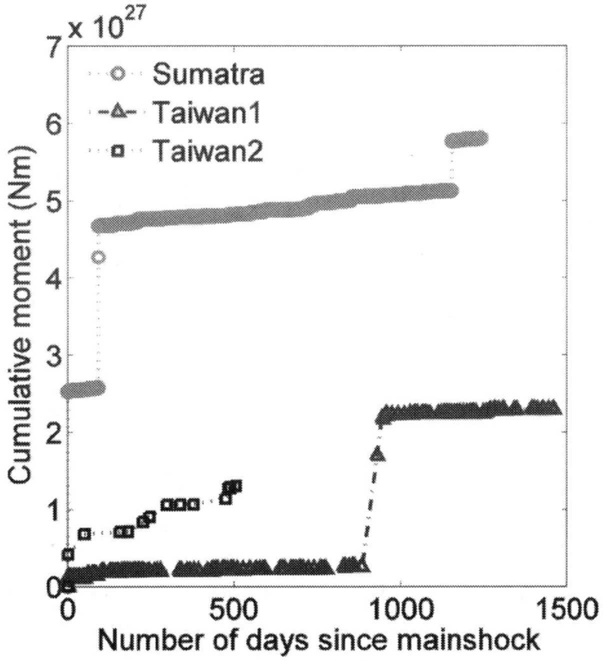

Figure 3.7: Plots of cumulative moment versus time since the mainshock for the datasets Sumatra, Taiwan 1 and Taiwan 2. The values for Taiwan 2 depicted here in the plot are 10 times the real values to ensure proper legibility of the figure (cf. [17]).

As foreshocks and aftershocks both happen on the same fault or fault system they are expected to yield the same slopes. Our claim, that the slope is a signature of the fault system, is further strengthened on analysis of the Alaska 2 aftershock sequence (see Figure 3.6 (C)). Here, the slope of the $Q(t)$ vs. t curve increases after about 754 days of the main event. One of the most significant events of the last century, the 1965 M_w 8.7 Rat Islands earthquake ruptured a 600 km long portion of the plate boundary to the west of the Amchitka Island. In the November 17, 2003 M7.7 earthquake, associated with the main shock or the first shock in the sequence we chose, the easternmost part of the 1965 zone failed again. On June 14, 2005, a series of moderate to strong earthquakes occurred in the Rat Islands region of the Aleutian Islands. The sequence started with a M 5.2 event at 08:03 UTC and the largest event (M 6.8) followed 9 hours later (at 17:10 UTC). The largest earthquake was situated 49 kilometers (31 miles) south-southeast of Amchitka. This new sequence of earthquakes re-ruptured the easternmost end of the 1965 rupture zone. This is the reason, we believe, for the increase in slope. The re-rupturing process meant that the earlier asperity distributions were changed and hence the region underwent a marked change in its seismicity pattern. The slope measures for the Alaska 2 sequence are given in Table 3.1 for the total inhomogeneous event listing i.e. $S_2 = 3.47$ for the earlier half and $S_2 = 4.08$ for the later half. The change in slope, as one can clearly observe is not within our error bounds. The slope change also occurs after about two years from the mainshock and hence should not be an artifact of change in minimum completeness level of the seismic network which is constant by now. Also the slope does not keep on changing and remains at $S_2 = 4.08$ for the rest of the sequence. The slope of $Q(t)$ definitely shows that some significant change occurred in the aftershock sequence (equivalently the fault system) due to re-rupturing.

However such cumulative statistics have already been attempted for the scalar seismic moment or Benioff stresses for aftershock sequences. We did a similar cumulative integral of scalar seismic moment for our sequences in Sumatra and Taiwan (the former was reliably converted to scalar seismic moment in [31] and the BATS CMT catalog for Taiwan was homogeneous and listed only broadband M_W values). The results are shown in the Figure 3.7. The resultant plots resemble a step function. Authors in [31] have tried to fit a power law and/or linear models piecewise to such data (in their case the cumulative Benioff stress). There seems to be no robust feature to this statistic, i.e. the cumulative moment versus time curve. Such cumulative curves have also been reported for theoretical models such as for the Critical Continuum-State Branching Model of Earthquake Rupture [50]. Precursory accelerating moment release before large earthquakes has been a widely discussed phenomenon until recent years, being regarded as observational evidence for the controversial critical-point-like model of earthquake generation [51, 52]. Another useful property of such seismic moment cumulants is that they help in monitoring the stress release modes for a given region and hence allow for discussions on the type of mechanisms underlying earthquake occurrences [53].

As mentioned earlier, the slope of $Q(t)$ is an estimate of the average magnitude of the aftershock sequence. We proceed on this line and try to obtain an expression connecting the slope and other parameters of aftershock statistics. For an aftershock sequence, apart from the first few aftershocks, let us assume that the event inter-occurrence times and their magnitudes are statistically independent of each other in the long term. Then for a large number of events we have,

$$Q(t) = \int_0^t m(t')dt' \approx \int_0^t \overline{m}(t')dt' \approx \overline{m}t, \tag{3.2}$$

where $\overline{m}$ is the average magnitude calculated from the GR distribution for the aftershock sequence (assuming that the b value is constant over time which is an observed fact [22]). Now this $\overline{m}$ can be calculated from the GR distribution as follows,

$$\overline{m} = \int_{m_{\min}}^{m_{\max}} m \frac{b \times \ln 10 \times 10^{-bm}}{(10^{-bm_{\min}} - 10^{-bm_{\max}})} dm. \tag{3.3}$$

If we assume that the GR law holds for all larger magnitudes then $m_{\max}$ tends to infinity and we have,

$$S = \overline{m} = m_{\min} + \frac{0.43}{b}. \tag{3.4}$$

This is the expression for the slope which connects it with the b value for the aftershock sequence. We have presented a comparison of the slopes obtained through fitting and $\overline{m}$ in Table 3.2 (except for the Alaska 2 sequence because of the complete inhomogeneity and change in slope midway). The results show a very good agreement. But the agreement is in general better for b as computed for the specific aftershock sequence rather than the global b value of unity. In some cases like in Loma Prieta, Taiwan 2 and Chamoli the error (i.e. $S - \overline{m}$) for $b \neq 1$ is an order of magnitude less than the errors obtained when $\overline{m}$ is computed using $b = 1$. Clearly, in general $S = \overline{m}(b)$ is a more accurate expression than $S = \overline{m}(b = 1)$. This brings us to an important realization. As can be seen in Table 3.1 and is also reported widely [22], the b-values for individual aftershock sequences are unique to the sequence in general and show wide variations. Moreover the b-value for a single aftershock sequence has been known to vary spatially between the extremities of the rupture [22].

However, temporally, the b-value is exceptionally robust except at times when we have rather large earthquakes. Now it is known that the b-value of a given aftershock sequence can be explained in terms of the surface similarity dimension of the causative fault network (the fault network being treated as a fractal) [54]. Also the Two Fractal Overlap model clearly demonstrates the GR law for the synthetic aftershock sequences and shows that the b-value

Event Tag	M_c	b	$\overline{m}(b=1)$	$\overline{m}(b)$	S	$S-\overline{m}(b=1)$	$S-\overline{m}(b)$
L	0.50	0.67	0.93	1.14	1.15	0.22	0.01
K	2.00	0.90	2.43	2.47	2.41	-0.02	-0.06
S	3.63	0.81	4.08	4.18	4.16	0.08	-0.02
M	3.33	0.79	3.78	3.90	4.05	0.27	0.15
C	-0.20	0.32	0.33	1.23	1.39	1.06	0.16
B	2.70	0.90	3.13	3.18	3.48	0.35	0.30
Z	2.80	0.99	3.23	3.24	3.47	0.24	0.23
Al 1	3.00	1.08	3.43	3.40	3.47	0.04	0.07
T 1	3.42	0.77	3.95	4.03	4.30	0.35	0.27
T 2	3.63	0.75	4.16	4.30	4.32	0.16	0.02

Table 3.2. The comparison of S and $\overline{m}$ computed from (3.4) using $b = 1$ (values listed as $\overline{m}(b = 1)$) as well as using b as computed by us from the sequences (values listed as $\overline{m}(b)$). $m_{\min} = M_c$ for all the sequences. The errors are listed as $S - \overline{m}$.

is purely dependent on the dimension and generation of the fractals involved in sticking and slipping (in our case we have the dependence on the fractal dimension). In the case of real aftershock sequences this would be equivalent to saying that the b-value is dependent on the fractal geometry of (a) the fault surfaces (b) the fault network. Therefore, the b-value might change after a large earthquake due to re-rupturing or extensive change in asperity distributions. Re-rupturing would affect the fractal geometry of the fault and, as discussed before, this should generally increase the slope of the $Q(t)$ statistic (not decrease). This is what we believe happened in Alaska 2. So as S is a function of the b-value it is characteristic of the fault geometry. Also one must keep in mind that our definition of the completeness magnitude is dependent on the roll-off of the frequency-magnitude distribution from the GR law curve towards the lower magnitude range. Our definition of a roll off or a fitting algorithm (like the one we used), based on creating synthetic distributions with the b-value obtained, depend on the value of b and a slightly smaller or larger b can alter our M_c calculations and hence affect our chosen $m_{\min}$. Theoretically, we also saw that the roll off can be defined as a log correction on the GR law and its onset is defined by the factor $n/3$, the same factor which decides the slope. Hence the roll off itself is a function of the generation number for the Cantor set or the geometry and extent of rupture of the fault. Thus the slope can be clearly called a definite characteristic of the fault zone as it is highly affected by the b-value. The usual temporal stability of the b-value explains the constancy in slope over long periods of time. The other question that we tried to address is: Is equation (3.4) an accurate expression for S or a good approximation? To check this we used the following observation. If we increase our $m_{\min}$ by some $\Delta m_{\min}$ above the minimum level of completeness then, we have, using (3.4),

$$\Delta S = \Delta \overline{m} = \Delta m_{\min} + \Delta \left(\frac{0.43}{b} \right). \tag{3.5}$$

Now as b is constant above the completeness level the second term on the right hand side goes to zero and we have,

$$\Delta S = \Delta m_{\min}. \tag{3.6}$$

We checked for this by recalculating $Q(t)$ for three different minimum magnitude thresholds for each sequence. We chose $m_{\min} = M_c + \Delta, \Delta = 0, 0.1$ and 0.2 as thresholds for each sequence. Nowhere did the results agree exactly to (3.6). Close agreement was seen only for Alaska 1. The results are reported in Table 3.3 and the plots for Muzzafarabad and Loma Prieta are shown in Figure 3.8. This implies that (3.4) might be an over-simplified expression for the slope and is not

Event Tag	M_c	$S(m_{\min} = M_c)$	$S(m_{\min} = M_c + 0.1)$	$S(m_{\min} = M_c + 0.2)$
L	0.50	1.15	1.17	1.21
K	2.00	2.41	2.49	2.57
S	3.63	4.16	4.21	4.28
M	3.33	4.05	4.08	4.10
C	-0.20	1.39	1.40	1.47
B	2.70	3.48	3.53	3.59
Z	2.80	3.47	3.48	3.55
Al 1	3.00	3.47	3.56	3.63
T 1	3.42	4.30	4.31	4.38
T 2	3.63	4.32	4.42	4.43

Table 3.3. Effect of the increment in $m_{\min}$ on the fitted slope of Q(t) for all the sequences in Table 3.2.

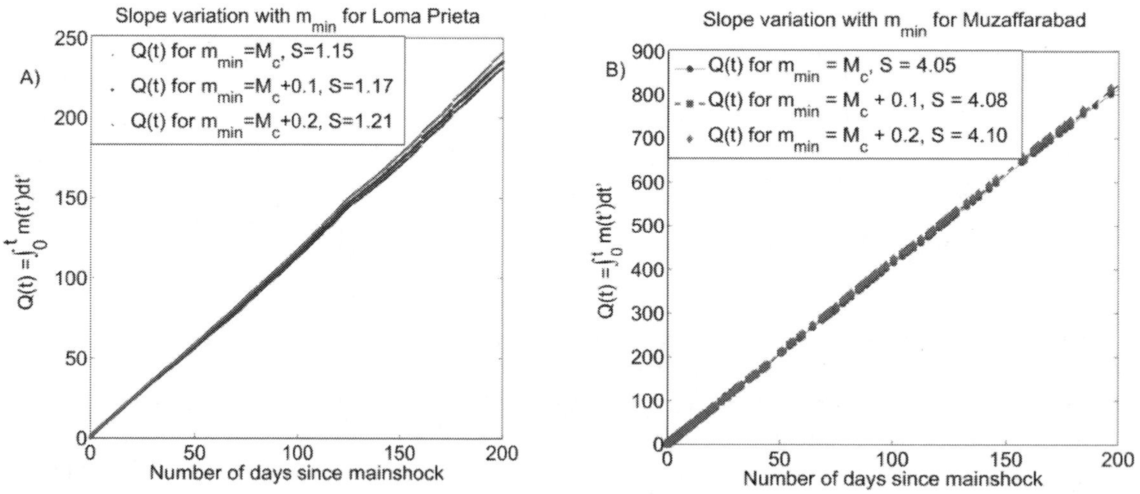

Figure 3.8: Slope variation in $Q(t)$ with $m_{\min}$ for the aftershock sequences in Loma Prieta and Muzaffarabad. First 200 days shown for the sake of clarity.

accurate. It explains the dependence of the slope on b reasonably whereas the dependence of the slope on $m_{\min}$ seems not to be straightforwardly linear. Hence the slope is probably not simply a manifestation of the lower magnitude cut-off. The failure of (3.4) and (3.6) to address the slope increment may also be an indication that aftershock magnitudes and inter-occurrence times may not be uncorrelated even long after the mainshock thus invalidating the line of mathematical argument used to arrive at (3.4).

3.3.2. Possible usefulness of the slope To end our discussion we check the nature of the linearity of the $Q(t)$ statistic: Is it deterministic or stochastic? We check this using the following expression for the numerical evaluation of the integral for $Q(t)$;

$$Q(t_i) = \int_0^{t_i} m(t')dt' \approx \sum_i m(t_i)\Delta t_i = Q(t_{i-1}) + m(t_i)(t_i - t_{i-1}) \tag{3.7}$$

which leads to the expression

$$t_{i+1} = \frac{m(t_{i+1})t_i - Q(t_i)}{m(t_{i+1}) - S}, \tag{3.8}$$

where we make use of the fact that $Q(t_{i+1}) = St_{i+1}$. For all the aftershock sequences, as a benefit of posteriori knowledge of the catalog sequence, we have with us $m(t_{i+1})$ as well as t_{i+1}. So starting with the mainshock and using the magnitudes of every subsequent aftershock we could make a synthetic occurrence time listing of t_{i+1} using (3.8) for each i. This synthetic t_{i+1} we will call t_{pred} while the actual catalog listed sequence of t_{i+1} we call t_{ac}. We calculated the quantity $t_{\text{pred}} - t_{\text{ac}}$ and created scatter plots against t_{ac}. The plots had very significant concentration of points around the line $y = t_{\text{pred}} - t_{\text{ac}} = 0$ but there were also some very large outliers. In general though the number of deviations were quite smaller. The concentration of most of the points around $t_{\text{pred}} - t_{\text{ac}} = 0$ clearly demonstrates the deterministic nature of the linearity. The details of the behavior of such plots will be discussed by us elsewhere. At the present we give two such plots for the Kobe and Muzaffarabad sequences to demonstrate this point. As pointed out earlier, the linearity in $Q(t)$ is maintained with the same slope if we shift our origin of integration to any aftershock in the sequence. As cumulative integrals are always subject to accumulation of errors, we also recalculated t_{pred} within various time windows within the series where the origin of the window is far from the mainshock. We observed lesser scatter within the time window than that observed within the same time window when the origin of integration was the mainshock indicating lesser error accumulation as expected. One usefulness of this result is that if we have apriori knowledge of (or we assume) the magnitude of the next aftershock in a sequence then (3.8) can be used in real time to predict time of occurrence by extrapolation (which makes perfect sense for this linear curve). Conversely we can have prediction tables where we can obtain a list of occurrence times for the next aftershock corresponding to a list of possible magnitudes of the event. We will discuss the prediction issue in more detail elsewhere. In real time calculations we can minimize errors by shifting our origin of integration or in other words putting $Q(t) = 0$ after every few aftershocks and recalculating all parameters in (3.8) again by again allowing $Q(t)$ to accumulate over the next few aftershocks and so on. This aspect of our study shows that the linearity of the $Q(t)$ statistic is deterministic and not stochastic. This affords it the unique ability to predict occurrence times of aftershocks given an estimate of the magnitude of the next event (which, as far as we know, is not possible by our current state of knowledge).

4. Conclusion

We have presented here some new results that have come out from our analysis of the Two Fractal Overlap model. The model is based on the fact that fault surfaces, both fresh and weathered, exhibit a fractal topography (Section 1.2). The model captures the stick-slip dynamics of overlapping fractal surfaces by using regular middle third removal Cantor sets (Section 2.1) wherein a Cantor set of a given generation slides over its replica with uniform velocity. Our model is essentially one of repeating seismicity on the same fault following a large magnitude event (or aftershocks). The statistical features of the synthetic earthquake time series thus produced are completely analytically tractable (Section 2.2). The model, as is evident from our

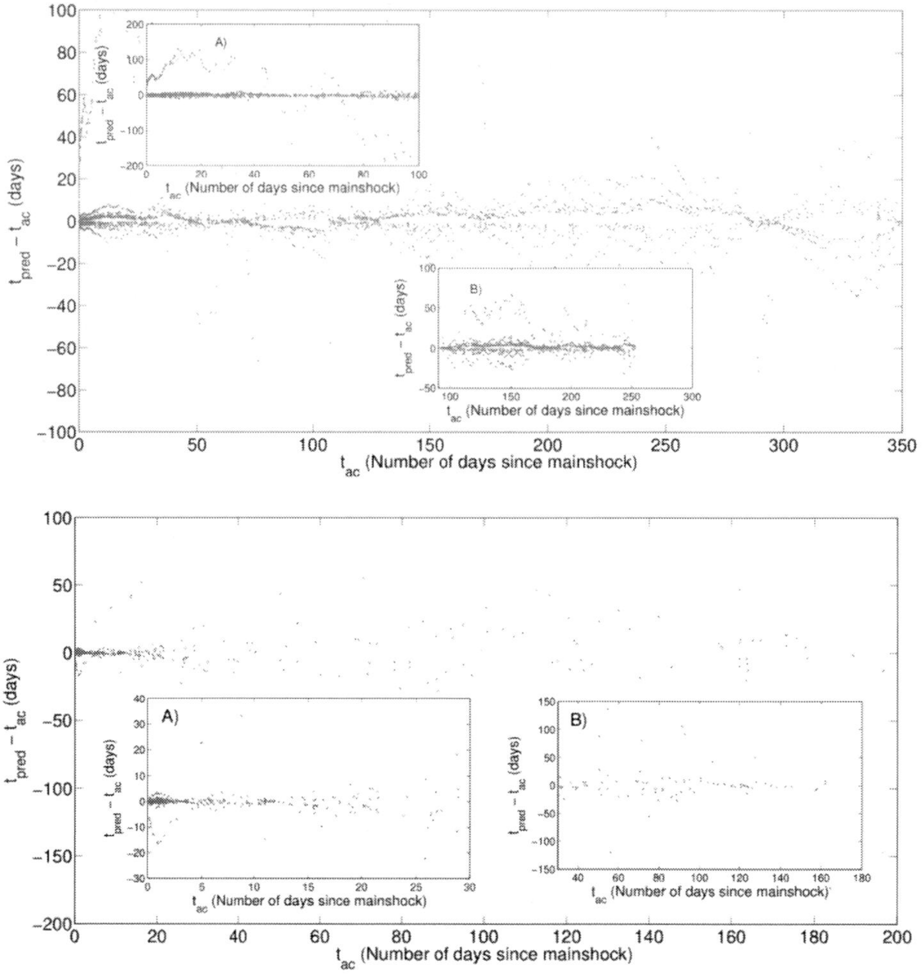

Figure 4.9: The $t_{\mathrm{pred}} - t_{\mathrm{ac}}$ vs. t_{ac} plots for the Kobe (top) and Muzaffarabad (bottom) sequences for first 350 and 200 days respectively. Some of the larger deviations from $y = t_{\mathrm{pred}} - t_{\mathrm{ac}} = 0$ lie beyond the extent of the y-axis shown here and are omitted to facilitate better viewing. Inset: different time windows within which $Q(t)$ was recalculated. For Kobe: Inset A) First 100 days only. Inset B) Between 90 and 250 days. Scatter lessened significantly with respect to the same time window when we integrate $Q(t)$ from the aftershock at the time origin of the window rather than the mainshock. For Muzaffarabad: Inset A) First 30 days. Inset B) Between 30 to 160 days. Again note the reduced scatter with respect to the same time window when we integrate $Q(t)$ from the aftershock at the time origin of the window rather than the mainshock.

analysis, captures the GR law (Section 2.3). If one agrees that aftershock time decay is given by the Omori law and not the modified Omori Law, then the model also captures the rudimentary features of this behavior (Section 2.4). It gives a hitherto unknown statistical feature of the temporal distribution of aftershock magnitudes which we have shown in Section 2.5. The model predicts that the time cumulant of magnitudes of synthetic seismicity produced by the model for a given generation number is a remarkable straight line with a slope that is characteristic of the fractal topography of the rupture surface. In Section 3 we have shown the proximal correspondence of the values of the model parameters with the b-value and the constant a in the GR law as observed for natural seismicity. The model also shows that there may be some physical basis behind the completeness magnitude M_c and the corresponding low magnitude roll off of the frequency magnitude statistics of earthquakes. There is also reasonable analogy between the behavior of the exponent p in Omori Law and the exponent which describes time decay of aftershocks in this model.

We have also shown that the new statistical law involving the time cumulant of magnitudes discussed in Section 2.5 is also observed in nature in the case of aftershocks and this might give us important information about the fractal geometry of the faults involved in producing an earthquake and its aftershock sequence. Specifically, we have shown that cumulative integral of magnitudes $Q(t)$ of a natural aftershock sequence over time t is linear, the slope S being characteristic of the fault zone. Hence some key features of the rupture zone may be extracted from such an analysis of the magnitude time series $m(t)$ of the earthquake aftershocks, in particular information about the fractal geometry of the fault. We have demonstrated the characteristic property of the slope and its relationship with the b value. We have also shown that though the slope depends on the lower magnitude cutoff, the relationship is not very simple or linear. More importantly we have demonstrated that the linearity of $Q(t)$ is deterministic and hence might very well provide methods to predict aftershock occurrence times in the future

Our focus here was on the Two Fractal Overlap Model which is a very simplistic model of earthquakes and do not claim in any way that this is the realistic physical scenario at the geological faults. But the fact that such a simplistic geometrical model mimics so much of nature is truly astonishing. The analysis, we reiterate, is one which requires very basic mathematics. The fractal overlap model tells us that the role of the fractal geometry of fault surfaces might play a significant role in controlling the statistical features of aftershock sequences.

References

[1] Gutenberg B and Richter C F 1944 *Bull. Seismol. Soc. Am.* **34** 185–188

[2] Omori F 1894 *J. Coll. Imp. Univ. Tokyo* **7** 111

[3] Utsu T, Ogata Y and Matsuura R S 1995 *J. Phys. Earth* **43** 1–33

[4] Burridge R and Knopoff L 1967 *Bull. Seismol. Soc. Amer.* **57** 341–371

[5] Carlson J M, Langer J S and Shaw B E 1994 *Rev. Mod. Phys.* **66** 657–670

[6] Olami Z, Feder H J S and Christensen K 1992 *Phys. Rev. Lett.* **68** 1244

[7] Bak P, Tang P and Weisenfeld K 1987 *Phys. Rev. Lett.* **59** 381

[8] De Rubeis V, Hallgass R, Loreto V, Paladin G, Pietronero L and Tosi P 1996 *Phys. Rev. Lett.* **76** 2599

[9] Hallgass R, Loreto V, Mazzella O, Paladin G and Pietronero L 1997 *Phys. Rev.* E **56** 1346

[10] Aochi H and Matsuura M 2002 *Pure Appl. Geophys.* **159** 2029–2044

[11] Sagy A, Brodsky E E and Axen G J 2007 *Geology* **35** 283

[12] Power W E, Tullis T E, Brown S R, Boitnott G N and Scholz C H 1987 *Geophys. Res. Lett.* **14** 29

[13] Brown S R and Scholtz S H 1985 *J. Geophys. Res.* **90** 12575

[14] Chakrabarti B K and Stinchcombe R B 1999 *Physica* A **270** 27

[15] Bhattacharyya P 2006 Geometrical models of earthquakes *Modeling Catastrophic and Critical Phenomena in Geoscience: A Statistical Physics Approach* ed Bhattacharyya P and Chakrabarti B K (Springer-Verlag) pp 155–168

[16] Pradhan S, Chakrabarti B K, Ray P and Dey M K 2003 *Phys. Scripta* **T106** 77

[17] Bhattacharya P, Chakrabarti B K, Kamal and Samanta D 2010 Fractal models of earthquake dynamics *Reviews of Nonlinear Dynamics and Complexity* vol 2 ed Schuster H G (Wiley-VCH Verlag GmbH & Co. KGaA) chap 4, pp 107–158 ISBN 9783527628001 URL `http://dx.doi.org/10.1002/9783527628001.ch4`

[18] Abramowitz M and Stegun I 1964 *Handbook of Mathematical Functions* 5th ed (New York: Dover)

[19] Okubo P G and Aki K 1987 *J. Geophys. Res.* **92** 345–356

[20] Scholz C H 1990 *The Mechanics of Earthquakes and Faulting* (Cambridge University Press, New York)

[21] Wiemer S and Wyss M 1997 *J. Geophys. Res.* **102** 15115–15128

[22] Wiemer S and Wyss M 2002 *Adv. Geophys.* **45** 259

[23] Schorlemmer D, Wiemer S and Wyss M 2005 *Nature* **437** 539–542

[24] Utsu T 1969 *J. Fac. Sci. Hokkaido Univ.* **3** 129

[25] Bhattacharya P, Phan M and Shcherbakov R 2011 *Bull. Seismol. Soc. Am.* In Press

[26] Hirata T 1986 *Zishin (J. Seismol. Soc. Japan)* **39** 478 in Japanese

[27] Kisslinger C and Jones L M 1991 *J. Geophys. Res.* **96** 11947–11958

[28] Wiemer S and Wyss M 2000 *Bull. Seismol. Soc. Am.* **90** 859–869

[29] Peng Z G, Vidale J E and Houston H 2006 *Geophys. Res. Lett.* **33** L17307

[30] Kanamori H 1977 *J. Geophys. Res.* **82** 2981

[31] Jiang C and Wu Z 2006 Pre-shock seismic moment release in different segments of a fault: the case of the 26 December 2004 Indonesia Mw = 9.0 earthquakes *Advances in Geosciences* vol 1 (Solid Earth) ed Ip W H (World Scientific, Singapore)

[32] Kamal and Mansinha L 1996 *Bull. Seismol. Soc. Am.* **86** 299
[33] Kamal and Chabak S K 2002 *J. Him. Geol.* **23** 63
[34] Mogi K 1962 *Bull. Earthquake Res. Inst.* **40** 125–173
[35] Das S and Scholz C H 1981 *J. Geophys. Res.* **86** 6039–6051
[36] Mikumo T and Miyatake T 1979 *Geophys. J. R. Astron. Soc.* **59** 497–522
[37] Scholz C H 1968 *Bull. Seismol. Soc. Am.* **58** 1117–1130
[38] Narteau C, Shebalin P and Holschneider M 2002 *J. Geophys. Res.* **107** 2359
[39] Nur A and Booker J R 1971 *Science* **175** 885–887
[40] Schaff D P, Beroza G C and Shaw B E 1998 *Geophys. Res. Lett.* **25** 4549–4552
[41] Dieterich J 1994 *J. Geophys. Res.* **99** 2601–2618
[42] Kagan Y Y 2004 *Bull. Seismol. Soc. Am.* **94** 1207–1228
[43] Perfettini H and Avouac J P 2004 *J. Geophys. Res.* -Solid Earth **109** B06402
[44] Shcherbakov R, Turcotte D L and Rundle J B 2004 *Geophys. Res. Lett.* **31** L11613
[45] Toda S, Stein R S, Reasenberg P A, Dieterich J H and Yoshida A 1998 *J. Geophys. Res.* **103** 24543
[46] PMD and Norsar 2007 Seismic hazard analysis and zonation for Pakistan, Azad Jammu and Kashmir Tech. rep. Pakistan Meteorological Department and Norsar, Norway
[47] Castellaro S, Mulargia F and Kagan Y Y 2006 *Geophys. J. Int.* **165** 913
[48] Sipkin S A, Person W J and Presgrave B W 2000 Earthquake bulletins and catalogs at the USGS National Earthquake Information Center Tech. rep. U.S.Geological Survey National Earthquake Information Center URL `www.neic.usgs.gov`
[49] Nalbant S S, Steacy S and McCloskey J 2006 *Geophy. J. Int.* **167** 309
[50] Kagan Y Y 2006 Why does theoretical physics fail to explain and predict earthquake occurrence? *Modeling Catastrophic and Critical Phenomena in Geoscience: A Statistical Physics Approach* ed Bhattacharyya P and Chakrabarti B K (Springer-Verlag)
[51] Sornette D and Sammis C G 1995 *J. Phys.* I **5** 607
[52] Jaume S C and Sykes L R 1999 *Pure Appl. Geophys.* **155** 279
[53] Fukuyama E, Kubo A, Kawai H and Nonomura K 2001 *Earth Planets Space* **53** 1021
[54] Turcotte D L 1997 *Fractals and Chaos in Geology and Geophysics* (Cambridge Univ. Press, Cambridge)

Journal of Physics: Conference Series **319** (2011) 012005 doi:10.1088/1742-6596/319/1/012005

Damage of silicate glasses during stress corrosion

F. Lechenault

[1] CEA-Saclay, IRAMIS, SPEC, F-91191 Gif-sur-Yvette, France
[2] LCVN, UMR 5587 CNRS-UM2, Place Eugène Bataillon, 34095 Montpellier Cedex 5, France

C. L. Rountree

CEA-Saclay, IRAMIS, SPCSI, F-91191 Gif-sur-Yvette, France

F. Cousin

CEA-Saclay, IRAMIS, Laboratoire Léon Brillouin, CNRS UMR12, F-91191 Gif-sur-Yvette, France

J.-P. Bouchaud

Science and Finance, Capital Fund Management, 6 Boulevard Haussmann, 75009 Paris, France

L. Ponson

Institut Jean Le Rond d'Alembert, UMR 7190, CNRS and Univ. Paris 06 UPMC, 75005 Paris, France

E. Bouchaud

CEA-Saclay, IRAMIS, SPEC, F-91191 Gif-sur-Yvette, France

E-mail: elisabeth.bouchaud@cea.fr

Abstract. We show that water penetrates into the silicate glass matrix during stress corrosion fracture by probing what is stored under the fracture surface using neutron reflection. The concentration profile determined for two different values of the external loading exhibits a region close to the fracture surface where the water content is fairly elevated, suggesting a high amount of damage.

1. Introduction

Although glass is the archetype of brittle elastic materials, it was shown by S. Wiederhorn that most of the mechanical energy released is not spent in the creation of free surfacess [1]. As a matter of fact, the measured fracture energies were shown to be seven to ten times larger than expected, and the nature of the "lost energy" has been since then the topic of passionate discussions [2, 3, 4, 5]. The first possibility one may think of as an origin to the observed discrepany between measured and expected energies is plasticity. However, although plastic deformation – a local densification – as high as $\sim 20\%$ can be observed when silicate glasses undergo shear or pressure or both [6], like under the tip of an indenter [7], such a contribution is not likely to arise at the tip of a crack submitted to mode I tension. As a matter of fact, it can be shown that the expected size of the plastic zone is of a few nanometers at most, so that the energy dissipated through plasticity is much smaller than the energy spent in surface creation.

It has also been suggested, first from Molecular Dynamics (MD) simulations [8, 9, 10, 11, 12] that energy dissipation might be due to the occurrence of "quasi-brittle" damage [13, 14, 15, 16]. Quasi-brittle damage is observed in very disordered materials such as wood [17, 18, 19], mortar or concrete [20, 21], which have negligible plasticity. In a perfect crystal, where atomic bond orientations and energies are δ-distributed, bonds at the crack tip will break first because stress concentration is maximum there. In contrast, because of disorder, the bonds which break first are likely to be at some distance from the one sitting exactly at the crack tip. As a result, small cracks nucleate, and subsequently grow in a zone ahead of the crack tip often referred to as the "Process Zone" (PZ), the extent of which depends on the location of the most brittle parts of the inhomogeneous specimen. For the materials quoted above, the PZ size may be of macroscopic dimensions. Does this effect arise in glass too, albeit at a much smaller scale? Glass is known to be quite homogeneous down to length scales of $\sim$10 nm. Hence, the nature of the disorder which might be at the origin of such quasi-brittleness is not completely clear. However, one can suggest that the local orientation of Si-O bonds with respect to the external applied stress may play an important role. In this case, the amorphous structure providing local random orientations, it can be a source of disorder sufficient to nucleate and grow damage at a distance from the crack tip. Indications of such damage can be seen on the fracture surfaces of metallic glasses [22, 23].

Direct observations of the mechanisms involved in the fracture of silicate glasses are more challenging, because of the very small length scales involved. Furthermore, following a dynamic crack during its propagation is inaccessible experimentally. Actually, in order to follow crack growth with an Atomic Force Microscope (AFM), with a scan size of $\sim$ 100nm, taking on the order of one minute to acquire, the crack velocity cannot exceed $\sim$ 1nm.s^{-1}. Depending on the degree of humidity, stress corrosion subcritical cracks in silicate glasses can progress at very slow velocities, i.e. typically 10^{-12} to 10^{-5}m.s^{-1} under very moderate external tensile stresses, thanks to a chemical reaction involving the water molecules in the surrounding environment. This is a complex phenomenon, which started to be studied in the sixties [24, 25] and is not yet fully understood (see [26] for a recent review).

In the classical picture, proposed first by Michalske and Bunker [27], water molecules break the Si-O bonds located exactly at the crack tip thanks to a hydrolysis reaction. For small enough external loads, the crack velocity is controlled by the rate of the chemical reaction, which depends both on the degree of ambient humidity and on the applied stress. This regime is traditionally referred to as "Stage I" [24, 28]. At higher applied loads, when the crack velocity reaches a value that depends on humidity, the slowest phenomenon (which imposes its kinetics to the crack velocity) is the diffusion of water molecules to the tip along the fracture surfaces. Since surface diffusion is not sensitive to the external applied load, the crack velocity in this "Stage II" does not depend on it either.

This classical picture, however, does not take into account the disordered character of the glass structure, while in glass, the complex arrangement of chemical bonds make the occurrence of disjunctions away from the tip most probable, which results in "quasi-brittle" [28, 16] damage. *In situ* Atomic Force Microscopy (AFM) experiments [2, 4, 29] were performed in order to follow a stage I stress corrosion crack during its propagation,and crack openings ahead of the main tip were claimed to be observed. These observations, however, are still very controversial [3, 30]. Because AFM observations are restricted to the free surface of the specimens, several artifacts can alter the measurements [5, 31]. And, as a matter of fact, there are several significant differences between the free surface and the bulk of the specimen. Of particular importance is the exposure to water in the case of stress corrosion fracture: while the free surface is in direct contact with the ambient humidity, water molecules have to diffuse within the material for Si-O bonds to break at a distance from the crack tip. Although experiments have been performed at high temperature only [32, 33], a rough extrapolation of Tomozawa et al's results [34] suggests a water diffusion coefficient in silica of the order of $\sim 10^{-21}$cm^2.s^{-1} at room temperature. This means that the penetration length of water molecules into unstrained glass should be approximately 3pm (respectively 0.3Å) during the time it takes for a crack moving at 10^{-6}m.s^{-1} (respectively at 10^{-8}m.s^{-1}) to cover 100μm.

However, because of the huge stresses concentrated at the crack tip, diffusion is enhanced by orders of magnitude in the vicinity of the tip during fracture [35, 36], as observed both in glass [37] and in several other materials [38, 39]. Therefore, water can penetrate into the glass and spread within a small zone around the crack tip. Because of the heterogeneity of the material mentioned above, water breaks bonds and creates microcracks ahead of the crack tip. This in turn increases further the diffusion of water, thereby creating more corrosion and potentially leading to a substantial damaged zone. Alternatively, and as suggested by recent experiments [40], water penetration can result in a volumic expansion of the zone saturated with water ahead of the crack, leading to effective cohesive forces that oppose the crack motion. This mechanism is similar to toughening by phase transformation in some ceramics, and does not require material deformation beyond its yield stress [41]. In both scenarii, a rather thick layer of water should remain trapped underneath the nominal fracture surface after the crack has propagated and stresses are relaxed. Since the diffusion constant is so small in unloaded silica glass (more than 100 days for travelling 1 nm), one should observe post-mortem a "fossil" water profile, essentially frozen-in at the time of its creation, with a thickness on the order of the size of the damaged zone. The aim of this work is to provide quantitative evidence for the above scenario using neutron reflectivity [42, 43] to measure the thickness of the water layer left behind the crack. We find that the penetration depth of water is on the order of a hundred ångströms, with a rather large volumic concentration in the vicinity of the free surface (of the order of 50%). This suggests that diffusion has been accompanied by a chemical degredation of the glass network. This observation strongly supports that the abnormally high fracture energy of silica glass results from stress corrosion damage within a nanometric volume surrounding the crack tip.

2. Fracture experiments

Fracture experiments were conducted in a highly controlled manner via Double Cleavage Drilled Compression (DCDC) samples. In this geometry, the stress at the crack tip naturally decreases, enabling us to conduct all our experiments in the sub-critical stress corrosion regime. DCDC samples used herein are cuboids of size 5 x 5 x 25mm^3 with a 1mm diameter hole drilled in the center. They are made of Corning 7980 pure silica. The fracture experiments were conducted in a glove box which had been saturated either with light or with heavy water. Two symmetrical precracks are first initiated from the hole of the sample as described in [4] (Fig. 1a). Subsequently the load is adjusted in order to reach a desired velocity [4]. Sample 1, broken in light water environment was submitted to an inital load of 1970N corresponding to a stress intensity factor K_I of 0.32MPa.m$^{1/2}$ (average velocity of 2.3^{-10}m.s^{-1}). For the sample 2, broken in a heavy water environment, Zone 1 (Fig. 2c) corresponds to a stress intensity factor K_I=0.61MPa.m$^{1/2}$. Zone 2 (Fig. 2c) corresponds to a stress intensity factor K_I=0.77MPa.m$^{1/2}$ (average velocity of 4.10^{-6}m.s^{-1}).

Note that heavy water has a coherent length density $b_w = 6.39\,10^{-6}$Å^{-2} higher than the one of silica ($b_s = 3.41\,10^{-6}$Å^{-2}), and of opposite sign to that of light water ($b_l = -0.53\,10^{-6}$Å^{-2}). If some water is trapped in the vicinity of the surface of the sample, the reflectivity of the sample should thus increase in the presence of heavy water (sample 2) whereas it would only weakly decrease with light water (sample 1).

Figure 1. Fracture experiment. (a) Scheme of the DCDC sample submitted to uniaxial compression. Two symmetric cracks progress under local mode I tension from the cylindrical hole drilled in the middle of the specimen. (b) Picture of the actual experiment. A strength gauge and a displacement gauge allow for the measurement of force and displacement during the experiment.

3. Neutron reflection experiments

Specular Neutron Reflectivity (SNR) measurements have been carried out on the horizontal time-of-flight EROS reflectometer (LLB, Saclay, France) with a fixed angle θ of 1.195, with a neutron white beam covering wavelengths λ from 4Å to 25Å, covering an accessible q-range (scattering wavevector $q = 2\pi \sin\theta/\lambda$) from 0.005 Å^{-1} to 0.032Å^{-1}.

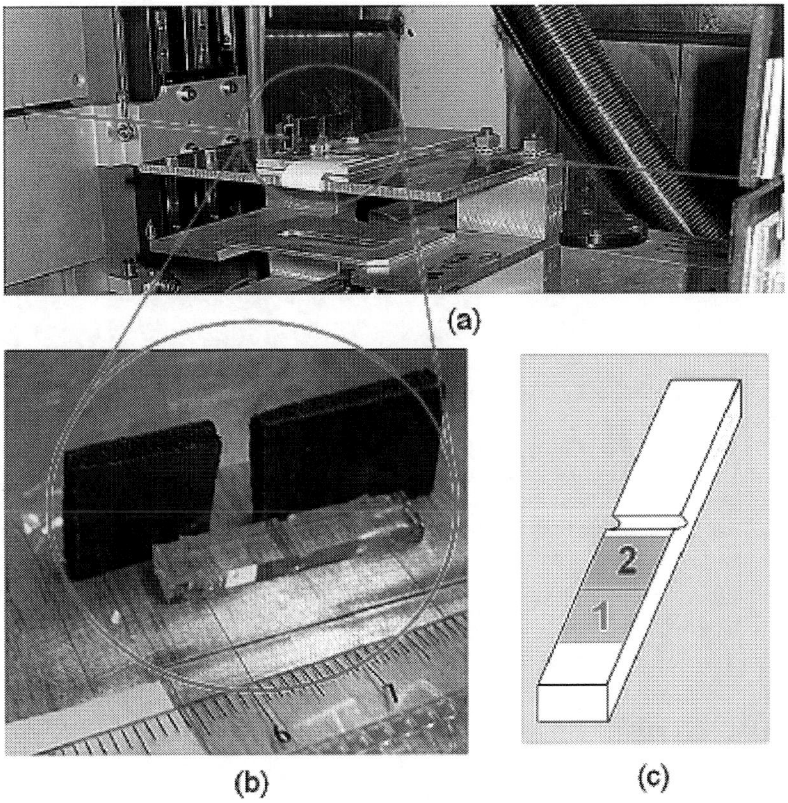

Figure 2. (a) Picture of the experimental setup. The neutron beam is schematized in red, incoming from the last slit of the collimator, reflecting on the sample and going within the slit in front of the detector. (b) Picture of the broken sample, showing the two black sheets of B_4C used to select one area of interest (see text). (c) Sketch of the broken sample, with the two parts described in the text.

In order to select Zone 1 or Zone 2 (see Fig. 2c)) in sample 2 described above, we used the following trick. The sample was almost completely hidden on the neutrons path by two black sheets of B_4C, a strong neutron absorber, to let the neutrons illuminate only the desired region (Fig. 2 a and b). In order to test that the selected region was flat enough to allow a correct measurement, we have checked that the full width at half max of the alignment rocking curve was lower than 0.25°. When this was not the case, the illuminated region was reduced until this condition was met. The resulting illuminated surfaces were very small, from $\sim 10\text{mm}^2$ to $\sim 25\text{mm}^2$. Because of this smallness, we used very long counting times to get a reasonable noise-to-signal ratio, up to 48 hours per illuminated region. In particular, we measured the background independently from the sample with high precision, enabling its subtraction with a good accuracy.

4. Results

Fig. 3 shows also that the reflectivity of the control specimen corresponds perfectly to the Fresnel reflectivity $R_F = (q - \sqrt{(q^2 - q_c^2)})^2/(q + \sqrt{(q^2 - q_c^2)})^2$ of a semi-infinite silica diopter for which the coherent length density is equal to b_s. The onset of total reflection is given by $q = q_c = \sqrt{4\pi b_s}$, where $b_s = 3.41\,10^{-6}\text{Å}^{-2}$ is the value found in the literature [43, 44].

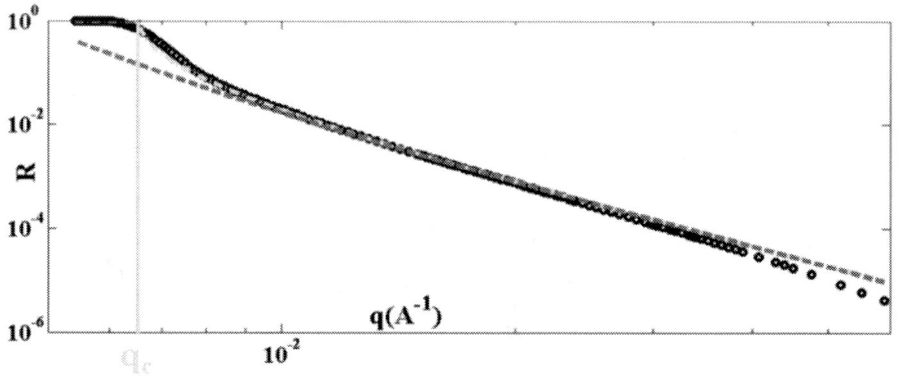

Figure 3. Experimental neutron reflectiviy of the control sample plotted as a function of the scattering wavevector q (open black circles). The theoretical Fresnel reflectivity (green dashed-dotted line) $R_F = (q - \sqrt{(q^2 - q_c^2)})^2/(q + \sqrt{(q^2 - q_c^2)})^2$ superimposes very well to the experimental curve up to $q \simeq 4.10^{-2}\text{Å}^{-1}$. The total reflectivity plateau takes place for $q < q_c$ with $q_c = \sqrt{4\pi b_s}$. The 2nd order Born approximation used for a slab of pure silica is plotted in dotted dark green, and it can be seen that it fits very nicely both the experiment and the theoretical expression as soon as $q > 9.10^{-3}\text{Å}^{-1}$, i.e. for $R \leq$some 10^{-2}.

While Fig. 4 shows that the reflectivity is unchanged when fracture is achieved in a light water environment, Fig. 5 shows a huge increase in the reflectivity of samples broken in a heavy water environment when compared to the reflectivity of the unbroken control specimen.

5. Theoretical analysis of the experimental observations

While the experimental curve presented in Fig. 4 does not show any measurable effect, the curves presented in Fig. 5 clearly show a huge change in the reflectivity of broken samples when compared to the reflectivity of the unbroken control specimen. Since the roughness of the fracture surfaces is larger than the roughness of the control specimen, one should *a priori* expect a decrease of the reflectivity of the broken samples. The difference seen in Fig. 5 is therefore underestimated.

In order to fit the neutron reflectivities measured in Zones 1 and 2 (Fig. 5), we have used a second order Born approximation, assuming that the heavy water concentration is not a constant, but that it decreases with the distance z from the free surface of the tested sample as $\phi(z)$. This concentration profile translates into a coherent length density profile $b_w(z) = b_w\phi(z)$. Since most of the workable signal is obtained in a region of large scattering vectors q, far enough from total reflexion, reflectivities are quite small, and hence the reference situation is the free case, when all the incident neutrons are transmitted.

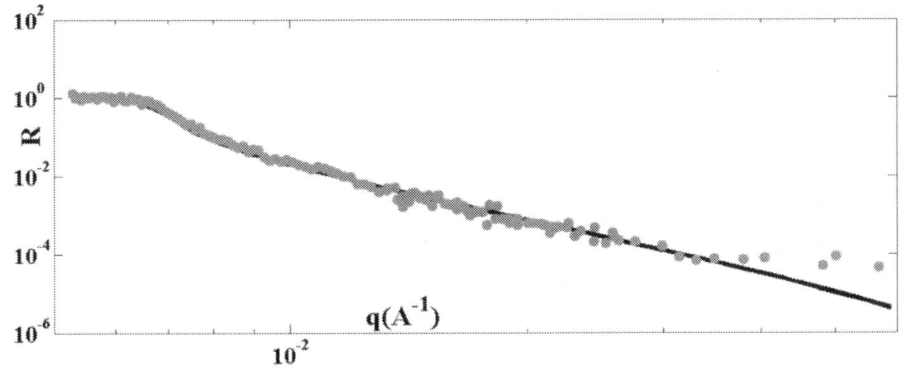

Figure 4. Experimental neutron reflectiviy of the light water stress corrosion fracture surface (red dots) compared to the measured Fresnel reflectivity (plain black line) of the control sample.

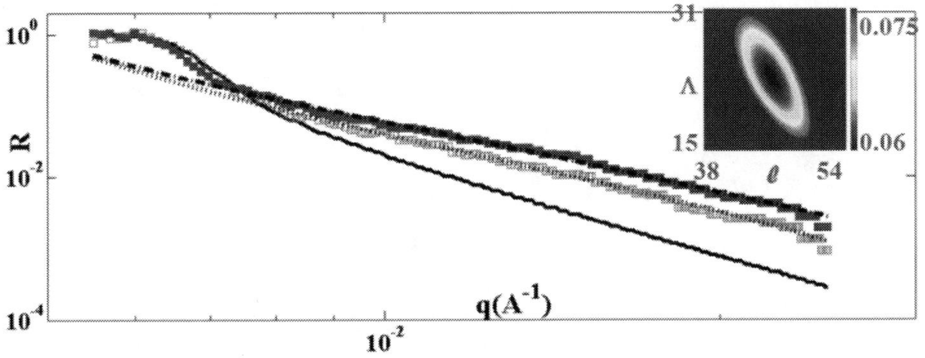

Figure 5. Experimental neutron reflectivities plotted as a function of the scattering wavevector q for: the control sample (plain black line),and for Zones 1 and 2 on the fracture surface of the specimen broken in a heavy water environment, corresponding respectively to stress intensity factors $K_I = 0.61\text{MPa.m}^{1/2}$ (empty red squares) and $K_I = 0.77\text{MPa.m}^{1/2}$ (full blue squares). The experimental results corresponding to Zone 1 and Zone 2 are fitted using Eqs. (10), (12) and (13). Fits are respectively a dotted black line and a black dashed line. The best fits correspond to $\phi_0^I = 0.348 \pm 0.003$, $\ell^I \approx 43\text{Å}$ and $\Lambda^I \approx 35\text{Å}$ in Zone 1, and $\phi_0^{II} = 0.567 \pm 0.003$, $\ell^{II} \approx 46\text{Å}$ and $\Lambda^{II} \approx 23\text{Å}$ in Zone 2. Inset: Contour lines of the fit root-mean square error in the plane (ℓ, Λ) in Zone 2, showing that while the combination $\ell_{eff} = \Lambda + \ell$ is rather well pinned down by the fit, $\Lambda - \ell$ is a "soft" direction. The relative experimental rms error per point is 0.075, whereas the minimum relative error achieved by the fit is 0.06.

The neutron wave function ψ obeys the following eigenvalue equation:

$$\frac{d^2\psi(z)}{dz^2} + (q^2 - V(z))\psi(z) = 0 \tag{1}$$

with $V(z) = 4\pi(b_s + b_w\phi(z))$ a small perturbation: $V(z) \ll q^2$.

We write:

$$\psi = \psi_0 + \psi_1 + \psi_2 + ... \tag{2}$$

with ψ_1 and ψ_2 the first and second order corrections (in V).

Adapting the calculation of [45], one finds:

$$\psi_{n+1}(z) = \int_0^\infty dz' V(z')\psi_n(z')G_0(z, z') \tag{3}$$

where the Green function $G_0(z, z' > 0)$ relevant for our boundary conditions refers to the free case:

$$\begin{array}{ll} G_0(z,z') = -\frac{1}{2iq}\exp\left(-iq(z'-z)\right) & \text{if } z<0 \\ G_0(z,z') = -\frac{1}{2iq}\exp\left(-iq|z'-z|\right) & \text{if } z>0 \end{array} \tag{4}$$

Hence:

$$\psi_1(z) = \frac{i}{2q}[\int_0^z dz'V(z')\exp -iqz + \int_z^\infty dz'V(z')\exp -2iqz'\exp iqz] \tag{5}$$

and:

$$\psi_2(z) = \int_0^\infty dz'V(z')\psi_1(z')G_0(z', z) \tag{6}$$

with $z' > 0$ and $z < 0$, i.e.:

$$\psi_2(z) = \frac{i}{2q}\int_0^\infty dz'V(z')\psi_1(z')\exp\left(-iq(z'-z)\right) \tag{7}$$

Adding the first and second order contributions, we get the total wave vector perturbation:

$$\begin{aligned} \psi_1(z)+\psi_2(z) &= \left(\frac{i}{2q}\int_0^\infty dz'V(z')\exp -2iqz'\right)\exp iqz \\ &-\frac{1}{4q^2}\int_0^\infty dz'V(z')[\exp -2iqz' \\ &\int_0^{z'} dz''V(z'') + \int_{z'}^\infty dz''V(z'')\exp -2iqz'']\exp iqz \end{aligned} \tag{8}$$

Using the above expression, we get the expression of the reflectance r to order V^2:

$$\begin{aligned} r &= \frac{i}{2q}\int_0^\infty dz'V(z')e^{-2iqz'} - \frac{1}{4q^2}\int_0^\infty dz'V(z') \times \\ &\times \left[e^{-2iqz'}\int_0^{z'} dz''V(z'') + \int_{z'}^\infty dz''V(z'')e^{-2iqz''}\right] \end{aligned} \tag{9}$$

The reflectivity R is $R = |r|^2$. In order to check the validity of our second order Born approximation, we first verify that our result in Eq. (9) tends to the Fresnel reflectivity for high values of q when V is a constant equal to $q_c^2 = 4\pi b_s$. This limit leads to the following reflectivity:

$$R = r_0^2 = \frac{q_c^4}{16q^4}\left(1+\frac{q_c^2}{2q^2}\right)^2 + O(\frac{q_c^8}{q^8}) \tag{10}$$

which coincides with the corresponding large q expansion of the Fresnel reflectivity (see Fig. 3).

We then tried to fit the reflectivities in Zones 1 and 2 (Fig. 5) using the simplest function involving a single length scale, i.e. $\phi(z) = \phi_0\exp\left(-z/\Lambda\right)$. Although this can be made to fit the Zone 1 results, the reflectivity increase in Zone 2 is too large to be accounted for using this simple function. Hence, guided by the idea that there might be a saturated layer of depth ℓ close to the surface that becomes more diffuse deeper in the sample, we posit that:

$$\begin{array}{ll} \phi_w(z) = \phi_0 & \text{if } z<\ell \\ \phi_w(z) = \phi_0\exp\left(-(z-\ell)/\Lambda\right) & \text{if } z>\ell \end{array} \tag{11}$$

This choice leads to a reflectance r that can be written as: $r = r_0 + r_1 + r_2$, with r_0 as the Fresnel reflectance (Eq. (10)) and:

$$r_1 = -\frac{2\pi b_w \phi_0}{q(2iq\Lambda + 1)} \exp\left(-2iq\ell\right) \tag{12}$$

$$\begin{aligned} r_2 &= -\frac{4\pi^2 (b_w\phi_0)^2}{q^2}\left[\frac{1}{2q^2}\left[\exp\left(-2iq\ell\right)(1+2iq\ell) - 1\right]\right. \\ &+ \left.\frac{\Lambda \exp\left(-2iq\ell\right)}{(2iq\Lambda+1)(iq\Lambda+1)}(\Lambda + 2\ell + 2iq\Lambda\ell)\right] \end{aligned} \tag{13}$$

Fig. 5 shows the best fits of the experimental measurements performed on the two fracture surfaces using Eqs. (10), (12) and (13):

$$R = |r_0 + r_1 + r_2|^2 \tag{14}$$

The best fit is achieved with $\phi_0^I = 0.348 \pm 0.003$, $\ell^I \approx 43$Å and $\Lambda^I \approx 35$Å in (slow) Zone 1, while in (fast) Zone 2, $\phi_0^{II} = 0.567 \pm 0.003$, $\ell^{II} \approx 46$Å and $\Lambda^{II} \approx 23$Å. Note that ϕ_0 is very accurately determined by the fit, although the error bar we quote only accounts for statistical uncertainty, and not systematic effects coming from the choice of the fitting function and of the interval over which the data is fitted. On the other hand, the quality of fit has a "soft direction" in the plane (ℓ, Λ), as represented in the inset of Fig. 5. As expected, the total effective width of the layer, $\ell_{eff} = \ell + \Lambda$ is better determined than ℓ and Λ separately. The statistical error bar on ℓ_{eff} is smaller than 1Å, but again systematic errors are much larger. As shown in the inset of Fig. 5, ℓ_{eff} can be varied by $\sim \pm 5$Å and still lead to an acceptable fit.

In order to evaluate the change in reflectivity due to the presence of light water under the fracture surface, we have computed R using Eq. (14), assuming that $\ell \simeq 46$ Ådoes not vary much with K_I, and $\Lambda \simeq 90$Å, since the penetration depth seems to increase when the external load is decreased. Because b_l is not only negative but approximately ten times smaller in absolute value than b_w, the reflectivity loss does not depend much on the amount of water stored under the surface, i.e. it does not depend much on ϕ_0. We have chosen to compute R in Eq. (14) for two values of ϕ_0, 0.05 and 0.5. The two results are actually indistinguishable in Fig. 6 (green dots). This means that our method is restricted to the use of heavy water as a corrosive agent.

Figure 6. Expected reflectivity (green dots) using Eq. (14) for light water, assuming a concentration profile of the type described by Eqs. (11), with ϕ_0, 0.05 and 0.5, $\ell \simeq 46$ Åand $\Lambda \simeq 90$ Å. The solid black line is the reflectivity of the control specimen, and the open red circles correspond to the measurements for sample 1, broken in a light water environment.

6. Discussion

Our results [46] clearly show that heavy water is present over $\sim$ 65 to 85Å under the stress corrosion fracture surface of pure silica in the conditions of our experiments. This penetration depth is much larger than what is expected using a room temperature extrapolation of the diffusion coefficient of light water in silica [34] (to our knowledge, there are no such results concerning heavy water, for which diffusion should be even smaller). The heavy water concentration within the first 40-50 Å under the fracture surface is found to be very high (35 and 57%) - it actually seems to be larger for a larger value of the applied load -. These values

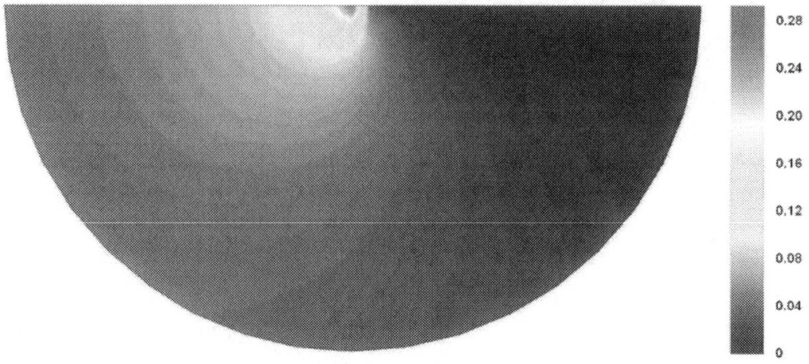

Figure 7. Finite element simulations of the dilation field $D(x, y) = \epsilon_{xx}(x, y) + \epsilon_{yy}(x, y)$ in the vicinity of the crack tip in the mid-plane of the specimen for an external loading $K_I = 0.61$ MPa.m$^{1/2}$. The crack propagates from right to left, and the color code indicates the amplitude of $D(x, y)$.

are too high to account for a situation where there would be no mechanical damage within the silica network. The resulting decrease of the concentration of silica may not have a negligible effect on the change of reflectivity, but it has not been taken into account in our interpretation of the data.

A likely explanation of our observations is as follows: diffusion enhancement in the vicinity of the crack tip, where huge tensile stresses are present ([39], [35], [47]), allows the water to penetrate inside the bulk and create a damaged zone which helps water progressing further still. Our neutron scattering experiment gives information about the water penetration depth in the direction perpendicular to the fracture surface. To investigate how far water penetrates parallel to the crack propagation direction, one would need a detailed self-consistent model for the coupled growth of the damaged zone and the diffusion of water. We grossly simplify the problem by computing the dilation field $D(x, y) = \epsilon_{xx}(x, y) + \epsilon_{yy}(x, y)$ in the vicinity of the crack tip within a purely elastic model, and postulating that the water penetrates in a region defined by $D(x, y) > D_c$, and infer the anisotropy of the water penetration from that of the iso-dilation lines. We analyze DCDC samples identical to those used in the experiments and compute the dilation field in the mid-plane of the specimen using finite element simulations, with element sizes decreasing exponentially as approaching the crack tip, so that the strain field is resolved at the nanometer scale within that region. As shown in Fig. 3 for an external loading $K_I = 0.61$ MPa.m$^{1/2}$, the domain at the crack tip with a high level of dilation extends deeper in the direction of propagation than in the perpedicular direction, by a factor $\sqrt{2}$. This result is consistent with the predictions of Linear Elastic Fracture Mechanics in the limit of very small distances from the crack tip [28]. Therefore, we estimate that water penetrates roughly $9-12$nm ahead of the crack tip, in the crack propagation direction. If one identifies the water-rich region with a damaged zone, our estimate is in agreement with the fact that the strain field observed in the vicinity of a stress corrosion crack tip is elastic only on scales larger than $\sim$ 10nm [48].

We observe an increase of the neutron reflectivity, not only with respect to the Fresnel case, but also with respect to the case where we assume a heavy water single-length scale, exponentially decreasing profile. This has prompted us to postulate the presence of an homogeneous layer of water, followed by an exponentially decaying profile. The width of the layer, $\ell \approx$ 4nm, is significantly larger than the height fluctuations on silica fracture surfaces which do not exceed 1nm [49, 50]. It is tempting to interpret this zone as a strongly damaged zone, with a density of microcracks that is larger when the stress intensity factor – and the crack velocity – is higher. This would naturally explain why $\phi_0^{II} > \phi_0^{I}$, although we have no clear explanation as why

$\ell^I \approx \ell^{II}$. This might be due to the cancellation of two opposite effects: a stronger stress enhances the diffusion of water in the bulk, but at the same time the crack speed is larger, leaving less time for the corrosion mechanism to operate. The latter mechanism in fact explains why the exponential region, which is probably more sensitive to diffusion, is wider in Zone 1 than in Zone 2 ($\Lambda^I > \Lambda^{II}$). In order to be more quantitative, one requires at this stage a detailed model for the formation of the damaged zone, dynamically coupled to the water profile. A more systematic study of the effect of the external applied stress on the heavy water content would be needed to provide a sound basis for such a quantitative model. Further experiments will be performed with larger specimens such as to increase the intensity of the reflected neutron beam.

Preliminary results have been obtained recently [51] using a Finite Element model which takes into account both the enhancement of water diffusion and damage formation in the vicinity of the crack tip: a concentration profile composed of a very damaged zone saturated in water followed by a rapid decrease of the water content was obtained, as well as an exponential dependence of the crack velocity with the applied stress intensity factor.

Acknowledgments
Part of this work was supported by ANR CORCOSIL. We acknowledge very interesting discussions with S. Chapuliot.

References

[1] Wiederhorn S 1969 *J. of the Am. Ceram. Soc.* **52** 99–105
[2] Célarié F, Prades S, Bonamy D, Ferrero L, Bouchaud E, Guillot C and Marlière C 2003 *Phys. Rev. Lett.* **90** 075504
[3] Guin J P and Wiederhorn S M 2004 *Phys. Rev. Lett.* **92** 215502
[4] Prades S, Bonamy D, Dalmas D, Bouchaud E and Guillot C 2004 *Int. J. of Sol. and Struct.* **42** 637–645
[5] Fett T, Rizz G, Creek D, Wagner S, Guin J P, López-Cepero J M and Wiederhorn S M 2008 *Phys. Rev. B* **77** 174110
[6] Rountree C L, Vandembroucq D, Talamali M, Bouchaud E and Roux S 2009 *Phys. Rev. Lett.* **102** 195501
[7] Kermouche G, Barthel E and Vandembroucq D 2008 *Acta Mater.* **56** 3222–3228
[8] Nakano A, Kalia R K and Vashishta P 1995 *Phys. Rev. Lett.* **75** 3138–3141
[9] Kalia R, Nakano A and Vashishta P 1997 *Phys. Rev. Lett.* **78** 689–692
[10] Rountree C L, Kalia R, Lidorikis E, Nakano A, Brutzel L V and Vashishta P 2002 *Annual Review of Materials Research* **32** 377–400
[11] Kalia R, Nakano A, Vashishta P and Rountree C 2003 *IEEE Computer Society* 36–39
[12] Rountree C L, Prades S, Bonamy D, Bouchaud E, Kalia R and Guillot C 2007 *J. of Alloys and Compounds* **434-435** 60–63
[13] Bažant Z and Cedolin L 1991 *Stability of Structures: Elastic, Inelastic, Fracture and Damage Theories* (Oxford University Press, New-York, USA)
[14] Bažant Z 1997 *Int. J. of Fract.* **83** 19–40
[15] Morel S 2007 *Int. J. of Sol. and Struct.* **44** 42724290
[16] Bonamy D and Bouchaud E 2011 *Physics Reports* **498** 1–44
[17] Morel S, Bouchaud E, Schmittbuhl J and Valentin G 2002 *Int. J. of Fract.* **114** 307–325
[18] S Morel E B and Valentin G 2002 *Phys. Rev. B* **64 (104101)** 1–8
[19] Morel S, Dourado N, Valentin G and Morais J 2005 *Int. J. of Fract.* **131** 385–400
[20] Bažant Z 2002 *Eng. Fract. Mech.* **69** 165–205
[21] Mourot G, Morel S, Bouchaud E and Valentin G 2005 *Phys. Rev. E* **71** 016136
[22] Xi X K, Zhao D Q, Pan M X, Wang W H, Wu Y and Lewandowski J J 2005 *Phys. Rev. Lett.* **94** 125510
[23] Bouchaud E, Boivin D, Pouchou J L, Bonamy D, Poon B and Ravichandran G 2008 *Europhys. Lett.* **83** 66006
[24] Wiederhorn S M 1968 *Int. J. of Fract. Mech.* **4** 171–177
[25] Wiederhorn S M 1969 *J. Am. Ceram. Soc.* **52**
[26] Ciccotti M 2009 *J. Phys. D: Appl. Phys.* **42, Special Issue "Fracture: from the atomic to the geophysical sclae", Guest Editors E. Bouchaud and P. Soukiassian** 214006
[27] Michalske T and Bunker B 1984 *J. Appl. Phys.* **56** 2686–2693
[28] Lawn B 1993 *Fracture of Brittle Solids- Second Edition* (Cambridge University Press)

[29] Rountree C L, Prades S, Bonamy D, Bouchaud E, Kalia R and Guillot C 2010 *Phys. and Chem. of Glasses-Europ. J. of Glass Sci. and Tech. Part B* **51** 127–132
[30] Lòpez-Cepero J, Wiederhorn S, Fett T and Guin J P 2007 *Int. J. Mater. Res.* **98** 1170–1176
[31] Lechenault F, Pallares G, George M, Rountree C, Bouchaud E and Ciccotti M 2010 *Phys. Rev. Lett.* **104** 025502
[32] Davis K and Tomozawa M 1996 *J. of Non-Cryst. Sol.* **201** 177–198
[33] Berger S and Tomozawa M 2003 *J. of Non-Cryst. Sol.* **324** 256–263
[34] Tomozawa M and Davis K M 1999 *Mat. Sci. and Eng. A* **272** 114119
[35] Larché F and Voorhees P 1996 *Defect and Diffusion Forum* **129-130** 31–36
[36] Greer A 1996 *Defects and Diffusion Forum* **129-130** 163–180
[37] Tomozawa M, Han W and Lanford W A 1991 *J. Am. Cer. Soc.* **74** 2573
[38] Aziz M J, Zhao Y, Gossmann H J, Mitha S, Smith S P and Schiferl D 2006 *Phys. Rev. B* **73** 054101
[39] Guery J, Baudry J, Weitz D A, Chaikin P M and Bibette J 2009 *Phys. Rev. E* **79** 060402(R)
[40] Wiederhorn S Private communication
[41] Kelly P M and Rose L R F 2002 *Prog. Mat. Sci.* **47** 463–557
[42] Bouchaud E, Farnoux B, Sun X, Daoud M and Jannink G 1986 *Europhys. Lett.* **2** 315–322
[43] Cousin F and Menelle A *La réflectivité de neutrons* (Les Editions du CNRS)
[44] http://wwwsfnassofr/Enseignement/Outils/Table/indexhtml
[45] Charmet J C and de Gennes P G 1983 *J. Opt. Soc. Am.* **73** 1777–1784
[46] Lechenault F, Rountree C L, Cousin F, Bouchaud J P, Ponson L and Bouchaud E 2011 *Phys. Rev. Lett.* **106** 165504
[47] Nazarov A V and Mikheev A A 2008 *J. Phys.: Condens. Matter* **20** 485203
[48] Pallares G 2010 *Analyse multi-échelle des mécanismes de propagation de fissure dans les verres d'oxydes* Ph.D. thesis Montpellier II University (France)
[49] Bonamy D, Ponson L, Prades S, Bouchaud E and Guillot C 2006 *Phys. Rev. Lett.* **97** 135504
[50] Ponson L 2007 *Annales de Phys.* **32** 1
[51] Chapuliot S and Bouchaud E Preliminary results

Spreading of Non-Newtonian and Newtonian Fluids on a Solid Substrate under Pressure

Moutushi Dutta Choudhury [1], **Subrata Chandra** [1], **Soma Nag** [1], **Shantanu Das** [2] **and Sujata Tarafdar** [1]

[1] Condensed Matter Physics Research Centre, Physics Department, Jadavpur University, Kolkata 700032, India

[2] Reactor Control Division, Bhabha Atomic Research Center, Trombay, Mumbai 400085, India

E-mail: `mou15july@gmail.com`

Abstract. Strongly non-Newtonian fluids namely, aqueous gels of starch, are shown to exhibit visco-elastic behavior, when subjected to a load. We study arrowroot and potato starch gels. When a droplet of the fluid is sandwiched between two glass plates and compressed, the area of contact between the fluid and plates increases in an oscillatory manner. This is unlike Newtonian fluids, where the area increases monotonically in a similar situation. The periphery moreover, develops an instability, which looks similar to Saffman Taylor fingers. This is not normally seen under compression. The loading history is also found to affect the manner of spreading. We attempt to describe the non-Newtonian nature of the fluid through a visco-elastic model incorporating generalized calculus. This is shown to reproduce qualitatively the oscillatory variation in the surface strain.

1. Introduction

Static and dynamic aspects of wetting and spreading are well studied classical text book problems. However, forced spreading of a fluid under an impressed load is not widely discussed, though the physics involved is challenging and the problem has important applications in technology as well. In real life many fluids are non-Newtonian, adding further complexity to the problem. General reviews concerning related problems are available [1, 2, 3], but there is scope for more experimental and analytical studies.

In the present work we report studies on spreading of non-Newtonian fluids between two glass plates, when the upper plate is loaded by a weight. We vary the weights, the fluids and the manner of loading.

We observe an interesting oscillation in the area of contact between the fluid and glass plate as a function of time. Earlier study of Newtonian fluids [3, 4] did not show such behavior. Based on the experimental results we try to explain this phenomenon using fractional calculus, which is known to be an appropriate technique to study non-Newtonian, visco-elastic materials [5, 6, 7]. Another remarkable observation is the appearance of a surface instability similar to viscous fingering. Viscous fingering under the condition of lifting, i.e. separating the plates is a well studied phenomenon [8, 9, 10]. In this case the pressure is lower within the fluid, compared to the air pressure outside, satisfying the Saffman-Taylor condition for instability [11]. However, in the present case the fingering develops during *compression*.

2. Materials and Methods

2.1. Materials

The fluids under study are two non-Newtonian fluids - arrowroot and potato starch gel and a Newtonian fluid ethylene glycol (GR) is also studied for comparison.

Starch granules suspended in water produce a whitish liquid. When heated, the granules of starch, which contain amylose and amylopectin, absorb water and swell up. Finally the starch grain collapses and long chained amylose molecules are leached out. The long chains join to form an amorphous mass with the water trapped inside. This is the gel phase. In the gel phase there are no granules with boundaries that scatter light, so it looks clear.

In order to get a homogenous gel of the starch in water the suspension has to be heated to temperature ($> 100\,^{\circ}$C). A preliminary set of experiments was done with arrowroot starch (Indo Moulders, India) available commercially for stiffening fabrics. 2.5gm of the starch is dissolved in 100ml of distilled water and it is heated up for 20 minutes and boiled for 2 minutes to get a clear gel. A pinch of dye is added to enhance the contrast and the solution is allowed to cool. Continuous stirring is necessary so that lumps are not formed and a smooth homogenous fluid is obtained.

The second set of experiments was done on a well characterized sample of potato starch ($(C_6H_{10}O_5)_m$) manufactured by Lobachemie Pvt. Ltd. (Mumbai), the procedure for preparing the gel is the same above but it has to be heated for 10 minutes. During heating and boiling, some water is lost due to evaporation. The ratio water:potato starch (w/w) changes from 40:1 before boiling to 35:1 after boiling and cooling.

The Newtonian fluid that we use is ethylene glycol GR,(E.Merck, India). Here also a little food colouring is added for enhancing contrast.

2.2. Methods

A droplet of the non-Newtonian fluid is placed on a smooth glass plate using a micro-pipette. The mass of the droplet varies from 0.04 gm to 0.06 gm.

The fluid drop is compressed by two different loading processes.

(i) A load of W kg. is placed on another glass plate (identical to the bottom plate and weighing 570 gm).Then the plate and load are placed together on the drop.

(ii) The upper glass plate is first placed on the droplet. Then the weight W is placed on top of the upper plate after an interval of 2 seconds.

Each of the non-Newtonian fluids is subjected to the two loading processes described above. Ethylene glycol is compressed by the second process. The change in area of contact between the fluid and plate is video recorded from below. The area is measured using Image Pro Plus software and plotted as a function of time.

3. Experimental Observations

The strain is defined as the fractional change in area per unit area at $t = 0$.

$$\epsilon(t) = \frac{a(t) - a_0}{a_0} \tag{1}$$

$a(t)$ and a_0 being respectively the area at time t and at the initial time $t = 0$. Time $t = 0$ is taken differently for the different loading processes. The results for the different sets of experiments are described below.

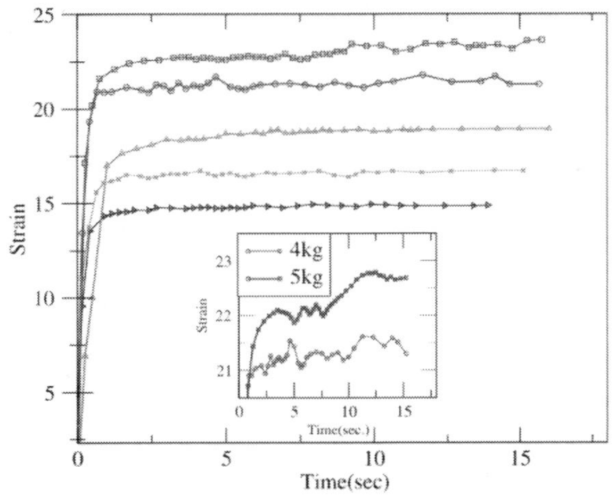

Figure 1. Variation in strain with time for arrowroot gel on glass, loaded by W 1 tp 5 kg by process (i) (black right triangle - 1 kg, red cross - 2 kg, green up triangle - 3 kg, blue square - 4 kg, magenta square - 5 kg) . The inset shows the erratic variations magnified for 4 and 5 kg respectively.

3.1. *Set (A) - Arrowroot gel*

- Loading process (i) Here the initial time is the instant when the total load - (plate + W) is placed in position. The variation in strain for the arrowroot gel is shown in figure(1) for different weights W. The area shows erratic noisy oscillations, more prominently displayed in the magnified inset.
- Loading process (ii) Here the initial time $t = 0$ refers to the instant when the weight W is placed, the upper plate already being in position. Figure(2) shows the variation in strain. Here the variation is less erratic, but all sets show that the area initially overshoots the final equilibrium value ϵ_∞, then approaches ϵ_∞ after one or two cycles of oscillation.
  In our experiments, the masses of the initial fluid drop for the different sets, could not be made exactly equal. This leads to some error which makes the data for 1 and 2 kg, as well as 3 and 4 kg appear very close to each other in figure(2).

This video is taken at 4 frames/sec. Though the oscillations are apparent, clearly, recording at higher speed is required. Instead of proceeding with the commercial arrowroot sample, which is not well characterized, we continued the experiments with potato starch.

3.2. *Set (B) - Potato Starch gel*

All the subsequent results show videos taken at 10 frames/sec.

- Loading process (i) Here again noisy variations, which may be superposed on oscillations are seen in the results (figure(3)).

- Loading process (ii) In this case, the erratic noise is much reduced and clear oscillations are observed in figure(4). The oscillations die down after a few cycles to the equilibrium value ϵ_∞. The amplitude of the oscillations increases with load W. The peak-to-peak differences for the first cycles for each load, were measured as: 0.13 cm^2 (2 kg.), 0.15 cm^2 (3 kg.), 0.18 cm^2 (4 kg.) and 0.27 cm^2 (5 kg.). The average frequency over the first 3 cycles was measured, but did not show a definite trend with load variation, the values ranging from

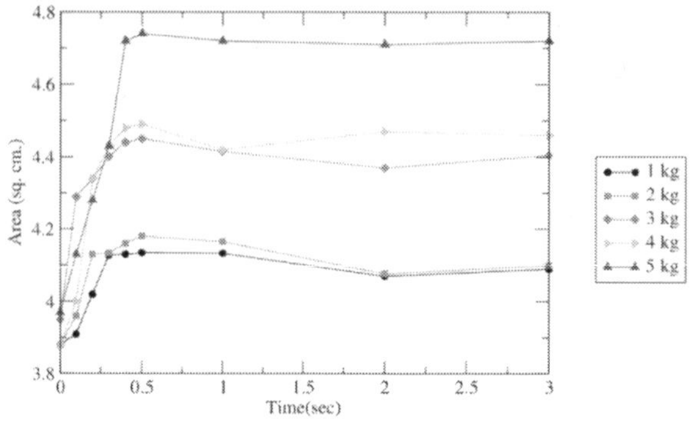

Figure 2. Variation in area of contact with time for arrowroot gel on glass, loaded by W 1 tp 5 kg by process (ii). Clear oscillations are seen, though not very sharp.

0.35 to 0.75 hz. For the load 5kg. however, all the measurements gave values close to 0.35 hz. There was a problem with the results for 1 kg. load and we have omited them in the present paper. More careful repeated measurements will be performed in future to verify all the observations.

The experiments are repeated for reproducibility and four such data sets for $W = 5$ kg are shown in figure(5).

3.3. Set (C) - Ethylene glycol

Ethylene glycol has been studied earlier, together with other Newtonian fluids [3, 4]. We have repeated the experiments, recording the video at 10 frames/sec and show the results here(figure(6)) for comparison with the gels . Loading process (ii) is employed here. The results are very clearly different from the previous sets for the non-Newtonian fluids. Here, the area increases smoothly and saturates to the equilibrium value for all loads W.

The results show convincingly that these non-Newtonian fluids show an oscillatory spreading, when compressed in a Hele-Shaw cell, unlike Newtonian fluids. It is to be noted that this is not a simple stick slip behavior, where the strain (here this is equivalent to area) would always increase, but in jumps. Here for the gels, the strain, i.e the area, actually decreases after a short time interval during loading and again increases. To establish that the film *does* shrink and expand in an oscillatory manner, we superpose two snapshots taken at an interval of 2 seconds, and show their difference in figure(7). The weight on the upper plate is 5 kg. Here the outer boundary corresponds to an instant of time *earlier* than the inner. So the oscillations in strain are genuine and not due to measuremental error. Further, the corrugated appearance of the boundary, demonstrates the Saffman-Taylor like instability developed.

In the subsequent sections we suggest a visco-elastic model to explain our experiments. At present, we focus on the less noisy data for loading process (ii). Presumably, loading process (i) needs a more complicated theory, because the weights are placed on the hemispherical drop directly. So change in shape i.e. flattening of the drop as well as change in area are to be considered in this case.

3.4. Fluid Characterization

It is well known that starch solutions are non-Newtonian [12, 13]. It is interesting that the suspension in water made without heating is shear thickening [14], whereas the gel made by boiling the solution is shear thinning.

Rheological studies of the arrowroot gel and potato starch gel, were done at Central Glass and Ceramic Research Institute(CGCRI, CSIR) by a Bohlin rotational Rheometer at 25°C temperature. No preshear was applied to the samples prior to measurement. Both gels show strongly shear thinning nature of the fluids with a yield stress. The apparent viscosity of potato starch gel is somewhat smaller than arrowroot for all strain rates.

A log-log plot of strain against shear rate shows a power law relation for both fluids over a wide range (figure(8)). An exponent 1.7 is obtained for arrowroot, while 1.5 is obtained for potato starch, as shown in the best fit power-law straight lines shown in figure(8).

The graphs indicate a non-zero yield stress, but the value of the yield stress was not measured precisely in the present set of experiments.

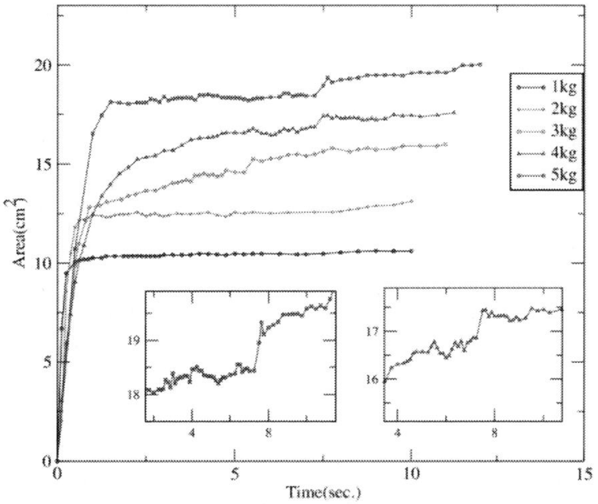

Figure 3. Variation in strain with time for potato starch gel on glass, loaded by W 1 tp 5 kg by process (i). The insets on the right and left show the erratic variations magnified for 4 and 5 kg respectively.

4. Mathematical Modeling

To analyze the stress-strain behavior we use a basic Kelvin-Voigt model [15, 16] of a spring and dash pot in parallel as a basic unit. The viscous behavior is represented by the dash pot and the elastic nature by the spring. The model is further generalized by taking the qth derivative of the strain, thus introducing non-Newtonian rheology through fractional calculus[5, 17]. Here q may be a fraction, for case $q = 1$ we have a Newtonian viscosity. The elastic term is assumed to be Hookean at present. The stress is then given by

$$\sigma(t) = \beta\tau^q \frac{d^q\epsilon}{dt^q} + E\epsilon \tag{2}$$

Here ϵ is the strain, E the elastic modulus and β a parameter characterizing the effective viscosity of the non-Newtonian fluid and τ a characteristic time of the system.

We assume a step function to represent the loading

$$\sigma(t) = \sigma \text{ for } t \geq 0, \text{ and } \sigma(t) = 0 \text{ for } t < 0$$

. The initial condition for strain is $\epsilon(t) = 0$ for $t < 0$.

The Laplace transform of equation(2) gives

$$\epsilon(s) = \frac{\sigma}{E}\left[\frac{1}{s} - \frac{s^{q-1}}{s^q + E/\beta}\right] \tag{3}$$

The inverse Laplace transform of equation(3) gives

$$\epsilon(t/\tau) = \frac{\sigma}{E}\left[1 - ML_q\Big(-\frac{E}{\beta}(t/\tau)^q\Big)\right] \tag{4}$$

where, $ML(-kt)$ is the one parameter Mittag-Leffler function defined by

$$ML_1(-kt) = e^{-kt}$$

$$ML_q(z) = \sum_{k=0}^{\infty}\frac{z^k}{\Gamma(qk+1)}$$

Visco-elastic systems typically have a 'memory', the strain at time t is determined by its previous loading history starting from $t = -\infty$. The Markovian system withour memory is a special case of general non-Markovian systems exhibiting correlation or anti-correlation. The general case can be treated naturally using generalized calculus methods, as follows.

One has to consider the Green's function for a general relaxation in equation(2), so we write the homogeneous equation with RHS equal to zero. The strain built up for any relaxation, as a function of a scaled dimensionless time t_r. may be treated as convolution integral of a strain variable with integral kernel $K_q(t)$, as [6, 7]. We use henceforth this reduced dimensionless time t_r, which is the time scaled by some characteristic time of the sytem.

$$\frac{d}{dt_r}\epsilon(\mathbf{t_r}) = -\int_0^{\mathbf{t_r}} \mathbf{K_q}(\mathbf{t_r} - \mathbf{t})\epsilon(\mathbf{t})\mathbf{dt} \tag{5}$$

We first consider two special cases, with no memory and infinite memory.

4.1. No memory

If the memory kernel is $K(t_r) = B_0\delta(t_r)$, we have the above system 2 without memory [6, 7] and the Green's function will be

$$\epsilon(t_r) = \epsilon_0 e^{-B_0 t_r}$$

that is the impulse response quickly decays to zero. Here ϵ_0 is initial strain of the system at $t_r = 0$.

This can be derived as follows:

$$K(t_r) = B_0\delta(t_r) \tag{6}$$

$$\frac{d}{dt_r}\epsilon(t_r) = -\int_0^{t_r} \delta(t - t_r)\epsilon(t)dt = -B_0\epsilon(t_r) \tag{7}$$

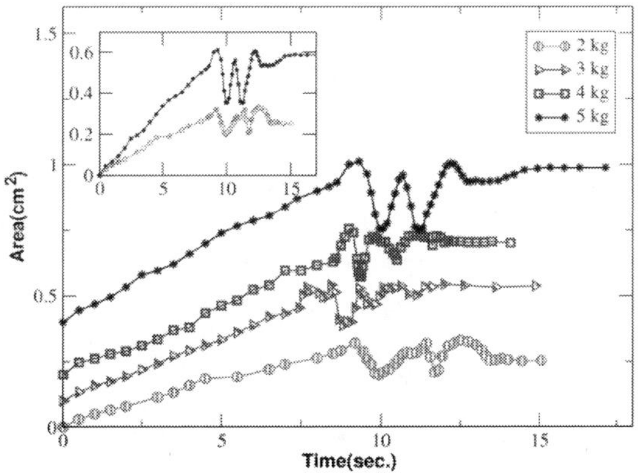

Figure 4. Variation in area of contact for potato starch on glass, loaded by W = 2,3,4 and 5 kg by process (ii). The curves are shifted along the y-axis with respect to each other for clarity. The inset shows the results for 2 and 5 kg. only, without the shift. Amplitude of the oscillations increases with load.

$$\epsilon(t_r) = \epsilon_0 e^{-B_0 t_r} \tag{8}$$

The homogeneous strain relaxation equation for no-memory case is a first order Ordinary Differential Equation i.e.

$$\frac{d}{dt_r}\epsilon(t_r) + B_0\epsilon(t_r) = 0 \tag{9}$$

4.2. Infinite memory

If the memory kernel is a constant say $K_2(t_r) = B_2$, then we will have oscillatory Green's function, which never decays to zero .

$$\frac{d^2}{dt_r^2}\epsilon(t) = -B_2\epsilon(t_r) \tag{10}$$

$$\epsilon(t_r) = \epsilon_0 cos(\sqrt{B_2}t_r) \tag{11}$$

4.3. Generalized case

If a generalized memory integral of the following form is taken

$$K(t) = B_q t_r^{q-2} \quad ; \quad 0 < q \leq 2 \tag{12}$$

one has

$$\frac{d}{dt_r}\epsilon(t_r) = -\frac{1}{\alpha^q}\left[\frac{d^{(1-q)}}{dt^{(1-q)}}\epsilon(t_r)\right] \tag{13}$$

where,

$$\alpha^q = \left[B_q\Gamma(q-1)\right]^{-1} \tag{14}$$

Integrating the equation(13), we have

$$\epsilon(t_r) - \epsilon_0 = -\frac{1}{\alpha^q}\left[\frac{d^{(-q)}}{dt^{(-q)}}\epsilon(t_r)\right] \tag{15}$$

Differentiating this to order q, we get the corresponding generalized differential equation.

$$\frac{d^q\epsilon(t)}{dt^q} - \epsilon_0\frac{t^{-q}}{\Gamma(1-q)} = -\alpha^{-q}\epsilon(t) \tag{16}$$

using the fact that the differentigral of order q of the constant ϵ_0 gives $\epsilon_0\frac{t^{-q}}{\Gamma(1-q)}$.

This is the equation for the system with the memory index entering as fractional order q of the Fractional Differential Equation with, $0 < q \leq 2$.

$q = 1$ corresponds to no memory and $q = 2$ corresponds to infinite memory with anti-correlation, or anti-persistence.

In equation(16) above, if the initial stress be ϵ_0 , assuming Heaviside's step function as the stress input, it modifies to equation(2), for $\alpha^{-q} \to B$.

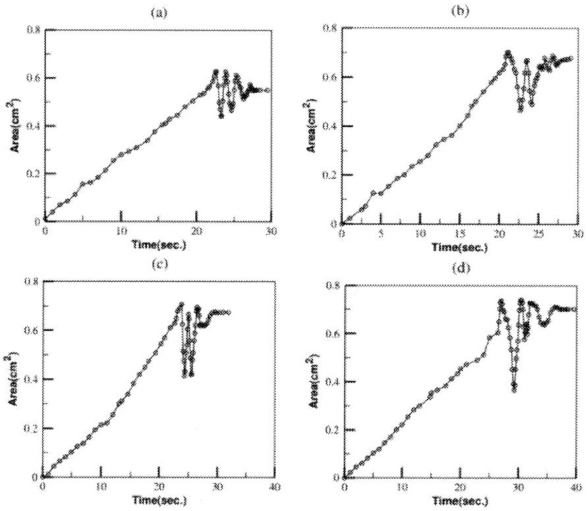

Figure 5. Four separate experiments (a) - (d) performed identically (by process 2), show the reproducibility of the nature of oscillations for potato starch gel loaded by 5 kg.

The non-Newtonian fluids without oscillatory behavior will have $0 < q < 1$, which leads to an equation also of fractional order, and the step-response will have monotonically increasing strain response, given by one argument Mittag-Leffler function. Its impulse response will be a function having a long tailed decay. In other words, the response will have long-range temporal correlation i.e. persistence. In equation(13) $q = 0$, 1 and 2 represent cases with respectively memory with complete correlation, no memory and memory with complete anti-correlation. These situations are analogous to a random walk with persistent memory, no memory and anti-persistent memory repectively.

The visco-elastic system with Newtonian viscous behavior can be modeled with a discrete ideal spring and a ideal dashpot. Whereas the more complicated case with non-Newtonian viscosity requires a different representation like a fractal chain of the ideal spring and ideal dashpot combination, or equivalently, a description using generalized calculus [5]. Our experiment clearly shows oscillatory strain and thus we infer the fractional order q of our system

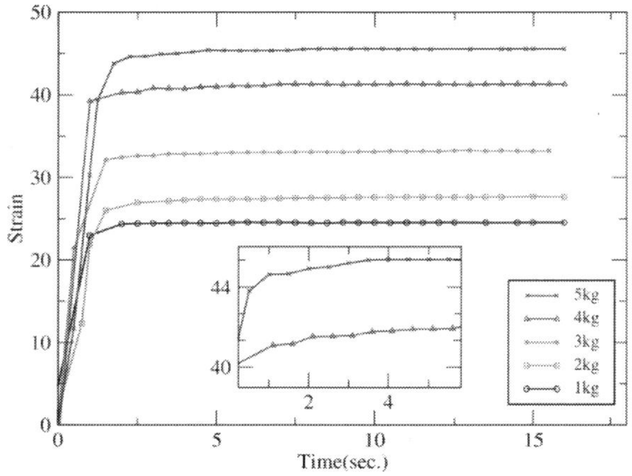

Figure 6. Variation of strain in time for ethylene glycol loaded by different W by process (ii). The inset shows that the increase is monotonic even on magnification.

to lie between 1 and 2. Neccessity for the use of a non-trivial memory kernel clearly demonstrates that results for strain at t should depend on earlier loading history, rather than the instantaneous load at t.

5. Results from the Model

We now plot the strain as a function of time using equation(4). For $q = 1$, the strain increases smoothly and saturates to the value $\epsilon_\infty = \sigma/E$, σ represents the load W in our experiment. For $q < 1$, a similar behavior is observed, with a slower variation in strain. For $q > 1$ however, we see an initial increase in strain overshooting ϵ_∞ followed by oscillations before saturating to ϵ_∞. The oscillations are more pronounced as q increases. These results are shown in figure(9). The amplitude depends on q and of course the magnitude of the strain changes proportionately to the load σ, also affecting the amplitude of oscillation.

Considering the parameter $B = \frac{E}{\beta}$ to represent the relative strengths of the elastic and viscous terms in equation(2), we may see how the system responds to changes in B. We find that variation of B changes the time period of the oscillations in strain, without affecting the amplitude. Variation of strain with W and B are shown in figure(10). The time period is smaller when elasticity dominates.

If we assume that the fractional change in area is equivalent to the strain in the system, the results for the starch solutions are reproduced qualitatively by the visco-elastic model with non-Newtonian rheology. Here the qth order derivative takes care of the non-linearity in the complex fluid. In the earlier paper [3] the Newtonian fluids were assumed incompressible and the change in film thickness was calculated from the constant volume of the fluid. Here the fluid may have a finite compressibility, so we refer only to the area which is measured directly. Determination of the exact value of q for our systems requires a more detailed analysis, with more information on the rheological properties. From the present study we may say q lies between approximately 1.5 and 1.8, because we observe 3 oscillation cyles. A value of q less than 1.5 would give less cycles, while q larger than 1.8 would give more. However, it is quite clear that q is not 1, i.e. the viscous contribution to the model is not Newtonian.

Figure 7. The figure shows the difference between two outlines of the blob of potato starch gel on glass, when compressed by a 5 kg. weight. The outer contour is of a snap taken 2 seconds *earlier* than the inner. This shows clearly that the compressed drop shrinks before increasing again,

6. Conclusion

To conclude, this work demonstrates the interesting phenomenon of oscillatory spreading of starch solutions on glass and illustrates how the approach of generalized calculus may be used to analyze it. The qualitative agreement of the variation in strain with time from the experiments and the theory is encouraging and further exploration along these lines promises to be rewarding.

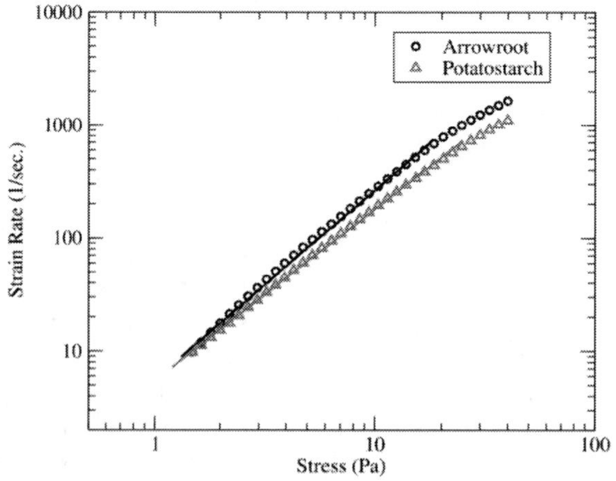

Figure 8. A double logarithmic plot of strain rate versus stress for arrowroot gel and potato starch gel, show that in both cases the fluids follow a power-law over a wide range. The straight lines are best power-law fits with exponents of 1.7 and 1.5 for arrowroot and potato starch respectively.

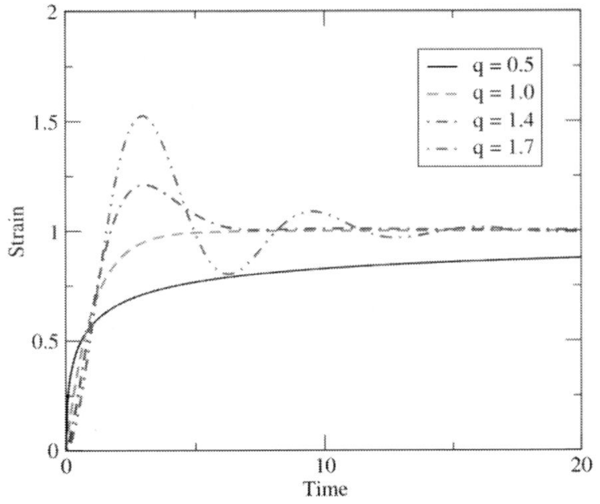

Figure 9. Variation in strain with time for different values of q from equation(4). An oscillatory behavior is seen, which dies down to the equilibrium value.

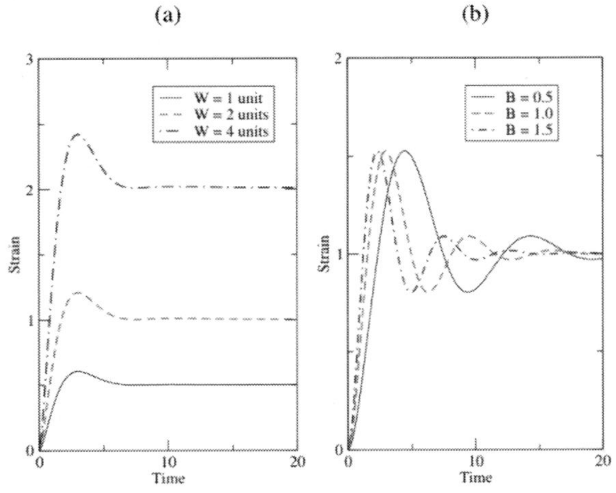

Figure 10. Variation in strain with time for different values of W and different values of B from equation(4). $q = 1.7$ in all cases.

Acknowledgments

Tapati Dutta and Prof. S.P. Moulik are gratefully acknowledged for helpful discussion. The authors thank UGC, Govt. of India for supporting this work and for providing a research grant to MDC.

References

[1] Engmann J, Servais C, Burbridge A S 2005 *J. Non-Newt. Fluid Mech.* **132** 1
[2] Bonn D, Eggers J, Meunier J, Rolley E, 2009 *Rev. Mod. Phys* **81** 739
[3] Nag S, Dutta T, Tarafdar S, 2011 *J. Colloid Int. Sci.* **356** 293
[4] Nag S, Dutta S, Tarafdar S, 2009 *Applied Surface Science* **256** 353
[5] Heymans N, Bauwens J C, 1994 *Rheologica Acta* **33** 210

[6] Das S, 2010 *Mathematico-Physics of Generalized Calculus* Limited edition, available at Jadavpur University and University of Calcutta
[7] Das S, 2007 *Functional Fractional Calculus for System Identification and Controls* **Berlin:Springer-Verlag**
[8] Sinha S, Dutta T and Tarafdar S, 2008 *Eur. Phys. J. E* **25** 267
[9] Ben Amar M and Bonn D, 2005 *Physica D* **209**, 1
[10] Poivet S, Nallet F, Gay C and Fabre P, 2003 *Europhys. Lett.* **62** 244
[11] Vicsek T, 1989 *Fractal Growth Phenomena* (Singapore: World Scientific)
[12] Moorthy S N, Larsson H and Eliasson A C, 2008 *Starch/Starke* **60** 233
[13] Singh J, Singh N, Saxena S K, 2002, *J. Food Eng.* **52**, 9
[14] Fall A, Huang N, Bertrand F, Ovarlez G. and Bonn D., 2008 *Phys. Rev. Lett.***100** 018301
[15] Christensen R M, 2003 *Theory of Viscoelasticity, 2nd ed.* **New York: Dover**
[16] Bland D R, 1960 *The Theory of Linear Viscoelasticity* **London: Pergamon Press**
[17] Glockle W G and Nonnenmacher T F, 1991 *Macromolecules* **24** 6426

Spectral renormalization group theory on networks

Eser Aygün[1] **and Ayşe Erzan**[2]

[1]Department of Computer Engineering, and [2]Department of Physics Engineering, Istanbul Technical University, Maslak, Istanbul 34 469, Turkey

E-mail: erzan@itu.edu.tr

Abstract. Discrete amorphous materials are best described in terms of arbitrary networks which can be embedded in three dimensional space. Investigating the thermodynamic equilibrium as well as non-equilibrium behavior of such materials around second order phase transitions call for special techniques.

We set up a renormalization group scheme by expanding an arbitrary scalar field living on the nodes of an arbitrary network, in terms of the eigenvectors of the normalized graph Laplacian. The renormalization transformation involves, as usual, the integration over the more "rapidly varying" components of the field, corresponding to eigenvectors with larger eigenvalues, and then rescaling. The critical exponents depend on the particular graph through the spectral density of the eigenvalues.

1. Introduction

Networks are ubiquitous in modeling discrete random media. For example granular media are characterized by contact or force networks [1, 2]. Networks of slow regions are found to percolate near the glass transition. [3, 4] Apart from interest in the network properties themselves, networks provide a convenient scaffolding on which one may describe the fluctuations in quantities that do not live on periodic lattices or are not necessarily embedded in a metric space.

This paper describes an attempt at developing a renormalization group treatment of a scalar field living on a complex network.

Critical phenomena on complex networks have been thoroughly studied. The field has been expertly reviewed by Dorogovsev et al. [5]. The Bethe approach gives rise to exact results for networks which are asymptotically tree like [6, 7]. Mean field like methods have been adopted to heterogeneous graphs [8]. A Landau-type phenomenological approach has been developed for arbitrary networks [9], where the explicit dependence of the critical behavior on the full degree distribution $P(k)$ is demonstrated. It is known [6, 7] that for networks with a divergent $\langle k^2 \rangle$, there is no phase transition except at infinite temperature in the "thermodynamic limit." The phase transition exists and is of infinite order for finite number of nodes N, and a special universal behavior is observed for all known models falling into this class. The critical behavior of the response function associated with the order parameter is mean field (exact) for scale free networks having degree distributions with convergent second moments [6, 10]. Exact real-space transformations have also been performed on scale free hierarchical networks, see, e.g. [11]. A Berezinskii-Kosterlitz-Thouless phase with power law correlations and a phase transition that is of infinite order has been observed on networks with long tailed degree distributions,

Journal of Physics: Conference Series **319** (2011) 012007 doi:10.1088/1742-6596/319/1/012007

e.g., $P(k) \sim k^{-\gamma}$ with γ sufficiently small [7, 12, 13, 14], leading to departures from mean field behavior. The presence of nodes with a diverging number of edges gives rise to Griffiths sigularities [15, 16, 17].

In spite of the rich literature on critical phenomena on arbitrary networks, the possibility of developing a "field theoretic" renormalization group on networks has not been explored so far, at least within a condensed matter context. Although some basic mathematical concepts needed to construct such a method are available, it turns out that there are also a number of unexpected challenges.

The paper is organized as follows. In Section 2 some basic mathematical tools are assembled, in Section 3 the field theoretic renormalization group, à la Wilson is very briefly reviewed. In Section 4 a renormalization group transformation is introduced for the relevant Landau-Ginzburg Hamiltonian density on a scale free network and the main difficulties are outlined. In section 5, a replica approach is used to perform quenched averages over stochastic spectral densities and a brief discussion of renormalization on deterministic hierarchical lattices is provided. In Section 6 we present some conclusions and pointers for future work.

2. The Graph Laplacian and the Laplace Spectrum

Let us define a graph as a collection of N vertices, or "nodes," labeled $i = 1, \ldots, N$ and connected to each other by edges. The adjacency matrix $\mathbf{A}$ completely specifies such a graph, with

$$A_{ij} \equiv \begin{cases} 1 & \text{if } (i,j) \text{ connected by an edge} \\ 0 & \text{otherwise} \end{cases} \tag{1}$$

If no self-interactions are allowed $A_{ii} = 0$. The "degree" k_i of the ith node is the number of edges that connect it to other nodes, which are then called its "neighbors," and $k_i = \sum_j A_{ij}$.

For an undirected graph, $\mathbf{A}$ is symmetric, by definition; in this paper we will deal only with undirected graphs. The invariants of the adjacency matrix are of course independent of the order in which the nodes are labeled, and so are all the different statistics on the graph in which one may be interested, such as the degree distribution (the probability distribution of the number of edges that the nodes have). On the other hand, unless the nodes are identified in the same way, it is not trivial to compare two graphs.

We will focus here on graphs that are not necessarily embedded in metric spaces, so that the edges do not carry information regarding the "distance" between neighboring nodes. Distance between an arbitrary pair of nodes (i, j) on the graph is simply defined as the least number of edges one has to traverse in going from node i to node j.

The formal analogue of the Laplace operator on an arbitrary graph, is the graph Laplacian [18, 19, 20] which is defined as

$$\mathbf{L} = \mathbf{D} - \mathbf{A} \tag{2}$$

where

$$D_{ij} = \delta_{ij} k_i \tag{3}$$

and k_i is the degree of the ith node.

Note that for any scalar field $\mathbf{f} = (f_1, \ldots, f_N)$ on a network of size N,

$$(\mathbf{Lf})_i = \sum_j L_{ij} f_j = k_i f_i - \sum_{j \in \mathcal{N}_i} f_j \ , \tag{4}$$

where $\mathcal{N}_i$ is the set of neighbors of the node i.

By construction, $\sum_j L_{ij} = 0$. For an undirected graph the eigenvectors

$$\mathbf{L}\, \mathbf{u}_\lambda = \lambda \mathbf{u}_\lambda \tag{5}$$

are orthogonal since $\mathbf{A}$ is symmetric, the eigenvalues λ are real and non-negative (please do not mistake the symbol λ for a wavelength!). If the graph consists of one connected component, $\lambda_0 = 0$ is non-degenerate. There is a finite gap between the zeroth and the first eigenvalue $\lambda_1 > 0$, which is bound away from zero for finite graphs. It is convenient to put the eigenvalues in ascending order with $0 < \lambda_1 \leq \lambda_2 \ldots \leq \lambda_{N-1}$, and define the largest eigenvalue as $\lambda_{N-1} \equiv \Lambda$.

The following useful properties are easy to show. The elements of the eigenvector belonging to a non-degenerate zero eigenvalue will be constant (which can be chosen $c = N^{-1/2}$ for normalization),

$$u_{\lambda_0}(i) = c \ \forall i \ . \tag{6}$$

If the graph consists of a number of connected components, the matrix $\mathbf{L}$ is of block diagonal form; the number of zero eigenvalues is equal to the number of connected components, and the corresponding eigenvectors each have elements that are constant over one of the connected components, and zero otherwise. The eigenvectors for $\lambda \neq 0$, on the other hand, satisfy

$$\sum_i u_\lambda(i) = 0 \ . \tag{7}$$

However, it should be noted that, unlike plane waves, in general $\sum_\lambda u_\lambda(i)\, u_\lambda(j) \neq \delta_{ij}$, and similarly, $\sum_\lambda u_\lambda(i) \neq 0$.

On a hypercubic lattice of dimension d, the degree of each node i is $k_i = 2d$, and Eq. (4) is simply the second order difference operator acting on $\mathbf{f}$. For $d = 1$, it is easy to see that if we divide the right hand side with the Euclidian distance between the neighboring nodes, $-\mathbf{L}$ goes over to the Laplacian in the continuum limit, i.e.,

$$-\nabla^2 f(x) = \lim_{\Delta \to 0} \{2f(x) - [f(x+\Delta) + f(x-\Delta)]\}/\Delta^2 \ . \tag{8}$$

Arguments for the convergence to continuous Laplace operators can be made rigorous [18, 19]. Moreover [19],

$$\mathbf{f}^\dagger \mathbf{L} \mathbf{f} = \frac{1}{2} \sum_{i,j}^{N} A_{ij} (f_i - f_j)^2 \tag{9}$$

converges to

$$(f(x), \mathcal{L} f(x)) = \int_\Omega |\nabla f|^2 dx \tag{10}$$

where the integral is over the support of the function $\mathbf{f}$.

Note that the right hand side of Eq.(9) essentially counts the number of edges over which $(f_i - f_j)^2$ is appreciably different from zero. One may use the Rayleigh-Ritz theorem [21],

$$\min_{||\mathbf{f}|| \neq 0} \frac{\mathbf{f}^\dagger L \mathbf{f}}{\mathbf{f}^\dagger \mathbf{f}} = \min \lambda \tag{11}$$

where λ are the eigenvalues of the operator $\mathbf{L}$, to partition the graph into two clusters which are connected by the fewest possible edges [19]. The non-trivial solution to the minimization problem is clearly found by setting $\mathbf{f} = \mathbf{u}_{\lambda_1}$ and min$\lambda = \lambda_1$. It can be verified by explicitly (by computing the eigenvector $\mathbf{u}_{\lambda_1}$) that its elements fall roughly into two comparable sized groups of positive and negative values. This clustering scheme can be generalized to larger numbers of clusters [19].

So far, it looks like the graph Laplacian provides us with an operator whose eigenvectors will be the analogue of the complex exponentials, so that we will be able to perform a

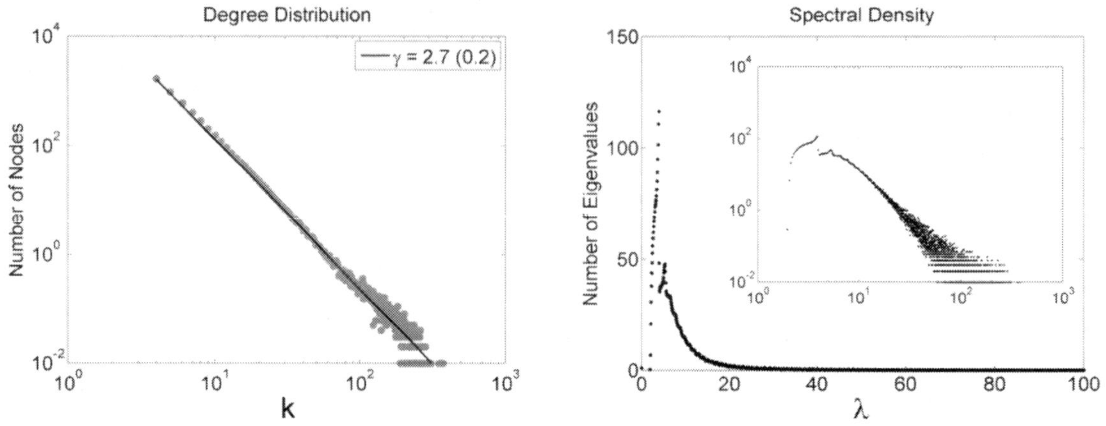

Figure 1. (Color on line.) The degree distribution (left panel) and Laplacian spectral density plotted against the eigenvalues λ (right panel), for a "scale free" network. Shown is the preferential attachment model[22] for the number of initial nodes $m_0 = 5$, and $m = 4$ new nodes added at each time step, grown to a size of $N = 5000$ and averaged over 100 realizations. The left panel has double logarithmic axes, and the degree distribution has been fitted to a power law $k^{-\gamma}$, with the root mean square error of the fit reported in parentheses (see Table 1). For this figure the (linear) k axis has been binned into 5000 intervals. The Laplace spectrum (right panel) is shown both with linear and double logarithmic (inset) axes. The van Hove singularity in the spectral density and the crossover from an exponential to a power law decay is quite evident. Both histograms have been obtained with 5000 bins on the linear scale. See text.

transformation on our function **f** and analyze fluctuations at different resolutions, in analogy with Fourier components of different wavenumbers. (As noted above, however, the kernel of the transformation is not self adjoint, unlike the Fourier transform.)

It is useful to take a look at the Laplace spectra of some sample networks. Clearly, a complex, scale free network would be of greater interest in this context. The degree distribution of a network generated using the Barabasi-Albert preferential attachment model [22] is shown in Fig. 1a. Although the degree distribution fits a power law $\sim k^{-\gamma}$ perfectly over the whole of its range, the Laplace spectral density (Fig. 1b) is not scale free; it displays a sharp peak followed by a discontinuity and then, in the upper part of its range, it crosses over from an exponential to a power law decay, $\rho(\lambda) \sim \lambda^{-\beta}$, followed by a sharp cutoff (see Fig. 2b). The behavior is very reminiscent of the density of vibrational states of a solid, exhibiting the van Hove singularity [23, 24].

Table 1. Numerical values for γ and β for the Barabasi-Albert[22] model. The number of initial nodes is $m_0 = 5$; the number of nodes added at each time step is m, and the total number of nodes is $N = 5000$. The ensemble averages of the degree and spectral distributions over 100 realizations have been fitted, and these numbers are reported below. For the reported exponents, the k and λ values have been log-binned into 500 and 50 bins respectively. The parentheses are root mean square error of the fits as illustrated in Fig. 2.

m	γ	β
3	2.7 (0.05)	1.9 (0.03)
4	2.7 (0.2)	1.9 (0.06)
5	2.9 (0.01)	1.9 (0.09)

We do not have an analytic relation between β and γ; it is quite possible that γ all by itself does not determine β and more work is needed to understand this dependence.

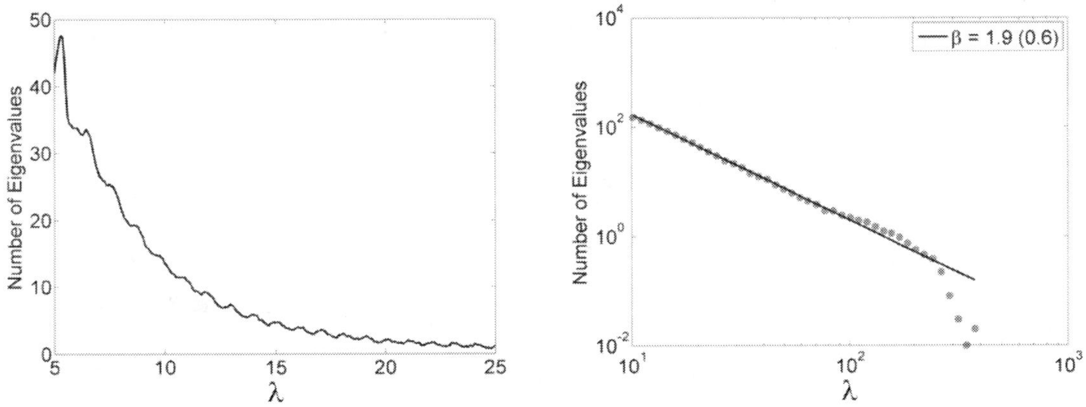

Figure 2. (Color online.) The tail end of the Laplace spectral density, ensemble averaged (over 100 realizations) for the same scale free network as shown in Fig. 1. The left panel is on a linear scale; the points have been connected by a line. The right panel shows a fit to $\rho(\lambda) \sim \lambda^{-\beta}$ on a log-log scale, with log-binning (see Table 1).

The ensemble averaged tail of the Laplace spectrum is shown in Fig. 2. The detail of Fig. 2 is astonishing. It displays a self-similar piling up of further and further decorations superposed on the power law decay. (For a BA model with aging of the sites [26], the Laplace spectrum displays even more pronounced periodic modulations.) In fact, if these decorations can be represented as a self-similar Weierstrass-Mandelbrot function [25],

$$W(\lambda) \sim W_0(\lambda) + \sum_{n=0}^{N} \frac{g(s^n \lambda)}{s^{\phi n}} \tag{12}$$

where g is a periodic function and $s > 1$ is a constant, then $W(\lambda)$ has the scaling form

$$W(s\lambda) \sim s^{\phi} W(\lambda) \ . \tag{13}$$

3. Order-Disorder transitions, fluctuations and the Field Theoretical Renormalization Group

The Ising model, with a scalar field $\sigma_i = \pm 1$ residing on the vertices of a lattice, is the basic paradigm for order-disorder transitions in statistical physics. The Ising Hamiltonian, or energy function for any given set $\{\sigma_i\}$, is, for short range interactions,

$$H = -J \sum_{<ij>} \sigma_i \sigma_j - h \sum_i \sigma_i \tag{14}$$

where $< ij >$ signifies that i and j are connected directly by an adge on the lattice. For $J > 0$, configurations in which the "spins" σ_i predominantly have the same sign as their neighbors lower the energy with respect to random configurations. In thermal equilibrium at sufficiently low temperatures and sufficiently high connectivity of the network (for the embedding dimension $d \geq 2$ on periodic lattices), the spins become "ordered," i.e., the "magnetization" which is equal to the thermodynamic expectation value $\langle \sum_i \sigma_i \rangle$ becomes non-zero,, even when the external magnetic field h is equal to zero. The magnetization can be regarded as an "order parameter." At the precise temperature T_c where this ordering sets in, a continuous phase transition takes place.

At the "critical" point $T = T_c$, $h = 0$, the system is characterized by fluctuations at all wavelengths and a diverging correlation length. As a result, the system becomes invariant under scale transformations on a Euclidean lattice. Slightly away from the critical point, changing the resolution at which the system is observed has the effect of changing the effective temperature (and magnetic field). The rate at which this change occurs determines the critical exponents which characterize the singular behavior of the free energy, order parameter and response functions of the system. [27, 28, 29] (In the rest of this section we mostly follow the presentation of Goldenfeld [27], Chapter 12.)

3.1. The Ginzburg-Landau Approach

The Ginzburg-Landau approach [29] to critical fluctuations involves a passage to a continuum description, which is based on a local averaging of the discrete variables such as $\{\sigma_i\}$ over volumes still very small compared to the total size of the system, so that one may eventually take them to be of infinitesimal size. In this way, a continuous variable corresponding to the local order parameter density is obtained. Let us call this variable $\psi(\mathbf{x})$ where $\mathbf{x}$ indicates the spatial coordinates. The effective Ginzburg-Landau (G-L) Hamiltonian in the absence of an external magnetic field is then given by,

$$H = \int_\Omega d\mathbf{x}\{\frac{1}{2}\left[r_0\psi^2(\mathbf{x}) - \psi(\mathbf{x})\nabla^2\psi\right] + v_0\psi^4(\mathbf{x})\} \quad . \tag{15}$$

where the coupling r_0 is assumed to be proportional to $(T - T_c)/T_c \equiv t$, and $v_0 > 0$. It is customary to absorb a factor of $1/k_BT$ into H, with k_B being the Boltzmann constant, so that H is dimensionless. This effective Hamiltonian can be expressed in terms of the Fourier coefficients of the order parameter,

$$\hat{\psi}(\mathbf{q}) = \frac{1}{2\pi}\int_\Omega e^{i\mathbf{q}\cdot\mathbf{x}}\psi(\mathbf{x})d\mathbf{x} \quad . \tag{16}$$

The $\mathbf{q}$ signify wavevectors, the components of which are $q_\alpha = 2\pi n_\alpha/|\Omega|^{1/d}$, with each n_α ranging over the positive and negative integers (periodic boundary conditions are assumed). For a d-dimensional hypercubic domain Ω, in the continuum limit, the density of wave vectors within a volume $d\mathbf{q}$ is given by $|\Omega|/(2\pi)^d$. Note that $\mathbf{q}$ are the eigenvalues of the gradient operator on the complex exponentials (which are the eigenfunctions); the eigenvalues of the Laplace operator on the same functions are simply $-||\mathbf{q}||^2$.

Let us write the effective Hamiltonian as $H = H_0 + H_{\text{int}}$, where H_0 is quadratic and diagonal in the $\psi(\mathbf{q})$,

$$H_0 = \frac{1}{2}\int_0^\Lambda \frac{q^{d-1}dq}{(2\pi)^d}[r_0 + q^2]\hat{\psi}^*(\mathbf{q})\hat{\psi}(\mathbf{q}) \quad . \tag{17}$$

Assuming isotropy in q-space, we have gone over to polar coordinates, obtaining the density (per unit volume $|\Omega|$) of wavevectors in the interval dq to be $S_d\,\frac{q^{d-1}}{(2\pi)^d}$, where S_d is the area of the unit sphere in d dimensions. It is important to note that this density is in the form of a power law in q. The upper cutoff Λ is given by $2\pi/l$ where l is the lattice spacing.

The interacting part of the Hamiltonian, H_{int}, involves couplings between Fourier components at different wavenumbers (and therefore different spatial scales) and is given by,

$$H_{\text{int}} = v_0\int_0^\Lambda d\mathbf{q}_1\ldots d\mathbf{q}_4\hat{\psi}(\mathbf{q}_1)\ldots\hat{\psi}(\mathbf{q}_4)\,\delta(\sum_{i=1}^4\mathbf{q}_i) \quad . \tag{18}$$

The partition function is,

$$Z = \int_{-\infty}^{\infty}\cdots\int_{-\infty}^{\infty}\prod_{0\le|\mathbf{q}|\le\Lambda} d\hat{\psi}_{\mathbf{q}}e^{-H} \quad . \tag{19}$$

3.2. *Renormalization Group à la Wilson*

Renormalization à la Wilson [30] is carried out by *i)* integrating out the relatively small wavelength (large q) components of the fluctuating field in the partition function [27, 28, 29]. Choosing a re-scaling parameter b, and dividing the range of q into two parts, the partition function can be written (exactly) as,

$$Z = \int_{-\infty}^{\infty} \cdots \int_{-\infty}^{\infty} \prod_{0\leq|\mathbf{q}|\leq\Lambda/b} d\hat{\psi}_{\mathbf{q}}^{\text{lower}} \prod_{\Lambda/b<|\mathbf{q}|\leq\Lambda} d\hat{\psi}_{\mathbf{q}}^{\text{upper}} e^{-H} \ . \tag{20}$$

where we have added an extra label on the functions $\hat{\psi}(\mathbf{q})$ to indicate the range (upper or lower)of wave vectors within which $\mathbf{q}$ lies. We would now like to do the integration over the fluctuations in the upper q range, $\hat{\psi}_{\mathbf{q}}^{\text{upper}}$. This can of course only be done exactly for $u_0 = 0$, and therefore the interacting part of the G-L Hamiltonian has to be taken care of perturbatively. *ii)* The next step involves rescaling the q to restore the original range of scales over which the fields fluctuate, and thus the original form of the Hamiltonian density. *iii)* Requiring that the coefficient of the diffusive (q^2) term in H_0 remain invariant under this operation fixes the rescaling factors acquired by the fields $\hat{\psi}(\mathbf{q})$ and the couplings r and v. It can easily be demonstrated that successive transformation with the parameter b satisfy all the *semi*group properties (there is no inverse).

The usefulness of the field theoretic approach of Wilson lies in the precise prescription for the Renormalization Group (RG) transformation, and the fact that although approximations have to be made for an interacting theory they can be systematically improved, as a power series in the coupling v_0, in contrast to the "Real Space" RG approach [27], where one has to perform uncontrolled approximate partial summations over the partition function Z. Moreover, such concepts as an upper critical dimension, beyond which critical behavior is exactly mean field, emerge only within the field theoretic RG formalism.

4. Field Theoretic RG on a Complex Network?

Having in mind the advantages of the Wilson approach to the renormalization group, we would like to construct the analogue of the field theoretic RG on complex networks. The idea is to expand order parameter fluctuations in eigenvectors of the graph Laplacian, write down the equivalent of a Ginzburg-Landau Hamiltonian, and then perform partial summations over the partition function, to eliminate the high-eigenvalue components. The development will closely follow the prescription in the previous section.

4.1. *Order parameter expansion in eigenvectors of the Laplacian for a Gaussian model*

Let us model our approach on the Ginzburg-Landau expansion of the Hamiltonian as in Eq. (15,17) and initially take just a Gaussian, or non-interacting model, with a field $\psi(i)$ residing on the vertices of a graph, with $i = 1, \ldots, N$. In the absence of a field (which can be included without any problem)

$$H_0 = \frac{1}{2} \sum_{i}^{N} \psi(i) \left[r_0 + L\right] \psi(i) \ . \tag{21}$$

Let us define the transformed fields $\hat{\psi}(\lambda)$ by

$$\psi(i) = \sum_{\lambda} \hat{\psi}(\lambda)\, u_\lambda(i) \ , \tag{22}$$

in terms of $\mathbf{u}_\lambda$, the normalized eigenvectors of the Laplace operator associated with the eigenvalues λ, and

$$\rho(\lambda) = \frac{1}{N}\sum_{i}^{N} \delta(\lambda - \lambda_i) \tag{23}$$

as the spectral density, i.e., the density of eigenvalues on the positive real line. Then we get,

$$H_0 = \frac{1}{2}\int_0^{\Lambda} \rho(\lambda)\, d\lambda\, \hat{\psi}(\lambda)\, [r_0 + \lambda]\, \hat{\psi}(\lambda)\ . \tag{24}$$

Note that, being the eigenvalue of the Laplacian, λ takes the place of the wavenumber squared, i.e., q^2 (it is actually dimensionless).

4.2. Naive renormalization of the Gaussian theory

The partition function can now be written as an integral over the different coefficients $\hat{\psi}_\lambda \equiv \hat{\psi}(\lambda)$ (we will use these notations interchangeably). In order to perform the partial integration of the partition function as in Eq. (20), we choose a scale factor s. We must now integrate out the coefficients $\hat{\psi}_\lambda$ for $\lambda > \Lambda/s$, where Λ is the largest eigenvalue of $\mathbf{L}$ on the given network.

Denoting the Gaussian partition function with Z_0, we have,

$$Z_0 = \int_{-\infty}^{\infty} \cdots \int_{-\infty}^{\infty} \prod_{0\le\lambda\le\Lambda/s} d\hat{\psi}_\lambda^{\text{lower}} \prod_{\Lambda/s<\lambda\le\Lambda} d\hat{\psi}_\lambda^{\text{upper}} e^{-H_0}\ , \tag{25}$$

where H_0 is now given by Eq. (24). Performing the set of integrals over the $\hat{\psi}_\lambda^{\text{upper}}$ yields,

$$Z_0 = e^{g(r_0)} \int_{-\infty}^{\infty} \cdots \int_{-\infty}^{\infty} \prod_{0\le\lambda\le\Lambda/s} d\hat{\psi}_\lambda^{\text{lower}} e^{-H_0^{\text{lower}}}\ , \tag{26}$$

where for future use we can define $Z_0^{\text{upper}} = e^{g(r_0)}$, where $g(r_0)$ is the free energy contributed by the high-λ degrees of freedom. Since Z_0 only involves doing Gaussian integrals, we easily calculate

$$Z_0^{\text{upper}} = \prod_{\Lambda/s<\lambda\le\Lambda} \sqrt{\frac{2\pi}{\lambda + r_0}}\ . \tag{27}$$

The Hamiltonian involving the remaining degrees of freedom is,

$$H_0^{\text{lower}} = \frac{1}{2}\int_0^{\Lambda/s} \rho(\lambda)\, d\lambda\, \hat{\psi}_\lambda^{\text{lower}}\, [r_0 + \lambda]\, \hat{\psi}_\lambda^{\text{lower}}\ . \tag{28}$$

The RG transformation will be complete when we rescale the λ with s so that the integral is once again over the range $(0, \Lambda)$, and we can drop the label "lower" from the Hamiltonian, while having to "renormalize" the coupling r_0.

Let us make the naive assumption that the spectral density is a homogenous function over its whole range, $\rho(\lambda) \sim \lambda^{-\beta}$, where β depends on the degree distribution, in particular on the exponent γ.

Now make a change of variables $\lambda' = s\lambda$ (so that $\lambda = \lambda'/s$). Define the rescaling factor for the fields via $\hat{\psi}^{\text{lower}}(\lambda'/s) = z\hat{\psi}'(\lambda')$. The resulting Hamiltonian is now going to be in the same form as Eq. (24), except that r_0 will acquire a multiplicative factor. Calling the renormalized Hamiltonian H_0', Eq. (28) becomes,

$$H_0' = \frac{1}{2}\int_0^{\Lambda} \rho(\lambda') d\lambda' s^{\beta-1} z^2 \hat{\psi}'(\lambda') \left[r_0 + \frac{\lambda'}{s}\right] \hat{\psi}'(\lambda') \tag{29}$$

We assume we can fix the renormalization factor z of the fields $\hat{\psi}'$ for all λ', by requiring the coefficient of the λ' term in the Hamiltonian to remain unchanged. This gives $z = s^{(2-\beta)/2}$. Substituting this back into Eq. (29) the various rescaling factors simplify and we find that we can define the renormalized coupling

$$r = sr_0 \tag{30}$$

Notice that as long as the spectral density is homogeneous, this result is inevitable for the Gaussian theory, independent of the scaling exponent β for the spectral density. (β should not be confused with the critical exponent of the order parameter!) Recall that the spectral density for the wavenumbers $\mathbf{q}$ in the Euclidean case reviewed in Section 3, was $\propto q^{d-1}$. It is useful, for later comparison, to define $-\beta = D - 1$. The fact that the spectral density decays, instead of growing with λ is going to have strong consequences later on.

In the Ginzburg-Landau approach we have assumed that $r_0 \propto t$, therefore the rescaling factor in front of r_0 is related to the temperature renormalization, i.e., we here have $t \rightarrow t' = s^y t$, with $y = 1$.

Recall that the eigenvalues of the Laplacian are "like" q^2, so that the scaling factor of the "length like" quantities in the problem is only $\sqrt{s}$. If we are very cavalier, we can identify the analogue of the correlation length exponent $\nu = (2y)^{-1} = 1/2$, as we would expect from a Gaussian theory.

Since we have not embedded our network in a metric space, ascribing to λ the dimensionality of a (length)$^{-2}$, and similarly making the identification of $(2y)^{-1}$ with the correlation *length* exponent is rather tenuous, just as it is somewhat problematic to define a correlation length on this non-metric space. (Note that the dimension d is also not defined for this system. However, for the product $d\nu$, the identification $2 - \alpha = d\nu$ can still be made, where α is the specific heat exponent.)

4.3. The ψ^4 theory

Consider adding an interaction term $H_{\rm int} = v_0 \sum_i \psi^4(i)$, to the non-interacting or Gaussian theory. In terms of the transformed fields this gives,

$$H_{\rm int} = v_0 \int_0^{\Lambda} \prod_{\mu=1}^{4} \left[\rho(\lambda_\mu) d\lambda_\mu\right] \hat{\psi}(\lambda_1) \ldots \hat{\psi}(\lambda_4) \Phi(\lambda_1, \lambda_2, \lambda_3 \lambda_4) \ . \tag{31}$$

where $\Phi(\lambda_1, \ldots, \lambda_4) \equiv \sum_{i=1}^{N} \prod_{\mu=1}^{4} u_{\lambda_\mu}(i)$.

We now encounter a property of the eigenvectors of the ordinary Laplacian on a Euclidean lattice (e.g., plane waves) which is not shared by the graph Laplacian: An element by element product of the eigenvectors (eigenfunctions) of the ordinary Laplacian yields yet another eigenvector since the exponentials add, modulo 2π. For the graph Laplacian,

$$u_{\lambda_k}(i) u_{\lambda_l}(i) \ldots u_{\lambda_m}(i) \neq u_{\lambda_n}(i) \ . \tag{32}$$

in general, i.e., no λ_n can be found such that the equality holds. If the equality were to hold, we would have had $\Phi = 0$ unless $\lambda_n = 0$, by Eq. (7). Recall that in the ordinary Euclidean case, the ψ^4 term in the Fourier transform representation gives rise to a Dirac delta function connecting the wavevectors, $\delta(\mathbf{q}_1 + \mathbf{q}_2 + \mathbf{q}_3 + \mathbf{q} + 4)$, i.e., a sum rule on the total incoming and outgoing "momenta," Eq. (18), while here, due to Eq. (32), no such rule applies. This gives rise to a markedly different scaling behavior as we will see later on.

The interaction term has to be handled perturbatively, and following the same prescription [30, 27] here we get,

$$Z(r_0, v_0) = \int_{-\infty}^{\infty} \cdots \int_{-\infty}^{\infty} \prod_{0 \leq \lambda \leq \Lambda} d\hat{\psi}_\lambda e^{-H}$$
$$= Z_0^{\text{upper}} \int_{-\infty}^{\infty} \cdots \int_{-\infty}^{\infty} \prod_{0 \leq \lambda \leq \Lambda/s} d\hat{\psi}_\lambda e^{-H_0^{\text{lower}}} \langle e^{V[\hat{\psi}^<, \hat{\psi}^>]} \rangle_0^{\text{upper}} \ . \tag{33}$$

where the interaction term now contains coefficients with λ in both the lower and the upper range,

$$V[\hat{\psi}^<, \psi^>] \equiv -H_{\text{int}} \left[\hat{\psi}(\lambda)^{\text{lower}}, \hat{\psi}(\lambda)^{\text{upper}} \right] \tag{34}$$

where we have implicitly defined $\hat{\psi}(\lambda)^> \equiv \hat{\psi}(\lambda)^{\text{upper}}$, etc., and the brackets mean,

$$\langle e^{V[\hat{\psi}^<, \psi^>]} \rangle_0^{\text{upper}} = \frac{1}{Z_0^{\text{upper}}} \int_{-\infty}^{\infty} \cdots \int_{-\infty}^{\infty} \prod_{\Lambda/s \leq \lambda \leq \Lambda} d\hat{\psi}_\lambda e^{-H_0^{\text{upper}}} e^{V[\hat{\psi}^<, \hat{\psi}^>]} \ . \tag{35}$$

At this point, it is standard to take a cumulant expansion, $\langle e^{-x} \rangle \simeq e^{-\langle x \rangle} e^{\frac{1}{2}[\langle x^2 \rangle - \langle x \rangle^2]}$. So now we have to compute $\langle V[\hat{\psi}^<, \psi^>] \rangle_0^{\text{upper}}$ and $\langle V^2[\hat{\psi}^<, \psi^>] \rangle_0^{\text{upper}}$. The angular brackets involve the same Gaussian measure as in Eq. (35) and the only terms that survive are those that are even in the $\hat{\psi}(\lambda)^>$, with

$$\langle \hat{\psi}(\lambda)^> \hat{\psi}(\lambda')^> \rangle_0^{\text{upper}} = \delta_{\lambda \lambda'} \frac{1}{\lambda + r_0} \equiv \delta_{\lambda \lambda'} G(r_0, \lambda) \ . \tag{36}$$

defining a "contraction" between two fields $\hat{\psi}(\lambda)^>$ and $\hat{\psi}(\lambda')^>$. Unless we refer specifically to the r_0 dependence, we will drop this argument and just write $G(\lambda)$ for the Green's function.

4.4. First order in perturbation theory

To keep track of these computations one makes use of Feynman diagrams, and the diagrams we need are given in Fig. B1 in Appendix B. As usual, the contribution from $\langle V[\hat{\psi}^<, \psi^>] \rangle$ to the quadratic interaction comes from the singly contracted term, Fig. B1a, with the combinatoric multiplicative factor 6. [27]. This term is given by

$$Q_2^{(1)} = 6v_0 \int_0^{\Lambda/s} d\lambda_1 d\lambda_2 \rho(\lambda_1) \rho(\lambda_2) \hat{\psi}^<(\lambda_1) \hat{\psi}^<(\lambda_2) \sum_i I_1(i) \, u_{\lambda_1}(i) u_{\lambda_2}(i) \ , \tag{37}$$

where

$$I_1(i) = \int_{\Lambda/s}^{\Lambda} d\lambda \, \rho(\lambda) G(\lambda) u_\lambda^2(i) \ . \tag{38}$$

$I_1(i)$ does not depend any more on any of the eigenvalues, and therefore is a constant under re-scaling, however it depends on r_0, s, and Λ. (see Appendix B)

If, for the moment, we neglect the dependence of the integrand on the eigenvectors $\mathbf{u}_{\lambda_\mu}$, $\mu = 1, 2$, the scaling behavior of $Q_2^{(1)}$ is obtained by counting powers of s which arise when we make the transformation to $\lambda' = s\lambda$. This is the same as counting the number of integrals over λ (each contributes a power $1 - \beta$) and counting the powers of $\hat{\psi}$, which contribute a factor of

z , i.e., a power of $(2-\beta)/2$ each. In total we have the rescaling factor $-2+2\beta+2-\beta=\beta$. Taking into account the forefactor of 1/2 appearing in H_0, Eq. (28) we get,

$$\frac{1}{2}Q_2^{(1)} = 12s^\beta v_0 \int_0^\Lambda d\lambda_1 d\lambda_2 \rho(\lambda_1)\rho(\lambda_2)\hat{\psi}(\lambda_1)\hat{\psi}(\lambda_2)\sum_i I_1(i)u_{\lambda_1}(i)u_{\lambda_2}(i) \ , \tag{39}$$

for the contribution from the interaction term to the quadratic coupling, to first order in v_0. This contribution is not diagonal in the eigenvalues λ, unlike the original Gaussian coupling in H_0, and moreover it depends on λ, so in principle the original H_0 also has to be modified. Note that, had the new quadratic term been diagonal, the renormalization factor would have been $s^{-1+\beta}z^2 = s$, rather than s^β.

Grouping together all the quadratic terms, absorbing any possible corrections to the scaling coming from $\sum_i I_1(i)\,u_{\lambda_1}(i)u_{\lambda_2}(i)$ into what we shall define as $\tilde{I}_1$ and using Eq. (30), we get,

$$r = sr_0 + 12v_0 s^\beta \tilde{I}_1 \ , \tag{40}$$

with the fixed point equation,

$$r = \frac{12v_o s^\beta \tilde{I}_1}{1-s} \ . \tag{41}$$

The uncontracted term in $\langle V\rangle$ with all the fields in the lower-λ range (Fig. B1b) gives, again by the same kind of power counting as above, a renormalized coupling constant

$$v = v_0 s^{2\beta}\tilde{\Phi} \ , \tag{42}$$

where possible corrections to scaling are included in $\tilde{\Phi}$. This equation has zero as fixed point, $v^* = 0$, leading to $r^* = 0$ via Eq. (41). Thus, to first order in v_0, the Gaussian fixed point is stable, analogously to the Euclidean case.

We can now make a generalization, using our definition $D-1=-\beta$. Consider a Feynman diagram with n legs. This will mean n integrals over λ, in the absence of "momentum conservation," and therefore a factor of s^{-nD}. Each leg carries a factor of $\hat{\psi}^<$, and therefore we get a factor of $z^n = s^{(D+1)n/2}$, resulting in a re-scaling factor of $s^{(1-D)n/2} = s^{n\beta/2}$. But $\beta > 0$ and the power of s is now always positive. We see that there is no analogue of the upper critical dimensionality, above which the ψ^4 interaction terms become irrelevant in the ordinary ψ^4 Euclidean theory for scalar fields. (One recovers the Ginzburg-Landau-Wilson result for $n=4$ by substituting d for D, taking only $n-1$ factors of s^{-D} and $z=s^{(d+2)/2}$ due to the q^2 diffusive term, rather than a linear term in λ.)

All the contributions to the renormalized Hamiltonian, to all orders in perturbation theory, will grow indefinitely under successive renormalizations, i.e., are "relevant" in the RG terminology. To first order, we are only saved by the fact that there exists a unique fixed point at $v^*=0$. We seem to be faced with an uncontrollable phenomenon of "proliferation" !

4.5. Second order in perturbation theory

Now we would like to see explicitly what happens when we take into account terms that are higher order in v_0.

The bubble diagram (Fig. B1c) is already familiar from the Euclidean case, and contributes to the renormalized 4-vertex. Using the results of Appendix B, we have

$$v = v_0 s^{2\beta}\left[\tilde{\Phi} - 36v_0(I_1)^2\right] \ . \tag{43}$$

There are further terms which contribute to the quadratic coupling, which would not survive in the ordinary case because of the "momentum conservation" rule (two "lower" momenta cannot be summed with two "upper" ones to give zero). These calculations are given in Appendix B, and here we will only report the results.

Performing the rescalings over the second order contributions and putting together all the results, we have,

$$r = sr_0 + 12v_0 s^\beta I_1 - v_0^2\left[(72+48)(I_1)^3\right] s^\beta \ . \tag{44}$$

The solution for the fixed point equations, besides the trivial one where $r^* = 0,\ v^* = 0$, are

$$v^* = \frac{\tilde{\Phi} - s^{-2\beta}}{(I_1)^2} \ , \tag{45}$$

and

$$r^* = s^\beta v^* I_1 \frac{12 - v^*\left[120\,(I_1)^3\right]}{1-s} \ . \tag{46}$$

Taking I_1 to lowest order in r_0 gives,

$$I_1 \sim \Lambda^{-\beta}\frac{(s^\beta - 1)}{\beta} \tag{47}$$

which leads to

$$v^* \propto \frac{\beta^2(1-s^{-2\beta})\Lambda^{2\beta}}{(s^\beta-1)^2} \ . \tag{48}$$

We see that, up to the approximations we have made, the s dependence does not drop out of the Eqs. (45,46). Apart from the fixed point at $v^* = 0$, there is no finite fixed point for v.

There is one other connected graph to second order in v_0, which is obtained by taking one field from each subgraph and contracting them (see Fig. B1f) obtaining a ψ^6 interaction. We have computed its scaling factor explicitly, and find, in accordance with the power counting scheme of the previous section, that it scales like $s^{3\beta}$, i.e., it is yet another relevant coupling.

The Laplace spectrum for a scale free network, shown in Figs. 1,2, is not a homogeneous function over the entirety of its range. We could think of taking the re-scaling parameter to be very close to unity, $s = 1+\delta$ with $\delta \ll 1$ but this does not help us find a non-trivial fixed point, and taking $\delta \to 0$ leads to a blow-up of Eq. (48), just as taking $\Lambda \to \infty$ does. Substituting (48) into (46) leads to the same problems. We must conclude that the only physical fixed point in this perturbative treatment is at $r^* = 0,\ v^* = 0$.

5. Stochastic and deterministic complex networks

5.1. Replica approach to a quenched average over the spectral density

In the previous subsections we have pushed forward the computations with the naive assertion that the spectral density scales simply as $\rho(s\lambda) \sim s^{-\beta}\rho(\lambda)$. However, it is clear from the Figs. 1 and 2 that the truth is somewhat more complicated than that. In particular, we proposed that the average over 100 realizations of the network can be represented by the Weierstrass-Mandelbrot function (12). Therefore, in order to really speak about a non-stochastic spectral density, we should take an average over different realizations. In the present case, this has to be a quenched average taken over the free energy of the system, since by assumption the network is fixed and does not fluctuate within the relaxation times of the fields (spins) living on the network.

It is standard to use the "replica method" [31, 32, 33] to be able to take the average over the logarithm of the partition function. The trick is to introduce n independent replicas of the system and to take the average over Z^n, finally using the identity $\lim_{n\to 0}(Z^n - 1)/n = \ln Z$.

To illustrate how the calculations have to be done let us take the non-interacting, Gaussian theory as a starting point. Then,

$$Z^n = \int_{-\infty}^{\infty} \prod_{\alpha=1}^{n} \prod_{z} d\hat{\psi}_z^{(\alpha)} e^{-\sum_\alpha H_0^\alpha} \ , \tag{49}$$

where $\alpha = 1, \ldots n$ is the replica index,

$$\sum_{\alpha=1}^{n} H_0^\alpha = \frac{1}{2} \int_0^\Lambda dz \frac{1}{N} \sum_i^N \sum_\alpha^n \hat{\psi}_z^{(\alpha)} (r_0 + z) \hat{\psi}_z^{(\alpha)} \delta(z - \lambda_i) \ . \tag{50}$$

and we have used Eq. (23).

If the distribution over different realizations of the spectrum $\lambda_0, \ldots, \lambda_{N-1}$ is denoted by $\mathcal{P}(\{\lambda_i\})$, we have

$$\langle Z^n \rangle = \int_{-\infty}^{\infty} \prod_{\alpha=1}^{n} \prod_{z} d\hat{\psi}_z^{(\alpha)} \int d\lambda_0 \ldots d\lambda_{N-1} \mathcal{P}(\{\lambda_i\}) e^{-\sum_\alpha H_0^\alpha} \ . \tag{51}$$

This expectation value now looks like a path integral over individual paths $\{\lambda_i\}$. Let us define

$$\langle \frac{1}{N} \sum_{i,\alpha} (r_0 + z) \delta(z - \lambda_i) \rangle \equiv \rho(z) \ . \tag{52}$$

A cumulant expansion up to second order gives,

$$\langle e^{-\sum_\alpha H_0^\alpha} \rangle \simeq e^{-C_1 + \frac{1}{2} C_2} \tag{53}$$

where

$$C_1 = \frac{1}{2} \langle \int_0^\Lambda dz \frac{1}{N} \sum_{i,\alpha} (r_0 + z) \delta(z - \lambda_i) (\hat{\psi}_z^\alpha)^2 \rangle =$$
$$\sum_\alpha \int dz (r_0 + z) \rho(z) (\hat{\psi}_z^\alpha)^2 \ . \tag{54}$$

The second cumulant gives rise to bi-quadratic terms which couple different replicas,

$$C_2 = \frac{1}{4} \int_0^\infty \prod_i d\lambda_i \mathcal{P}(\{\lambda_i\}) \int dx dy \frac{1}{N^2}$$
$$\times \sum_{i,j} \sum_{\alpha,\beta} [(r_0 + x)(r_0 + y) \ \delta(x - \lambda_i) \ \delta(y - \lambda_j)] \ (\hat{\psi}_x^\alpha)^2 (\hat{\psi}_y^\beta)^2 - C_1^2$$
$$= \frac{1}{4} \int dz (r_0 + z)^2 \rho(z) \sum_{\alpha,\beta} (\hat{\psi}_z^\alpha)^2 (\hat{\psi}_z^\beta)^2 \} \ , \tag{55}$$

where the off-diagonal terms in i, j have been cancelled by C_1^2. Thus it turns out that averaging the spectral density over different realizations, one ends up with a bi-quadratic term in the Hamiltonian, which could have been included from the start. We get, for the effective Hamiltonian,

$$H^{\text{eff}} = \frac{1}{2} \int_0^\Lambda d\lambda \rho(\lambda) \ (r_0 + \lambda) \ \sum_\alpha [\hat{\psi}_\lambda^{(\alpha)}]^2$$
$$- \frac{1}{4} \int_0^\Lambda d\lambda \rho(\lambda) (r_0 + \lambda)^2 \sum_{\alpha,\beta} [\hat{\psi}_\lambda^{(\alpha)}]^2 [\hat{\psi}_\lambda^{(\beta)}]^2 \ . \tag{56}$$

Note, however that there is no small parameter which allows us to terminate the cumulant expansion or in which to expand the biquadratic term. (For treatment of this problem via functional RG methods see [34].

The first term in the Hamiltonian, up to the summation over the replica indices, is essentially what we started off with in Section 4.2 and in the $n \to 0$ limit will give the same free energy. The renormalization of the quadratic term will therefore be the same, with $r = sr_0$. The bi-quadratic terms with the new interaction $U_4 \equiv r_0^2 + 2r_0\lambda + \lambda^2$ now have to be treated in a perturbative fashion relative to the Gaussian term, using another cumulant expansion. The different terms in the uncontracted vertex acquire the re-scaling factors $s^{3-\beta}$, $s^{2-\beta}$ and $s^{1-\beta}$, in this order. Although this has to be checked in detail, we conjecture that they are thus irrelevant for $\beta > 3$, and therefore to first order the theory remains Gaussian. The new quadratic couplings coming from the once-contracted graphs, however, depend on λ and are not trivial.

Inclusion of a $v_0\psi^4$ interaction, under the quenched averaging, gives rise to a bi-quadratic and a quartic term where the number of λ integrals are reduced to 2 and 1 respectively, with further simplifications now due to Gaussian contractions in the cumulant expansion in v_0. Power counting suggests that these terms are irrelevant for $\beta > 3/2$ and $\beta > 5/3$ to first order in v_0.

5.2. Deterministic hierarchical lattices and the matrix extension method

For some hierarchical networks obtained by successive decorations of a seed graph, the spectrum of the normalized Laplacian can be computed iteratively.[35] Under graph decoration, the"matrix extension transformation" yields all the new eigenvalues, in terms of the existing ones [36, 37]. It has been shown that most of the spectrum is given by the pre-images of the so called "spectral decimation" transformation $\lambda = R(\lambda')$, and converges to the Julia set of R^{-1} as $N \to \infty$ [35]. Moreover, one can easily check that the attractor (the Julia set) is chaotic, with the preimages of many eigenvalues jumping back and forth between different intervals $0 < \lambda < 1$ and $\lambda > 1$. We conjecture that similar properties may also hold for the un-normalized Laplace spectrum which we have been treating in this paper. Thus there is no smooth way in which to rescale existing eigenvalues to restore the integrated-out ones; although the network is deterministic, the rescaling transformation on the spectrum itself could just as well be stochastic.

It is interesting to recall a chaotic "real space" renormalization group transformation [38] encountered in a frustrated Potts model on a hierarchical lattice. This example shows that a finite interval below the critical temperature may become densely populated with critical points (singularities of the free energy) leading to BKT like behavior, similar to that found by Andrade and Herrmann [11].

6. Conclusions

In this paper we have tried to explore the possibilities offered by the eigenvectors and eigenvalues of the graph Laplacian to develop a field theoretic renormalization group (FTRG) approach to order-disorder phenomena on complex networks. We have taken a tutorial approach which intends to acquaint the reader with the ideas of FTRG, and then to build upon these ideas in order to develop the analogous machinery on a complex network. We have been able to carry over most of the basic concepts and to implement many of the usual procedures.

The proliferation of higher order terms in the renormalized Hamiltonian was brought under control by going over to a quenched average over different realizations of the stochastic network using a replica approach. We found that for $\beta > 3$, the first order perturbation expansion in the bi-quadratic couplings introduced by the quenched average were irrelevant, and that the theory remained Gaussian. Inclusion of a quartic interaction can also be shown to be irrelevant for $\beta > 3/2$ up to first order in v_0. Further work is in progress.

The λ-dependence acquired by the effective renormalized couplings persist, and in principle would call for a nonlinear transformation $\lambda' = T_s(\lambda)$ parameterized by the scale factor s.

This, in turn, calls for an initial λ-dependent temperature like coupling constant, $r(\lambda)$. The λ-dependent transition temperature, also indicated by the calculation in section 5.1, may be a way to understand the Berezinskii-Kosterlitz-Thouless (BKT) power law behavior of correlations obtained over a range of temperatures for $T < T_c$ found in Real Space RG calculations on scale free networks. [11]

Acknowledgments
AE would like to thank Zehra Çataltepe for having introduced her to the graph Laplacian, Lucilla Arcangelis and Anita Mehta for a careful reading of the manuscript. She also acknowledges partial support by the Turkish Academy of Sciences.

Appendix A. Normalized Laplacian

It is possible to define the "normalized" graph Laplacian via

$$\tilde{\mathbf{L}} \equiv \mathbf{D}^{-1}\mathbf{L} = \mathbf{I} - \mathbf{D}^{-1}\mathbf{A} \ , \tag{A.1}$$

where $\mathbf{I}$ is the identity matrix.

Eigenvectors of the normalized Laplacian are no longer orthogonal to each other, but one has,

$$\mathbf{v}_\mu^\dagger \mathbf{D} \mathbf{v}_{\mu'} = \delta_{\mu,\mu'} \ . \tag{A.2}$$

An advantage to using the normalized Laplacian in this context is that the eigenvalues are confined to a fixed interval $[0, 2]$ such that $\mu_0 = 0$, $0 < \mu_1 \le \mu_2 \ldots \mu_{N-1} \le 2$. In this way the eigenvalue spectrum gets denser and smoother as the graph becomes larger.

For many properties of normalized and unnormalized Laplacian spectra, see Ph. D. thesis of A. Banerjee. [20]

Appendix B. Second order contributions to the renormalized Hamiltonian

For compactness of notation we have defined $d_\mu \equiv d\lambda_\mu \rho(\lambda_\mu)$, where λ_μ appears in a subscript it will be just represented by μ, and $\int_{\Lambda/s}^{\Lambda} \equiv \int^{>}$.

The second order contribution from the perturbation expansion (Fig. B1c) to the ψ^4 interaction in the renormalized Hamiltonian is given by the doubly contracted term,

$$36 v_0^2 \int^{<} \ldots \int^{<} \prod_{\mu=1,\ldots,4} \left[d_\mu \hat{\psi}_\mu^{<} \right] Q_4^{(2)} \tag{B.1}$$

and

$$Q_4^{(2)} = \sum_{i,j} \{ u_1(i) u_2(i) u_3(j) u_4(j) \int^{>} \int^{>} \prod_{\nu=a,b} [d_\nu G(\lambda_\nu)] \, u_a(i) u_a(j) u_b(i) u_b(j) \} \tag{B.2}$$

where we have re-numbered lines so that the indices $\mu = 1, 2, 3, 4$ belong to the external legs and $\nu = a, b$ belong to the internal lines; i, j index the elements of the eigenvectors. Defining, after Eq. (38),

$$I_2(i,j) \equiv \int^{>} d\lambda \rho(\lambda) \, G(\lambda) \, u_\lambda(i) u_\lambda(j) \ , \tag{B.3}$$

we may re-write,

$$Q_4^{(2)} = \sum_{i,j} \{ u_1(i) u_2(i) u_3(j) u_4(j) \, [I_2(i,j)]^2 \} \ . \tag{B.4}$$

The effective coupling $36 v_0^2 Q_4^{(2)}$ depends on the eigenvectors $\lambda_1, \ldots \lambda_4$. The re-scaling factor for this 4-vertex can be obtained by power counting to be $s^{2\beta}$, up to the corrections which might come from this dependence. In the best RG tradition [27] we will neglect the λ dependence of the 4-vertex in what follows. If we also neglect the (i,j) dependence of I_2 and I_1 (which then become identical) we simply have

$$Q_4^{(2)} \propto (I_1)^2 \ . \tag{B.5}$$

Expanding I_1 in r_0 gives, from Eq. (38),

$$I_1 = \simeq \Lambda^{-\beta} \left[\frac{1}{\beta}(s^{\beta} - 1) - r_0 \frac{\Lambda^{-1}}{\beta + 1}(s^{\beta+1} - 1) \right] \ . \tag{B.6}$$

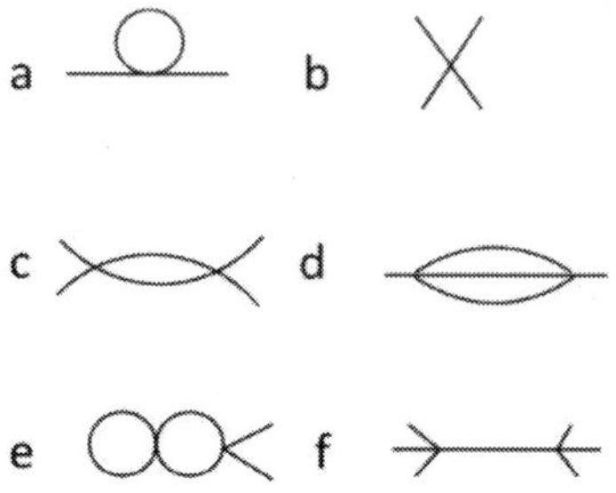

Figure B1. The Feynman diagrams to which reference is made in the text. In the context of the ψ^4 theory, the diagrams a), e) and d) contribute to the renormalized quadratic couplings; b) and c) to the quartic coupling, and f) to a sixth order coupling. As usual, a vertex (where four lines meet) is proportional to the coupling constant v_0, the "internal" lines (which are connected to a vertex at both ends) denote the Green's function $G(r_0, \lambda)$, Eq. (36), and each "external" leg carries a factor $\hat{\psi}_\lambda^{\text{lower}}$.

One contribution (Fig. B1d) the quadratic coupling to the second order is, with 1,2 denoting the external legs,

$$\begin{aligned} Q_{2,1}^{(2)} &= 72 v_0^2 \sum_{i,j} u_1(i) u_2(i) \int^{>} \int^{>} \int^{>} \prod_{\mu=3,4,5} d_\mu G(\lambda_m u) u_3(i) u_3(j) u_4(i) u_4(j) u_5(j)^2 \\ &= 72 v_0^2 \sum_{i,j} u_1(i) u_2(i) \left[I_2(i,j) \right]^2 I_1(i) \ , \end{aligned} \tag{B.7}$$

Note that, to the same approximation as above, we have

$$Q_{2,1}^{(2)} \propto (I_1)^3 \ . \tag{B.8}$$

The other contribution is given by is (Fig. B1e)

$$Q_{2,2}^{(2)} = 48 v_0^2 \sum_{i,j} u_1(i) u_2(j) \left[I_2(i,j) \right]^3 \propto (I_1)^3 \ . \tag{B.9}$$

Journal of Physics: Conference Series **319** (2011) 012007 doi:10.1088/1742-6596/319/1/012007

References

[1] Mehta A, Barker GC and Luck J-M 2009/5 *Physics Today* 40
[2] Mehta A 2007, *Granular Physics* (Cambridge: Cambridge University Press)
[3] Bennemann C, Donati C, Baschnagel J and Glotzer SC 1999 *Nature* **399** 246
[4] Yılmaz Y, Erzan A and Pekcan Ö 2002 *Eur. Phys. J E* **9** 135
[5] Dorogovtsev SN, Goltsev AV 2008 *Rev. Mod. Phys.***80** 1275
[6] Dorogovtsev SN, Goltsev AV and Mendes JFF 2002 *Phys. Rev. E* **66** 016104 ;
[7] Dorogovtsev SN, Goltsev AV, Mendes JFF and Samukhin AN 2003 *Phys. Rev. E* **68** 046109
[8] Bainconi G 2002 *Phys. Lett. A* **303** 166
[9] Goltsev AV, Dorogovtsev SN and Mendes JFF 2003 *Phys. Rev. E* **67** 026123
[10] Bradde S, Caccioli F, DallAsta L and Bianconi G 2010 *Phys. Rev. Lett.* **104** 218701
[11] Andrade RFS and Herrmann HJ 2005 *Phys.Rev.E* **71** 056131.
[12] Bauer M, Coulomb S and Dorogovtsev SN 2005 *Phys. Rev. Lett.* **94** 20060
[13] Bauer M, Coulomb S and Dorogovtsev SN 2006 *Phys. Rev. Lett.* **96** 109901
[14] Costin O, Costin RD and Grünfeld CP 1990 *J. Stat. Phys.* **59** 1531, cited in the previous Reference.
[15] Griffiths RB 1969 *Phy. Rev. Lett.* **23** 17
[16] Kaufman M and Griffiths R B 1982 *J. Phys. A* **15** L239
[17] Hinczewski M 2007 *Phys. Rev. E* **75** 061104
[18] Chung FRK 1997, *Spectral graph theory*, (Washington DC: American Mathematical Society)
[19] von Luxburg U 2007 *Stat. Comput.* **17** 395
[20] Banerjee A 2008 "Spectrum of the Graph Laplacian as a Tool for Analyzing Structure and Evolution of Networks," Ph. D. Thesis, University of Leipzig.
[21] `http://planetmath.org/encyclopedia/RayleighRitzTheorem.html, 10 June 2011.`
[22] Barabasi BA and Albert R 1999 *Science* **286** 509
[23] Ilyin I, Procaccia I, Regev I, Shokef Y 2009 *Phys. Rev. B* **80** 174201
[24] Karmakar S, Lerner E, Procaccia I 2011 "Direct Estimate of the Static Length-Scale Accompanying the Glass Transition," arXiv:1104.1036v1.
[25] Mandelbrot BB 1983, *The Fractal Geometry of Nature*, (New York: W. H. Freeman)
[26] Dorogovtsev SN and Mendes JFF 2000 *Phys. Rev. E* **62** 1842
[27] Goldenfeld N 1992, *Lectures on Phase Transition and the Renormalization Group*, (New York:Benjamin)
[28] Kadanoff LP *Statistical Physics - Statics, Dynamics and Renormalization* (Singapore: World Scientific)
[29] Kardar M 2007, *Statistical Physics of Fields*, (Cambridge: Cambridge University Press)
[30] Wilson KG and Kogut J 1974 *Phys. Rep.* **83**, 75
[31] Kac M (1968) *Trondheim Theoretical Physics Seminar* (Nordita Publ., 286;, cited in De Dominicis C and Giardina I 2006 *Random Fields and Spin Glasses*, (Cambridge: Cambridge Univesity Press)
[32] Edwards S F 1972 Proceedings of the Third International Conference on Amorphous Materials Douglass RW and Ellis B eds., (Wiley, New York); cited in De Dominicis C and Giardina I 2006 *Random Fields and Spin Glasses*, (Cambridge: Cambridge Univesity Press)
[33] Edwards S F and Anderson P W 1975 *J. Phys. F* **5** 965
[34] Zinn-Justin J 2007 *Phase Transitions and Renormalization Group* (Oxford: Oxford University Press), especially Chapters 14, 16.
[35] Molazemov L and Teplyaev A 2003 *Mathematical Physics, Analysis and Geometry* **6** 201
[36] Bajorin N, Chen T, Dagan A, Emmons C, Hussein M, Khalil M, Mody P, Steinhurst B, Teplyaev A 2008 *J. Phys. A***41** 015101
[37] Zhang Z, Qi Y, Zhou S, Lin Y and Guan J 2009 *Phys. Rev. E* **80** 41 015101
[38] Derrida B, Eckmann J-P and Erzan A 1983 *J. Phys. A* **16** 893

The stationary flow in a heterogeneous compliant vessel network

Marcel Filoche
Physique de la Matière Condensée, Ecole Polytechnique, CNRS, 91228 Palaiseau, France

Magali Florens
CMLA, ENS Cachan, CNRS, UniverSud, 61 av. du Président Wilson, Cachan, France

Abstract. We introduce a mathematical model of the hydrodynamic transport into systems consisting in a network of connected flexible pipes. In each pipe of the network, the flow is assumed to be steady and one-dimensional. The fluid-structure interaction is described through tube laws which relate the pipe diameter to the pressure difference across the pipe wall. We show that the resulting one-dimensional differential equation describing the flow in the pipe can be exactly integrated if one is able to estimate averages of the Reynolds number along the pipe. The differential equation is then transformed into a non linear scalar equation relating pressures at both ends of the pipe and the flow rate in the pipe. These equations are coupled throughout the network with mass conservation equations for the flow and zero pressure losses at the branching points of the network. This allows us to derive a general model for the computation of the flow into very large inhomogeneous networks consisting of several thousands of flexible pipes. This model is then applied to perform numerical simulations of the human lung airway system at exhalation. The topology of the system and the tube laws are taken from morphometric and physiological data in the literature. We find good qualitative and quantitative agreement between the simulation results and flow-volume loops measured in real patients. In particular, expiratory flow limitation which is an essential characteristic of forced expiration is found to be well reproduced by our simulations. Finally, a mathematical model of a pathology (Chronic Obstructive Pulmonary Disease) is introduced which allows us to quantitatively assess the influence of a moderate or severe alteration of the airway compliances.

1. Introduction

Numerous transport networks, especially biological networks, can be modeled as a complex assembly of flexible pipes. This is for instance the case of the cardiovascular system in vertebrates (arterial and venous), of the lung airway system in mammals, and of the venation in plants (see figure 1). In all these networks, the overall flow is the result of the complex interplay between the applied pressures at the extremities and the non linear hydrodynamic resistances of the pipes. When large pressures are applied, the fluid-structure interaction induces non linearities that may lead to important inhomogeneities in the flow distribution. Local characteristics of the vessels then drastically affect the global flow distribution pattern across the network. The total flow is therefore a direct signature of both the topology of the network and the compliance of the pipes.

For large or inhomogeneous networks where analytical calculations are impossible to carry out, numerical simulations are a invaluable tool to obtain qualitative and quantitative predictions

Figure 1. Left: Cast of the small airways of the human lung (E. R. Weibel). Right: venation of a leaf (S. Bohn, Rockefeller University, USA).

of the flow distribution for various working conditions, and to have a better understanding of the local and global features of these flexible networks [1]. Unfortunately, computing the flow in such systems requires in theory to solve the coupled equations governing the flow in the entire network of flexible pipes, and the 3D fluid-structure interaction in each pipe. This is in most cases out of reach for today's computing facilities.

We present here an alternative mathematical model that describes the flow into network through a large set of coupled equations, one for each flexible tube. First, we introduce a 1D model of the flow going through a single pipe. The fluid-structure interaction is modeled by state laws relating the local pipe diameter to the local transmural pressure. We then show that the 1D differential equation describing the flow in each pipe can be exactly integrated, provided one is able to estimate the average Reynolds number along the pipe. The 1D ODE in the pipe is then reduced to a scalar non-linear equation which relates the pressures at both ends of the pipe and the flow through the pipe. After adding Kirchoff laws at each node of the network to account for flow conservation, the entire flexible network is likened to an electrical network in which the unknowns are the pressures at the nodes of the network and the flow across the branches. Finally, this new model is applied to compute the flow in a specific case of particular interest, namely the lung airway system under forced expiration condition.

2. The 1D model of the flow through a flexible pipe

2.1. A differential equation for the transmural pressure

First, we assume a steady flow in the collapsible tube and steady pressures at both ends. Even in the case of an oscillatory flow, this simplification remains valid when the typical duration of a transient is small compared to the period of the cycle. The existence and the stability of solutions for the stationary flow in a collapsible tube has been examined and discussed in details in [2, 3, 4]. This assumption is for instance valid for most of the respiratory flow in the lung airway system [5].

Second, we consider that the flow is mostly one-dimensional along each pipe of the structure. The longitudinal coordinate along the tube is noted x. All the information about the cross-section is summarized into a position-dependent hydrodynamic diameter of the pipe, denoted $D(x)$.

Third, the fluid-structure interaction is modeled by a state law which relates the pipe diameter $D(x)$ to the local pressure difference across the pipe wall, also called *transmural pressure* P_{tm}.

This pressure is the difference between the local pressure P inside the pipe and the local pressure P_{out} outside the pipe: $P_{tm} = P - P_{out}$. The compliance of the tube is defined as the derivative of the state law with respect to the pressure. In most cases, this is a positive quantity which means that if the inside pressure exceeds the outside pressure, the transmural pressure is positive and the pipe opens. On the contrary if the transmural pressure is negative, the pipe tends to close. Figure 2 shows an example of a typical compliant behavior of a flexible pipe, i.e. two cross-sections of a pulmonary airway respectively under positive and negative transmural pressure.

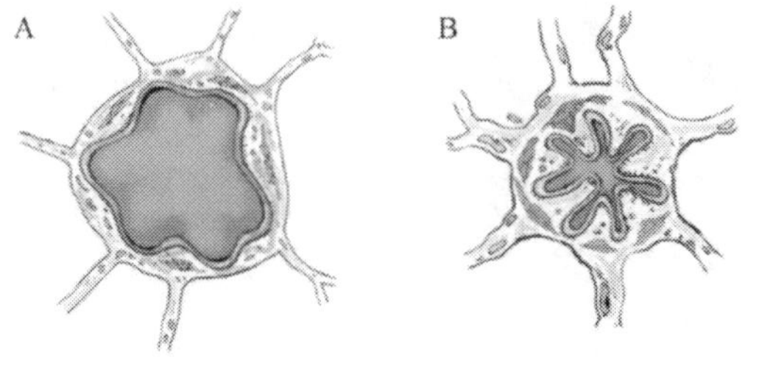

Figure 2. Typical behavior of a flexible pipe (here a bronchus) for two different values of the transmural pressure [6]. A: cross-section of the bronchus during inhalation, under positive transmural pressure. B: cross-section of the same bronchus during exhalation, under negative transmural pressure. One can observe the drastic reduction of the cross-section area hence the collapse of the hydrodynamic diameter.

The pressure profile is computed by solving in each pipe a 1D differential equation. For a steady and incompressible flow, the pressure at location x along the pipe is deduced from the pressure P_A at the pipe entrance (corresponding to the coordinate $x = 0$) using Bernoulli equation:

$$P(x) = P_A + \frac{1}{2}\rho\, v_A^2 - \frac{1}{2}\rho\, v^2(x) - \int_0^x f(y)\, dy \tag{1}$$

where P_A and v_A are respectively the pressure and the velocity of the flow at the pipe entrance, $v(x)$ is the velocity of the flow at point x, ρ is the fluid density, and $f(x)$ accounts for the energy loss per unit of distance. This last term always includes the viscous pressure loss per unit of distance and can also include a gravity term. The latter case is achieved by adding a constant term, equal to $(-\rho g \cos\theta)$ to $f(x)$, g being the acceleration due to gravity and θ the angle between the direction of gravity and the axis of the pipe.

If one assumes a uniform pressure in the medium surrounding the vessel network, then the transmural pressure P_{tm} also satisfies (1). The gradient of transmural pressure can be computed just by differentiating the above equation with respect to the coordinate x, and express the fluid velocity as a function of the flow rate Φ and the hydrodynamic diameter $D(x)$:

$$\frac{dP_{tm}}{dx} = -\rho v(x)\frac{dv}{dx} - f(x) = -\rho v(x)\frac{d\left(\frac{\Phi}{\pi D^2(x)/4}\right)}{dP_{tm}}\frac{dP_{tm}}{dx} - f(x) \tag{2}$$

This develops into:

$$\frac{dP_{tm}}{dx} = \rho v^2(x)\frac{2}{D(x)}\frac{dD}{dP_{tm}}\frac{dP_{tm}}{dx} - f(x) \tag{3}$$

Finally, the gradient of transmural pressure sums up:

$$\frac{dP_{tm}}{dx} = \frac{-f(x)}{1 - \frac{v^2(x)}{c^2(x)}} \quad \text{with} \quad \frac{1}{c^2(x)} = \frac{2\rho}{D(x)}\frac{dD}{dP_{tm}} \tag{4}$$

where $c(x)$ is the local wave speed.

For instance, the energy loss per unit of distance $f(x)$ can be evaluated in case of the bronchial network from Reynolds et al. [7]:

$$f(x) = \frac{128\eta\Phi}{\pi D^4(x)} \left(a + b \, Re(x)\right) - \rho g \cos\theta \tag{5}$$

where η is the fluid viscosity, $Re(x)$ is the local Reynolds number, and a and b are two phenomenological constants. The first term corresponds to viscous pressure loss per unit of distance; it is directly deduced from the experimental work of Reynolds et al [7]. The first part of this term corresponds to the viscous pressure loss for a Poiseuille flow. The constant a, which should be 1 for an infinite straight pipe, accounts for the fact that the Poiseuille profile needs some time to set up in the branches of the network. The second part of this term, characterized by a constant b, is a phenomenological term accounting for the loss when inertial effects are present, i.e. at higher velocities. In the following, we assume that the function $f(x)$ representing the energy loss per unit distance takes the general form (5) in our networks.

The denominator that appears in (4) corresponds to a negative feedback loop in the fluid-structure interaction that leads to flow limitation. It can be understood as follows. The dissipative pressure loss in the pipe tends to collapse the pipe which in turn generates an additional loss of transmural pressure by kinetic acceleration of the fluid. When the flow velocity is close to the wave propagation speed of a pressure disturbance along the pipe wall, the pressure loss increases causing the pipe to narrow even more. The resulting increase of the hydrodynamic resistance of the pipe automatically limits the flow rate. This mechanism has been described in details by Lambert et al. [8].

To sum up, the flow distribution in the network can be mathematically described by a set of differential equations, one for each pipe similar to (4). These equations are coupled by the continuity equations (Kirchoff laws) at each node of the network. Therefore, computing the flow in the entire network requires to solve in a coupled way as many differential equations as there are pipes in the network. Since biological networks usually correspond to several thousands of branches, this resolution is in most cases numerically unrealistic. To our knowledge, similar approaches have considered at most a few tens of flexible pipes in a network. Simulating large networks such as the lung airway system have been made possible only by assuming some sort of symmetry in the system. This allows one to solve only one differential equation per generation, this equation representing a large number of identical branches [8, 9, 5, 10, 11].

What we propose here is an alternative model based on an quasi analytical solution of the differential equation 4. After integration, each differential equation is transformed into a non linear scalar equation in which the unknowns are the transmural pressures at both ends of the pipe and the flow going through it. This will allow us to compute the flow distribution in the entire network, even in the case of large or non heterogeneous systems. We now describe this step.

2.2. Integrating the differential equation along a pipe

Starting from (4) one multiplies on both sides the equation by $\left(1 - v^2/c^2\right) D^4$ and then integrate between both ends of the pipe, noted A and B:

$$\int_A^B \left(1 - \frac{v^2(x)}{c^2(x)}\right) D^4(x) \, \frac{dP_{tm}}{dx} \, dx = -\int_A^B D^4(x) \, f(x) \, dx \tag{6}$$

This also expresses, using (5):

$$\int_A^B \left(1 - \frac{v^2(x)}{c^2(x)}\right) D^4(x) \, dP_{tm} = -\frac{128\eta\Phi}{\pi} \left[a \, L + b \int_A^B Re(x).dx\right] + \rho g \cos\theta \int_A^B D^4(x) \, dx \tag{7}$$

where L is the length of the pipe. Figure 3 shows a schematic representation of the flexible pipe and displays the various unknowns involved in the integration.

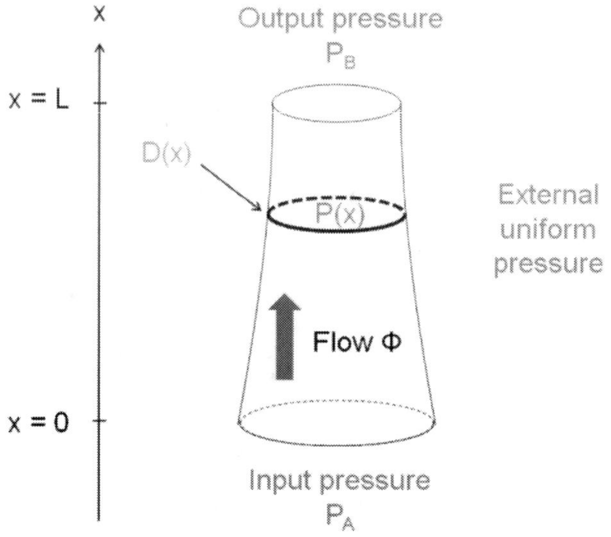

Figure 3. Schematic representation of the flow in one elementary flexible pipe of the network. L is the pipe length, $D(x)$ is the local diameter of the pipe at longitudinal coordinate x, P_A and P_B are the inner pressures at both ends of the pipes, and Φ is the flow rate going through the pipe. We recall here that the external pressure is assumed uniform.

The left hand side of (7) can be exactly integrated:

$$\int_A^B \left(1 - \frac{v^2(x)}{c^2(x)}\right) D^4(x)\, dP_{tm} = h(P_{tm,B}) - h(P_{tm,A}) - \frac{32\rho\Phi^2}{\pi^2} \ln\left(\frac{D_B}{D_A}\right) \tag{8}$$

where $P_{tm,A} = P_A - P_{out}$ (resp. B) is the local transmural pressure and h a function defined by: $h(P) = \int_0^P D^4(p)\, dp$. In other words, h is the primitive of the fourth power of the state law of the flexible pipe. The function $h(P)$ can be obtained from tabulated data or, when $D(P)$ is described by an analytical law as it is the case for instance in the airway system [8], it can even be an analytical function.

The right hand side of (7) writes:

$$-\frac{128\eta L\Phi}{\pi}\left[a + b\langle Re\rangle\right] + \rho g \cos\theta \int_A^B D^4(x)\, dx \tag{9}$$

where $\langle Re\rangle$ stands for the average Reynolds number along the pipe. The last term which requires to integrate D^4 can be rewritten using:

$$\Phi = \frac{\pi D^2(x)}{4}\, v(x) = \frac{\pi D^2(x)\eta}{4D(x)\rho}\, Re(x) \qquad \text{thus} \qquad D(x) = \left(\frac{4\rho\Phi}{\pi\eta}\right)\frac{1}{Re(x)} \tag{10}$$

which yields

$$\int_A^B D^4(x)\, dx = \left(\frac{4\rho\Phi}{\pi\eta}\right) L\ \langle Re^{-1}\rangle \tag{11}$$

Using (8) and (11), (7) finally writes:

$$h(P_{tm,B}) - h(P_{tm,A}) - \frac{32\rho\Phi^2}{\pi^2}\ln\left(\frac{D_B}{D_A}\right) =$$
$$-\frac{128\eta L\Phi}{\pi}\Big[a + b\langle Re\rangle\Big] + \rho g\cos\theta\left(\frac{4\rho\Phi}{\pi\eta}\right) L\,\langle Re^{-1}\rangle \tag{12}$$

As a result, in order to transform (12) into a scalar equation in which the only unknowns are the flow and the pressures at both ends of the pipes, one needs to compute the averaged quantities $\langle Re\rangle$ and $\langle Re^{-1}\rangle$ along the pipe. In most practical cases, both averages can be evaluated with a good precision using the average diameter of the pipe:

$$\langle Re^{\alpha}\rangle = \left(\frac{4\rho\Phi}{\eta\pi\left(\frac{D_S+D_E}{2}\right)}\right)^{\alpha} \tag{13}$$

One has to stress here that this approximation only affects the calculation of the dissipation in the tube and does not at all implicitly assume that the fluid velocity is constant along each pipe. We will see in the results section of the paper that this approximate calculation of the dissipation losses is justified. When applied to the lung airway system, it allows us to compute the stationary flow in the entire tree in a very good approximation at each stage of the breathing cycle.

In summary, we have transformed each differential equation describing the fluid-structure interaction in each pipe into a non linear scalar equation (12). Computing the flow through the network is now a feasible goal.

3. Computing the flow in the entire network

3.1. Continuity at the nodes of the network

When the entire network is considered, the scalar equations (12) describing the fluid-structure interaction in each pipe are coupled through Kirchhoff laws at each node of the network. Due to mass conservation at each node, the sum of the flow rates reaching a node is equal to zero. Also, the pressures at the ends of every pipe reaching the same node are supposed to be identical. This is equivalent to consider that the pressure losses are neglected in the branching region as compared to the pressure losses along the pipes.

$$\sum_{\text{branches } i} \Phi_i = 0 \qquad \text{and} \qquad \Delta P|_{branching} = 0 \tag{14}$$

The last assumption has been discussed for instance in the lung airway system by several authors [12, 13]. In the branching region, one can assume that the dissipative pressure losses are negligible since the transition from one branch to the next occurs on a short distance. Only the convective pressure losses could significantly contribute in the branching region. Furthermore, studies from the literature [14, 5] have shown no significant convective pressure drop occurring at the junctions. They are therefore neglected in our approach. These losses could nevertheless be taken into account in further studies by introducing a non-zero loss coefficient at the junctions as suggested by recent works [15, 1].

3.2. Boundary conditions and numerical scheme

In our model, the mathematical system describing the flow in the entire network consists in one scalar equation (12) per pipe, coupled by the Kirchoff laws described above at each node of the network. The closure of this system is achieved by imposing boundary conditions at all exit nodes and branches of the network. The system can be pressure driven which means that the pressures are imposed at each exit node of the network. This is for instance the case for the lung airway system under forced expiration that we will present in the next section. The system can also be flow driven which means that the flow rates are imposed in all branches exiting the network. It can also be a combination of these two boundary conditions.

For a given set of boundary conditions, the non linear system is solved by an iterative Newton-Raphson scheme. As a result one obtains the flow rates and the pressures at both extremities for each pipe of the network, i.e. the flow distribution in the entire network. This also allows us to access the diameter and the pressure at any location x of each pipe, just by integrating (8) between one end of the pipe and this location.

The time-dependent response of a network such as the lung airway system is obtained using a quasi-static approach. At each time step of the expiration varying boundary conditions are applied at the extremities of the network. In the case of the lung airway system, the atmospheric pressure is applied at the mouth (which corresponds to the entrance of the system), and time-dependent alveolar pressures are applied at the distal ends of the tracheobronchial tree. The steady flow corresponding to these boundary conditions is then computed in the entire network. Finally, the time-varying response of the network is obtained from the sequence of these successive steady states.

4. Simulation results: the lung airway system

Several examples of compliant vessel network have been quoted in the introduction. We have chosen here to focus on a particular network, the lung airway system. The lung airway system in humans is a complex branched distribution system whose structure exhibits highly hierarchical features and scale invariance [16, 17, 18]. Ideally, this transport system is intended to deliver as uniformly as possible the flux of fresh air into the gas exchange units located in the distal regions. During inhalation the thoracic cage is lifted up and out by the respiratory muscles, leading to a natural opening of the pulmonary airways. On the contrary during exhalation the pressure exerted by the diaphragm and the elastic energy stored in the respiratory muscles tend to close the airways. The flow pattern in the tree is the result of a complex interplay between the flexible airway structure and the applied pressure distribution. In extreme conditions, as in forced expiration, the system exhibits non linearities that may lead to important inhomogeneities in the flow distribution. As a consequence, measuring the total flow rate out of the lung has become a routine clinical examination as a signature of the mechanical properties of the compliant bronchial geometry.

In the following, we study the lung airway system at forced expiration. First, the parameters of the model are detailed. Then, the results of the model are analyzed and compared to classical spirometric measurements made on patients.

4.1. Parameters of the model

The state laws giving the relationship between the local airway diameter D and the local transmural pressure P_{tm} are taken from the works of Lambert et al. [8]. The state law of any bronchus is defined with two analytical functions:

$$\begin{aligned} D(P_{tm}) &= D_{max}\sqrt{\alpha_0\left(1-\frac{P_{tm}}{P_1}\right)^{-n_1}} && \text{if } P_{tm}<0 \\ D(P_{tm}) &= D_{max}\sqrt{1-(1-\alpha_0)\left(1-\frac{P_{tm}}{P_2}\right)^{-n_2}} && \text{if } P_{tm}\geq 0 \end{aligned} \qquad (15)$$

where D_{max} is the maximal diameter of the airway and α_0, n_1, n_2, P_1 et P_2 are adjustable parameters defining the flexibility or compliance of the airway. These parameters are tabulated in the paper of Lambert et al [8].

The energy loss per unit distance is defined by (5) in which the parameters a and b are respectively equal to 1.5 and 0.0035 in the lung airway system [7]. Since the fluid is air, gravity effects are negligible compared the viscous losses.

The lung airway system at exhalation is a pressure driven system. At the upper end (the mouth) the pressure is assumed to be constant and equal to the atmospheric pressure. At the extremities of the tracheobronchial tree, the pressure built by the respiratory muscles and the diaphragm output is called the alveolar pressure P_{alv}. Its time dependency is of exponential type, representing the relaxation of the lung towards equilibrium:

$$P_{alv}(t) = P_m(1 - e^{-t/\tau})\left(\frac{V_L(t) - V_R}{C_V}\right) - R_T\Phi_0(t) \tag{16}$$

where P_m is the maximal pressure of the expiratory muscles, τ is the time constant of the expiratory muscles, V_L is the lung volume, V_R is the residual volume (the minimal functional lung volume), C_V is the vital capacity, R_T is the lung tissue resistance and Φ_0 is the flow rate in the first pipe of the network, here the trachea. All parameters in (16) are taken from the literature, tabulated for an healthy young adult [19, 20, 21, 22].

4.2. Spirometry of forced expiration

Forced expiration is simulated using the quasi-static approach described above. At each time step, a steady state of the lung airway system is computed: the flow, pressures and diameters of each airway are determined for a given set of pressure boundary conditions. In particular, the expiratory flow rate and the lung volume are calculated, allowing to obtain flow-volume loops. These loops plot the flow rate as an implicit function of the exhaled volume. Figure 4 displays flow-volume loops obtained for different patient's efforts. From A to C, the patient's effort decreases. This decrease is accounted for by changing the alveolar pressure P_{alv} through the parameters P_m and τ [23]. The computation time to retrieve a complete flow-volume loop is about one hour on a standard laptop for a lung airway system consisting in about 30,000 different bronchi.

First, the curves obtained from our simulations match real flow-volume loops measured on patients [24]. In the case A, the peak flow is about 10 L/s and the expired volume during the first second is about 4.6 L, which is in good agreement with typical measurements in healthy adults. Moreover, we recover the two characteristic parts of the flow-volume curve. The first part of the curve depends on the patient's effort: the peak flow increases for larger effort. The second part of the curve is independent of the patient's effort and linear with respect to the expired volume: air flow does not increase when the alveolar pressure increases. Recovering this characteristic twofold behavior is an important result that underlines the quality of the numerical model.

Using our set of simulations, one can try to assess the validity of the approximation made in (13). To that end, at each time step and for each airway we derive the local diameter $D(x)$ at each location along each airway using equation 12. From these data an *a posteriori* estimate of average Reynolds number in the airway is computed:

$$\langle Re \rangle = \frac{1}{L}\int_A^B \frac{4\rho\Phi}{\eta\pi D(x)}.dx \tag{17}$$

Our computations shows that the differences between the approximated and the exact average Reynolds numbers are most of the time about a few percents. They never exceed 20% in any of the airways, a value that is reached for a very short duration in only a small number of airways. This justifies in a self-consistent way the integration scheme chosen in (12).

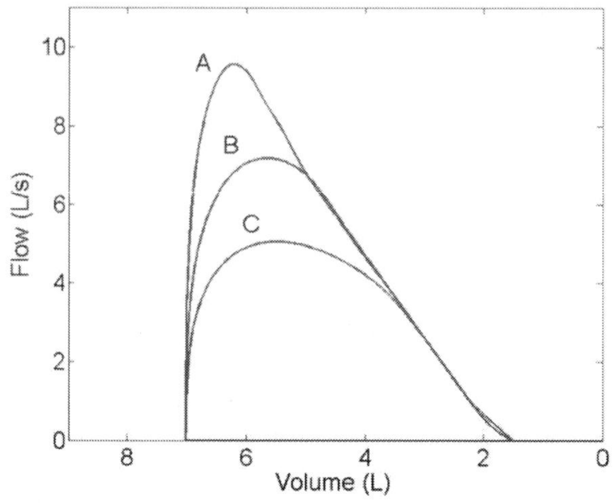

Figure 4. Flow-volume loops computed using a quasi-static computation of flow in the lung airway system. A: $P_m = 24$ kPa and $\tau = 0{,}2$ s; B: $P_m = 12$ kPa and $\tau = 0{,}3$ s; C: $P_m = 6$ kPa and $\tau = 0{,}4$ s [23].

4.3. Simulated pathology

After validation on healthy subjects, we now investigate the behavior of a pathological lung airway system under forced expiration. Several pathologies have been explored. The example presented here corresponds to a Chronic Obstructive Pulmonary Disease (COPD). COPD is a pulmonary disease characterized by an obstruction of the small airways. The alteration induces local and global modifications of the pulmonary mechanics. It is modeled in our case by introducing a local modification of the mechanical properties of the small airways. They become much more compliant due to a partial destruction of the surrounding parenchyma holding the airways. The elastic recoil forces also decrease so that the smallest airways are not completely open and tend to have a smaller diameter [25]. The global effects correspond to an increase of lung volumes [24], lung tissue resistance [21], and pulmonary compliance [26, 27].

Figure 5 displays the flow-volume loops simulated for an healthy airway system (A) and for two alteration levels (B and C). Both altered airway systems exhibit a reduced peak flow and a flow collapse after the peak flow as reported by West [24], and Pellegrino et al [28]. First, the decrease of the maximal diameters in small airways generates an increase of the airway resistance, which in turn induces a decrease of the expired flow. Secondly, the increase of both local and global pulmonary compliance leads to a more rapid collapse of the small airways. The dynamic compression of the small airways occurs faster and the expired flow drops sharply after the peak flow. One can observe in Figure 5 that this effect becomes even more important in the case of a severe alteration. All these simulation results are in very good agreement with the clinical observations.

5. Conclusion

In conclusion, we have derived a mathematical model which is capable of computing the flow distribution into a large heterogeneous network of flexible tubes. The fluid-structure interaction is introduced through state laws that relate the pipe diameter to the pressure difference across the pipe wall. After integration, the one-dimensional differential equations that govern the flow in each pipe of the network are transformed into a non linear system of scalar equations where

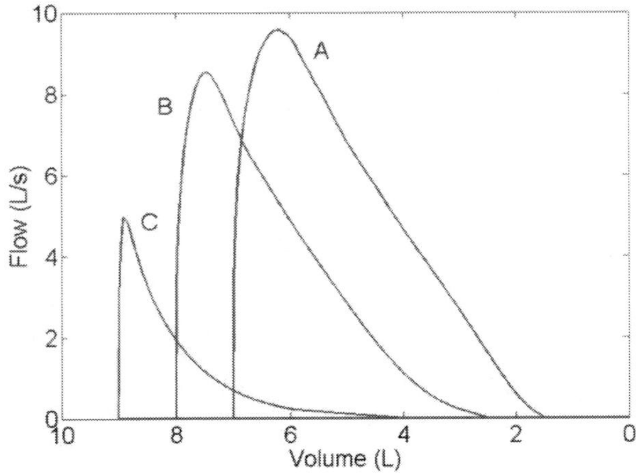

Figure 5. Flow-volume loops computed using our quasi-static model of the lung airway system. A: healthy airway system, B: moderate alteration of the airway system (moderate COPD), C: severe alteration of the airway system (severe COPD).

the unknowns are the pressures at each node of the network and the flow rates in each branch of the network. The system is closed by imposing boundary conditions at the exit nodes or in the exit branches of the network.

To assess its efficiency, this model has been applied to the human lung airway system. The topology of the vessel network and the compliance laws have been taken from the literature. This model allowed us to compute for the first time the response of a realistic asymmetric lung airway system at forced expiration, solving individually in each branch the flow rate during the entire expiration maneuver. The interest of this approach is that it permits to simulate large systems in a reasonable time, and to easily introduce morphometric or physiological modifications of the network in order to evaluate their influence on the global flow going through the structure. This approach is particularly fitted to understand the consequences of obstructive pathologies such as asthma or COPD at the local and global level during the breathing cycle in realistic bronchial trees. This could be used as a tool in a medical perspective to validate physiological hypothesis or to design new pulmonary function tests.

Finally, in the limit of a very dense network, one can think of combining the discrete equations (12) together with the mass conservation at each node to generate a continuous model of the flow through compressible porous media.

References

[1] Fullana J-M and Zaleski S 2009 *J. Fluid. Mech.* **621** 183-204
[2] Jensen O E and Pedley T J 1989 *J. Fluid. Mech.* **206** 339-374
[3] Elad D, Kamm R D and Shapiro A H 1989 *J. Fluid Mech.* **203** 401-418
[4] Kamm R D and Pedley T J 1989 *J. Biomech. Eng.* **111** 177-179
[5] Polak A G 1998 *Comput. Biol. Med.* **28** 613-625
[6] Tiddens H A W M, Donaldson S H, Rosenfeld M and Pare P D 2010 *Pediatr. Pulmonol.* **45** 107-117
[7] Reynolds D B 1982 *J. Biomech. Eng.* **104** 153-158
[8] Lambert R K, Wilson T A, Hyatt R E and Rodarte J R 1982 *J. Appl. Physiol.* **52** 44-56
[9] Isabey D 1982 *J. Biomech.* **15** 395-404
[10] Polak A G and Lutchen K R 2003 *Ann. Biomed. Eng.* **31** 891-907
[11] Barbini P, Brighenti C, Cevenni G and Gnudi G 2005 *Ann. Biomed. Eng.* **33** 518-530
[12] Pedley T J, Schroter R C and Sudlow M F 1970 *Respir. Physiol.* **9** 371-386

[13] Douglass R W and Munson B R 1974 *J. Biomech.* **7** 551-557
[14] Pardaens J, Van de Woestijne K P and Clement J 1972 *J. Appl. Physiol.* **33** 478-490
[15] Olufsen M S et al. 2000 *Ann. Biomed. Eng.* **28** 12811299
[16] Weibel E R and Gomez D M 1962 *Science* **137** 577-585
[17] Grotberg J B 1994 *Ann. Rev. Fluid Mech.* **26** 529-571
[18] Florens M and Sapoval B and Filoche M 2011 *J. Appl. Physiol.* **110** 756-763
[19] Agostini E and Fenn W O 1960 *J. Appl. Physiol.* **15** 349-353
[20] Weibel E R 1984 *The Pathway for Oxygen, Structure and Function in the Mammalian Respiratory System* (Cambridge: Harvard University Press)
[21] Verbeken E K, Cauberghs M, Mertens I, Lauweryns J M, Van de Woestijne K P 1992 *J. Appl. Physiol.* **72** 2343-2353
[22] Aguilar X, Fiz J A, Texido A, Vilalta P, Abad J, Richart C and Morera J 1996 *Respir. Med.* **90** 231-233
[23] Florens M and Sapoval B and Filoche M 2011 *Comput. Phys. Commun.* **189** 1932-1936
[24] West J B 2008 *Pulmonary Pathophysiology, The essentials, Seventh Edition* (Lippincott Williams & Wilkins)
[25] Guenard H and Rouatbi S 2002 *Rev. Mal. Respir.* **19** 230-240
[26] Greaves I A and Colebatch H J H 1980 *Am. Rev. Respir. Dis.* **121** 127-136
[27] Verbeken E K, Cauberghs M, Mertens I, Clement J, Lauweryns J M and Van de Woestijne K P 1992 *Chest* **101** 800-809
[28] Pellegrino R et al. J 2005 *Eur. Respir. J.* **26** 948-968

Continuum Models and Discrete Systems Symposia (CMDS-12) IOP Publishing
Journal of Physics: Conference Series **319** (2011) 012009 doi:10.1088/1742-6596/319/1/012009

Two-phase flow in porous media: power-law scaling of effective permeability

Morten Grøva, Alex Hansen

Department of Physics, NTNU, NO-7491 Trondheim, Norway

E-mail: Morten.Grova@ntnu.no, Alex.Hansen@ntnu.no

Abstract.
A recent experiment has reported power-law scaling of effective permeability of two-phase flow with respect to capillary number for a two-dimensional model porous medium. In this paper, we consider the simultaneous flow of two phases through a porous medium under steady-state conditions, fixed total flow-rate and saturation, using a two-dimensional network simulator. We obtain power-law exponents for the scaling of effective permeability with respect to capillary number. The simulations are performed both for viscosity matched fluids and for a high viscosity ratio resembling that of air and water. Good power-law behaviour is found for both cases. Different exponents are found, depending on saturation.

1. Introduction

Two immiscible fluids (such as oil and water) sharing a common pore space within a porous medium (such as sandstone) self-organize under flowing conditions [1, 2]. Viscous and capillary forces are thought to dominate this process. Some transients are known to display power-law behaviour, and have attracted significant attention for several decades.

Two-phase flow in porous media is not only important as an interesting application and testing-ground for theoretical models. With global oil reservoirs dwindling, the problem of recovering the 20 to 60 percent of oil which remains at the end of traditional recovery is of great importance. An improved understanding of the connection between pore-level dynamics (where the menisci sit) and the reservoir scale is crucial to the field of enhanced oil recovery.

This system therefore provides a unique challenge for physicists and others with an interest in complex systems and self-organized criticality [3]. Two models have been successful in capturing extreme limits of the parameter space for displacement processes: diffusion-limited aggregration [4] describes displacement of a more-viscous fluid by a less-viscous fluid (or gas) when capillary forces can be neglected while invasion percolation [5] describes displacement of a wetting fluid by a non-wetting fluid (or gas) when viscous forces can be neglected.

The displacement of one phase by another in the limits of high viscosity ratio or strongly dominant capillary forces leads to power-law scaling for, e.g., pressure-burst size distributions [6]. Also, the displacement structures have fractal dimensions [7, 8, 9]. For this reason we believe that in these limits the internal self-organization of fluid-fluid interfaces (menisci) are the result of critical behaviour giving rise to scale-free structures.

It is important to use a parameter space which is large enough to capture the relevant physics, but not larger. Viscosity ratio M and capillary number Ca are generally thought to be the two most significant parameters. Saturation S or fractional flow F, of which one should be kept fixed to achieve steady-state, is a third parameter. In this context, steady-state implies that macroscopic observables only fluctuate around well-defined, reproducible average values. In our simulations we control saturation, and find that fractional flow eventually reaches an average value, around which it fluctuates, and this average value does not depend on initial conditions or the realization of the porous medium disorder. If fractional flow is controlled, we expect the saturation to fluctuate around some well defined average when steady-state is achieved.

Depending on the applied flow-rate and saturation, both phases or only one of the phases may be mobile. In this work we consider the case where both phases are mobile under conditions of fixed saturation. We call this the regime of free fractional flow. The transition where one of the phases becomes immobilized will be referred to as the freezing transition. If we are below the freezing transition, such that only one phase would be mobile under conditions of fixed saturation, but both phases are forced to flow due to, e.g., simultaneous injection of both phases, we have a regime of forced fractional flow.

Together with F (or S if F is fixed), the effective permeability κ_{eff} is the main macroscopic quantity of interest. When these two are known, relative permeabilities are easily calculated.

Some earlier numerical work has explored steady-state properties. Pioneering work was done by Constantinides and Payatakes [10]. A similar model to the one used here has been described and discussed for simulations focusing on both transients and steady-state [11, 12]. The need for a statistical mechanics approach to describe steady-state has been suggested by work finding a relationship between the applied pressure-drop and fractional flow [13, 14] and first order phase-transition-like behaviour in the freezing transition [15]. An approach to a thermodynamics of two-phase flow in porous media has been skethched [16].

Little experimental work has been done under steady-state conditions. Rather, the focus has been on transients: flooding a sample by one of the phases until this is the only phase that flows. A notable exception is the work done by the Payatakes group [17, 18], which were etched-glass experiments performed under steady-state conditions. To our knowledge they did not explore scaling of effective permeability.

Progress was made in a recent experiment by Tallakstad *et al* [19, 20] where a power-law was found for the scaling of effective permeability with respect to capillary number,

$$\kappa_{\text{eff}} \sim \text{Ca}^{\beta}. \tag{1}$$

In their set-up, they find $\beta = 0.54 \pm 0.08$.

This scaling indicates that self-organized criticality persists over a wide range of parameters, i.e., the system does not only approach a critical point in some limit of the parameter space. In the experiment there was a large viscosity ratio due to one phase being air and the other a mixture of water and glycerol. Also, fractional flow was fixed by simultaneous injection of the two phases. Seven of fifteen injection syringes contained the non-wetting phase (air), so $F = 7/15$.

Two immediate questions present themselves. Does power-law scaling also happen in the absence of a large viscosity ratio? And what happens when we control saturation instead of fractional flow? Some progress has been achieved recently [21].

This work explores the dependency of effective permeability on saturation, viscosity ratio and capillary number, using a disordered transport network as a model for the porous medium.

2. The network simulator

An impression of the model is provided by Fig. 1.

A porous medium can be considered to be a transport network, consisting of tubes connected at pores, which in a network may be identified as links and nodes respectively. The network

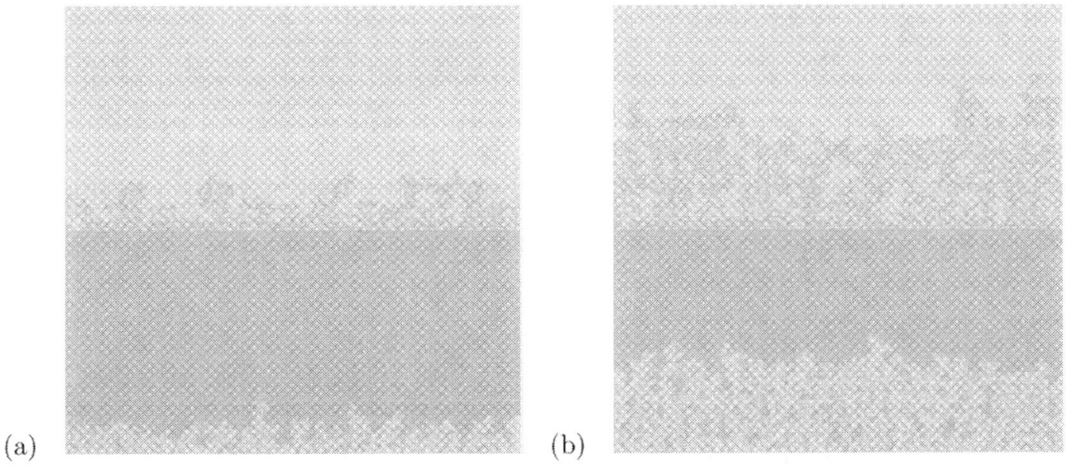

Figure 1. Two snapshots of the spatial distribution of non-wetting bubbles (red). The overall direction of flow is upwards. The initial condition was a horizontal band of non-wetting fluid. Bi-periodic boundary conditions are used. Capillary fingering and imbibition is seen along the middle and bottom mixing front, respectively. System size is 256×256. $\mathrm{Ca} = 10^{-4}$ and $S = 0.5$.

considered in this work is a square lattice inclined at 45° relative to the main direction of flow. Boundary conditions are bi-periodic (the network may be mapped onto a torus). Disorder is introduced by assigning random radii to tubes, from a flat distribution between 10-40% of the tube length. All tubes are assumed to be equally long, with length ℓ.

For our model, we consider the nodes to have no properties other than that of connecting links, while links are assigned a volume and radius r. Links also need to model the narrowing geometry of pore necks. For this purpose, links are assigned an hourglass-shape with respect to the capillary pressure of menisci.

A modified version of the Young-Laplace relation then gives the capillary pressure p_c as

$$p_c(x) = \frac{2\gamma}{r}\left(1 - \cos(\frac{2\pi x}{\ell})\right), \tag{2}$$

where x is the position of the meniscus in the interval $[0, \ell]$ and γ is surface tension.

The Hagen-Poiseulle relation for cylindrical tubes gives us the flow rate of a tube for a given pressure drop,

$$q = -\frac{\pi r^4}{8\ell\mu}\left(\Delta p + \sum p_c\right), \tag{3}$$

where μ is the viscosity of the fluid contained in the tube and Δp is the difference in pressure of the two nodes connected by the tube. The sum runs over all menisci within the tube. If the viscosities of the two fluids are different, the viscosity of the above equation is replaced by the volume-weighted effective viscosity, averaged over the volume of the tube.

A constant flow-rate Q is achieved by solving the pressure field for two applied pressure-drops ΔP and extrapolating linearly to the desired Q using the relationship

$$Q = a\Delta P + b, \tag{4}$$

where a is the single-phase mobility and b is a flow-rate induced by the capillary pressures. This approach is valid due to the flow in each tube being a linear function of Δp.

The effective permeability is given by

$$\kappa_{\text{eff}} = \frac{Q}{A\Delta P}, \tag{5}$$

which is unity if there are no macroscopic effects due to capillaries, and smaller than unity if the flow-rate is smaller than it would be without the effect of capillaries.

The effective pressure drop P_{eff} is the pressure drop normalized by the pressure drop that would be required in the absence of capillaries, and is nothing but the reciprocal of κ_{eff}.

The main parameter of our model is the relative strength of capillary and viscous forces, parametrized by the capillary number

$$\text{Ca} = \frac{\mu Q}{\gamma \phi A}, \tag{6}$$

where A is the cross-sectional area of the sample and ϕ is its porosity. If the viscosities of the two fluids are different, the viscosity of the above equation must be replaced by the volume-weighted effective viscosity, averaged over the entire system.

The viscosity ratio M is defined by

$$\text{M} = \frac{\mu_{nw}}{\mu_w}, \tag{7}$$

where the subscripts denote non-wetting and wetting viscosity.

Saturation S is defined by the ratio of non-wetting to total volume,

$$S = \frac{V_{nw}}{V}. \tag{8}$$

Fractional flow F is defined by the ratio of non-wetting to total flow,

$$F = \frac{Q_{nw}}{Q}. \tag{9}$$

The system is moved forward by solving the pressure field and timestepping with the timestep chosen by the largest flow-rate within the system. Each configuration consists of a set of bubbles, each with two menisci. At all timesteps the Kirchhoff equations are satisfied by solving these for the pressure and flow field. This is done using a customized version of the conjugate gradient method with an incomplete cholesky preconditioner [22].

Fig. 2 provides an impression of the saturation, flow and pressure spatial distributions for snapshots of a displacement process. Fig. 3 provides snapshots of the saturation and flow spatial distribution for a steady-state configuration.

3. Bubble mechanics – coalescence and snap-off

Menisci are created and destroyed as they are transported across nodes in the network. For each timestep the total volume of non-wetting liquid entering each node is distributed to the links transporting fluid away from the nodes, weighted according to its flow-rate. Within each link, there is a piston-like movement of bubbles, with the exception of the case when a maximum number of bubbles is exceeded, in which case the two nearest bubbles are found and merged.

An alternative mechanism of preventing the total number of bubbles from growing indefinitely is to set a minimum distance between menisci. When two menisci come closer than this distance they are merged. In the results presented here a minimum distance of 10 percent of the tube length was used. The effect of varying this distance, and also of changing to the method of a maximum number of bubbles, varied from 2 to 8, has been explored. No effect was found with respect to κ_{eff}, but $F(S)$ and the Ca value for the freezing transition are sensitive to this detail.

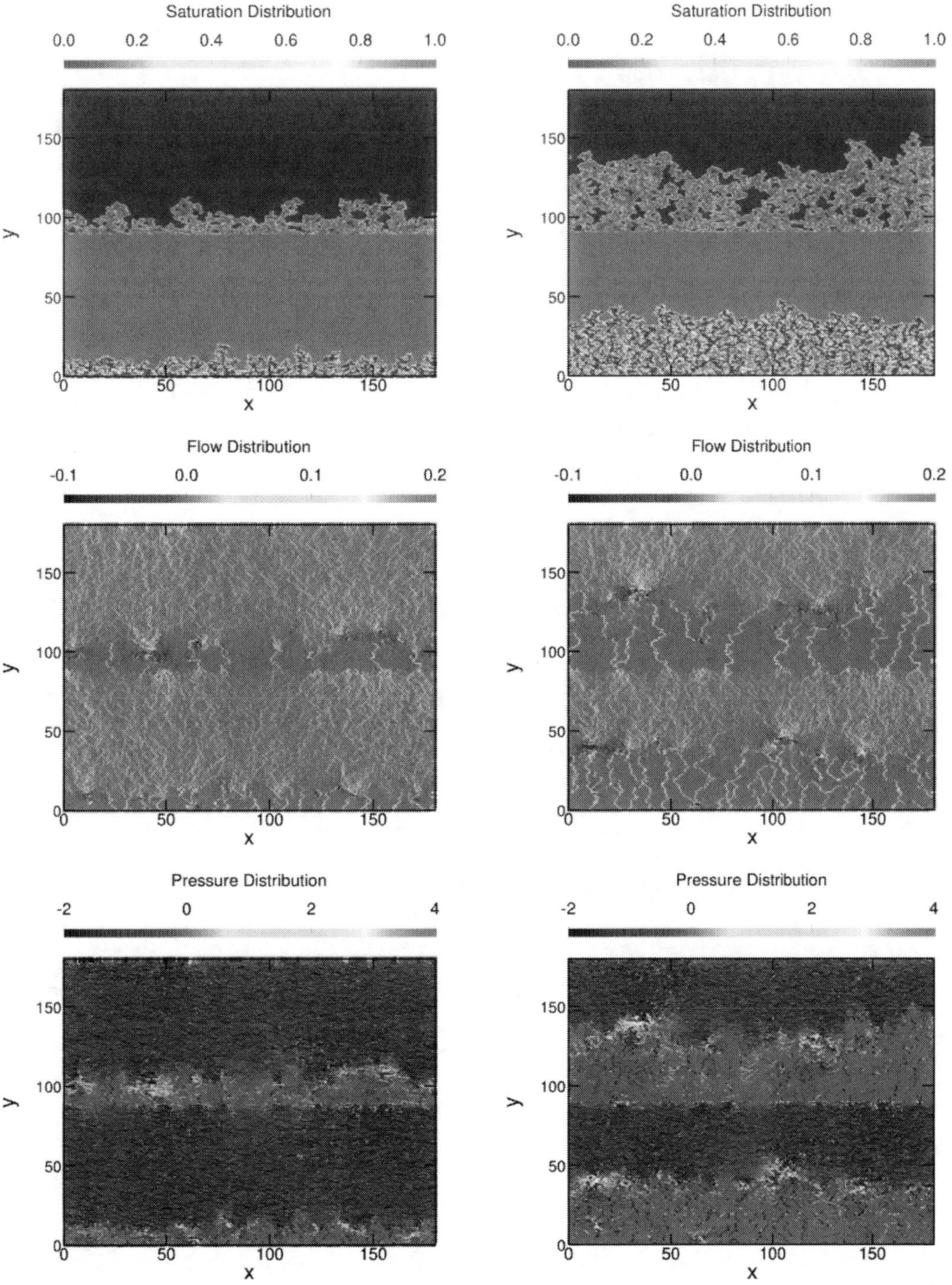

Figure 2. Saturation, flow and pressure distribution for the configuration shown in Fig. 1a (left) and Fig. 1b (right). System size is 256×256. $\mathrm{Ca} = 10^{-4}$ and $S = 0.5$.

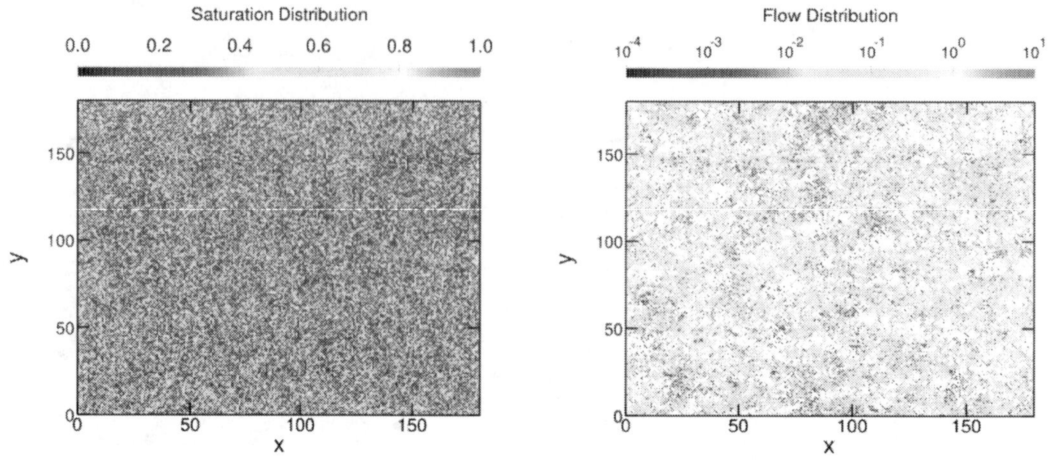

Figure 3. Saturation and flow distribution for a snapshot of a steady-state configuration. The flow distribution is plotted on a logarithmic color scale. Negative flow rates are not plotted and appear as white patches. System size is 256×256. $\mathrm{Ca} = 10^{-3}$ and $S = 0.5$.

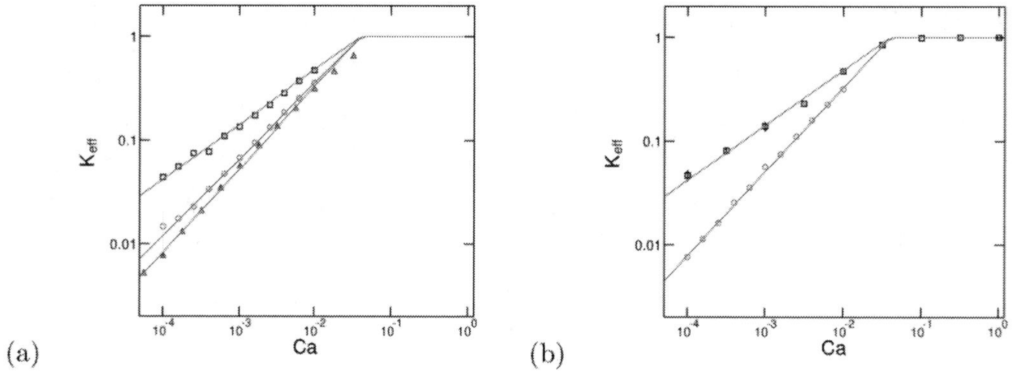

Figure 4. Effective permeabilities and capillary numbers plotted on logarithmic scales, with power laws fitted below a transition. Scaling exponents are given in table 1. (a) $S = 0.2$ (black squares), $S = 0.3$ (red circles) and $S = 0.4$ (blue triangles). $\mathrm{M} = 10^{-4}$. (b) $S = 0.2$ (black squares) and $S = 0.4$ (red circles). $\mathrm{M} = 1$.

Certainly, the way in which bubbles are transported across nodes, created and destroyed is of significance generally. However, the significance is greatest for low capillary numbers. Also, when the number of bubbles becomes large the sensitivity to the bubble mechanics appears to be reduced. In particular, a large number of bubbles reduces the effect of coalescence. We consider an understanding of this simplified system to be of importance in order to improve our understanding of the more involved problem which includes dependence on the detailed bubble mechanics.

Table 1. The scaling exponent β for different values of M and S. Determined by best-fits to the data points of Fig. 4.

M	S	β
10^{-4}	0.2	0.53
10^{-4}	0.3	0.74
10^{-4}	0.4	0.80

M	S	β
1	0.2	0.53
1	0.4	0.81

4. Simulation results

Flow-rates are chosen to explore the regime of free fractional flow, where both fluids move. For each run the saturation and flow-rate is kept fixed. System sizes of $L_x \times L_y = 256 \times 256$, measured in number of tubes, are easily accessible, however we find good enough statistics at smaller sizes. For the results of Figs. 4 and 5, 32×32 and 64×64 systems were used. Initial conditions consist of a random distribution of non-wetting bubbles which gives the desired saturation. The initialization conditions of these simulations are different than those used in [21], the significance of which will be discussed below. We run several individual realizations of disorder and initial conditions for each set of parameters, and present averaged results.

The freezing transition depends strongly on the effects of bubble snap-off, bubble coalescence and possibly film flow. Staying well above this transition simplifies the system considerably, as details of bubble mechanics and films are not expected to affect the qualitiative picture. As we wish to stay above the freezing transition, and this transition occurs at quite large values of Ca when S is very large or small, or $\mathrm{M} \gg 1$, the parameter space which we choose to explore becomes somewhat restricted.

For selected values within the parameter space we find that the effective permeability κ_{eff} appears to follow power-law scaling with respect to Ca below a threshold value. Figs. 4 and 5 and table 1 collect the results of some simulations.

We find that power-law scaling persists for capillary numbers up to $\mathrm{Ca} \approx 2 \cdot 10^{-2}$. Above this threshold it appears that all effects due to capillaries vanish, with both the effective permeability becoming unity and fractional flow becoming equal to saturation.

Below the threshold κ_{eff} decreases with decreasing Ca. This is caused by the menisci whose capillary pressure opposes the applied pressure outnumbering the rest, at any given time. As individual bubbles, consisting of a pair of menisci, are transported through hourglass-shaped tubes, the sum of capillary pressures acting on a single bubble changes direction as it is transported across the narrowest point of the tube. If there were only a single capillary tube, or if the porous medium were modelled as a bundle of capillaries, symmetry and a constant flow-rate would necessitate a perfect cancelling out of capillary effects for long times under steady-state conditions. Hence, the decrease of permeability can only be understood as a collective effect, where the ability of the system to choose between several flow paths yields correlations in both the direction and magnitude of capillary pressures, which conspire to produce the observed power-law behaviour.

For an example saturation of $S = 0.7$, shown in Fig. 5a, we have $\kappa_{\text{eff}}(\mathrm{Ca}) \sim \mathrm{Ca}^{0.67}$. For the same saturation, the dependence of fractional flow F on Ca is given in Fig. 5b. We see that $F(\mathrm{Ca}) = S$ above the threshold, while the non-wetting fluid sees a larger permeability than the wetting fluid below the threshold.

From Fig. 4 and table 1, we see that there is an abrupt change in β when S increases from 0.2 to 0.3. For $S = 0.3$ and above, β is somewhere in the range of $0.67 - 0.81$. Below this threshold, we find $\beta = 0.53$, close to that of the experiment.

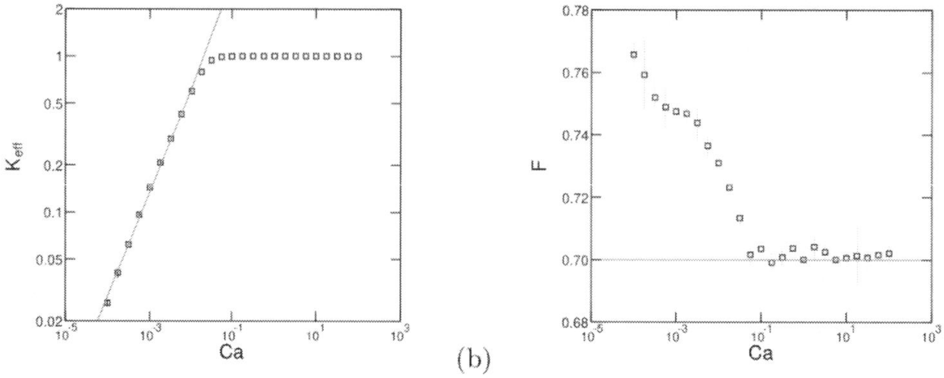

(a) (b)

Figure 5. (a) Effective permeability as a function of capillary number, with logarithmic scales. A power-law (red) is fitted with an exponent of 0.67. $S = 0.7$. $M = 1$. (b) Fractional flow as a function of capillary number. A horizontal line (red) of $F(\text{Ca}) = 0.7$ is added to emphasize the departure from $F(\text{Ca}) = S$. Parameters are the same as in (a).

The observation of two distinct regimes with respect to a sudden change in scaling exponent is surprising. It can be seen quite clearly when we plot $P_{\text{eff}}(S)$ for low Ca, as in Fig. 6, that there are indeed two distinct regimes. The onset of this transition also coincides with a sharp increase in $F(S)$.

Close inspection of individual samples reveals that samples with $S \approx 0.2$ contain stable, immobile clusters, as opposed to samples where $S \geq 0.3$. In the former case some residue of the initial conditions remains even after steady-state is reached, while in the latter case any initialization-dependent clustering is destroyed as the pressure builds up beyond some point where immobile clusters destabilize.

In [21] simulations were initialized with a gradual increase of surface tension while the system was flowing. After the desired value was reached the system continued to flow until it reached steady-state. For $S \geq 0.3$ the two initialization procedures give the same results, but for $S \approx 0.2$ and below the difference is significant. The initialization used here leads to two regimes, a small-exponent and a large-exponent regime, while the smooth increase in surface tension results in only a single regime, the large-exponent regime.

We have found little sensitivity to the viscosity ratio M. Even if heterogeneities in the saturation distribution develops at low Ca, capillary forces increase in importance relative to viscous forces in the same limit, and it does not appear that a viscous contrast is ever sufficient to have a significant impact.

As previously noted, the experiment of Tallakstad *et al* uses constant fractional flow $F = 7/15$ instead of constant S. Also, they are at capillary numbers below the freezing transition $\text{Ca} < 10^{-5}$. By our terminology, they are studying the regime of forced fractional flow. If they did not force both phases to flow by simultaneous injection, only the wetting phase (glycerol-water) would flow and the clusters of air would be immobile. We believe that this picture is compatible with our small-exponent regime, which also contains stable immobile clusters. With this assumption, the exponents of experiment and simulation are in excellent agreement.

Another difference between the experiment and our simulations is that the experiment uses a compressible gas instead of our incompressible non-wetting liquid. A close inspection of the experiment suggests that the effect of compressibility might not be important as this would introduce a sensitivity to absolute pressure which decreases along the direction of flow. Such a dependence is not observed.

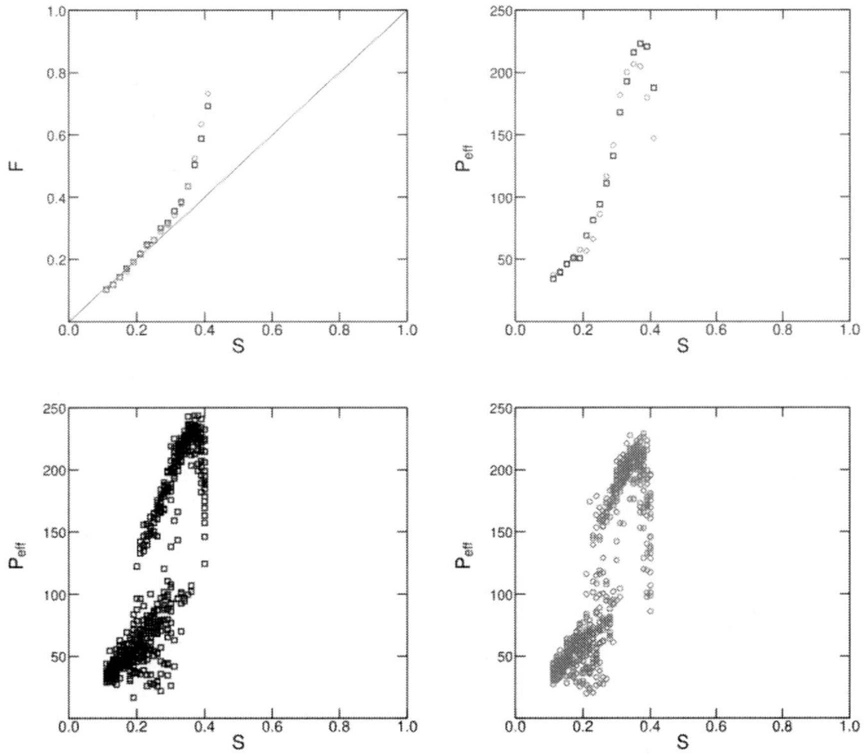

Figure 6. Top row: F and P_{eff} as functions of S, averaged over several realizations of the simulation, for $\text{M} = 1$ (black squares) and $\text{M} = 10^{-4}$ (red circles). Bottom row: P_{eff} as a function of S. The results of individual realizations are shown. This reveals the existence of two distinct regimes. All simulations have $\text{Ca} = 10^{-4}$.

5. Conclusion

Using a two-dimensional network simulator with fixed saturation we have explored a system which is similar, but not identical to the experiment of Tallakstad *et al* [19, 20]. This has been done for both viscosity matched fluids and for a viscosity ratio $\text{M} = 10^{-4}$ which is close to that of the experiment.

We have found power-laws both for viscosity matched fluids and for the high viscosity ratio. We find that there are several values for the exponent β, which depends on the non-wetting saturation S. The power-laws cover more than three decades in Ca. At the large Ca limit of the law the permeability becomes the single-phase permeability of the system, i.e., effects due to capillaries cancel out and do not affect macroscopic observables. The power-law is lost when Ca reaches a low value, beneath which κ_{eff} is independent of Ca. This transition occurs when one of the two phases becomes immobilized, which can be for Ca as low as $\approx 10^{-5}$, depending on S and M.

We have found that two regimes coexist within the parameter space of free fractional flow, a large-exponent well-mixed regime and a small-exponent regime with stable, immobile clusters and hysteresis. The experimental exponent indicates that forced fractional flow belongs to the latter regime.

We thank Santanu Sinha for valuable discussions.

References

[1] Dullien F 1992 *Porous Media: Fluid Transport and Pore Structure, 2nd ed.* (Academic Press, San Diego)

[2] Sahimi M 1995 *Flow and transport in porous media and fractured rock* (VCH Verlagsgesellschaft)

[3] Bak P, Tang C and Wiesenfeld K 1987 *Physical Review Letters* **59** 381–384

[4] Witten T A and Sander L M 1981 *Physical Review Letters* **47** 1400–1403

[5] Wilkinson D and Willemsen J F 1983 *Journal of Physics A: Mathematical and General* **16** 3365

[6] Furuberg L, Feder J, Aharony A and Jøssang T 1988 *Physical Review Letters* **61** 2117–2120

[7] Lenormand R, Touboul E and Zarcone C 1988 *Journal of Fluid Mechanics* **189** 165–187

[8] Måløy K J, Feder J and Jøssang T 1985 *Physical Review Letters* **55** 2688–2691

[9] Chen J D and Wilkinson D 1985 *Physical Review Letters* **55** 1892–1895

[10] Constantinides G and Payatakes A 1996 *AIChE Journal* **42** 369–382

[11] Aker E, Måløy K, Hansen A and Batrouni G 1998 *Transport in Porous Media* **32** 163–186

[12] Knudsen H, Aker E and Hansen A 2002 *Transport in Porous Media* **47** 99–121

[13] Knudsen H and Hansen A 2002 *Physical Review E* **65** 56310

[14] Knudsen H and Hansen A 2004 *EPL (Europhysics Letters)* **65** 200–206

[15] Knudsen H and Hansen A 2006 *The European Physical Journal B* **49** 109–118

[16] Hansen A and Ramstad T 2009 *Computational Geosciences* **13** 227–234

[17] Avraam D, Kolonis G, Roumeliotis T, Constantinides G and Payatakes A 1994 *Transport in Porous Media* **16** 75–101

[18] Avraam D and Payatakes A 1995 *Journal of Fluid Mechanics* **293** 207–236

[19] Tallakstad K, Knudsen H, Ramstad T, Løvoll G, Måløy K, Toussaint R and Flekkøy E 2009 *Physical Review Letters* **102** 74502

[20] Tallakstad K, Løvoll G, Knudsen H, Ramstad T, Flekkøy E and Måløy K 2009 *Physical Review E* **80** 36308

[21] Grøva M and Hansen A 2011 *preprint*

[22] Meijerink J and van der Vorst H 1977 *Mathematics of Computation* **31** 148–162

Melting of microgel colloidal crystals

Yi Peng, Ziren Wang and Yilong Han

Department of Physics, Hong Kong University of Science and Technology, Clear Water Bay, Hong Kong SAR, China

E-mail: `yilong@ust.hk`

Abstract. We experimentally studied the melting of colloidal crystals composed of diameter-tunable microgel spheres by bright-field and confocal video microscopies at the single-particle level. Thick (> 4 layers) and thin ($\leq$ 4 layers) films exhibited dramatically different melting kinetics. Thick films melted heterogeneously from grain boundaries in polycrystals and from surfaces in single crystals, while thin films melted homogeneously even in polycrystals. A novel heterogeneous melting at dislocation was observed in 5- to 12-layer films. The equilibrium phase behaviors in three thickness regimes were all different: thick films had a crystal-liquid coexistence regime which decreased with the film thickness and vanished at 4 layers, thin crystalline films melted into liquids in one step, while monolayers melted in two steps with an intermediate hexatic phase.

1. Introduction

Two-dimensional (2D) and three-dimensional (3D) crystals have qualitatively different melting behaviors. 2D melting often exhibits two continuous phase transitions with an intermediate hexatic phase as predicted by the KTHNY theory [1, 2], while 3D melting is a one-step first-order transition. The crossover between 2D and 3D melting, i.e. the thin-film melting, has not been well studied because it is difficult in theory and challenging in simulation and experiment. In particular, the strong long-wavelength fluctuations in low dimensions make the finite-size effect difficult to estimate [2]. Recent simulations show that at least 10^6 particles are necessary for measuring the melting behaviors of hard disks in 2D [3]. For multilayer systems, such large-scale simulations become challenging. Thin-film melting has been measured experimentally in some molecular systems, but these films simply melted heterogeneously from film-vapor interfaces as expected [4, 5]. In contrast, the non-trivial melting of thin films fully wrapped by solid substrates has not been explored. Recently we studied the melting of colloidal thin films confined between two flat glass walls [6, 7]. Here we review those results and discuss in detail the annealing techniques used and the surface melting observed in single crystals.

Besides the dimensionality, melting behavior is strongly affected by defects and surfaces [8]. For 3D crystals, melting always starts heterogeneously from surfaces and progresses into the bulk. If a 3D crystal is fully covered by flat solid substrates, then melting will only start from grain boundaries [9] which suppresses the melting from other defects. Here we explore the role of grain boundaries, dislocations and film-wall interfaces in 2D, thin-film and 3D melting. In fact, there lacks a quantitative theory about heterogeneous melting from defects in thin films. For example, all of the 2D melting theories including the KTHNY theory [1, 2], the grain-boundary-mediated melting theory [10] and the geometrical-defect-mediated melting theory [11] describe the homogenous melting of a single crystal. The defect-mediated melting theory for 3D crystals also focuses on the precursor defects in the homogenous melting of a single crystal without preexisting surfaces or defects. In real systems, however, crystals almost always melt heterogeneously from surfaces and preexisting defects. Hence the measurement of heterogeneous melting at the single-particle level is of practical interest and theoretical importance.

2. The experimental system

The colloidal crystals consisted of monodispersed N-isopropylacrylamide (NIPA) microgel spheres in aqueous solution. As temperature increases, NIPA polymers become less hydrophilic so that the diameter of the microgel sphere would decrease [12]. The pair potentials, $u(r)$, were directly measured from the liquid structure of a dilute (area fraction $\sim 10\%$) monolayer of spheres [13, 14], see Fig. 1. The effective diameter σ is defined as $u(\sigma) = 1k_BT$. For a typical batch of colloids, σ varies linearly from 1.27 μm at 24.1°C to 1.14 μm at 28.0°C.

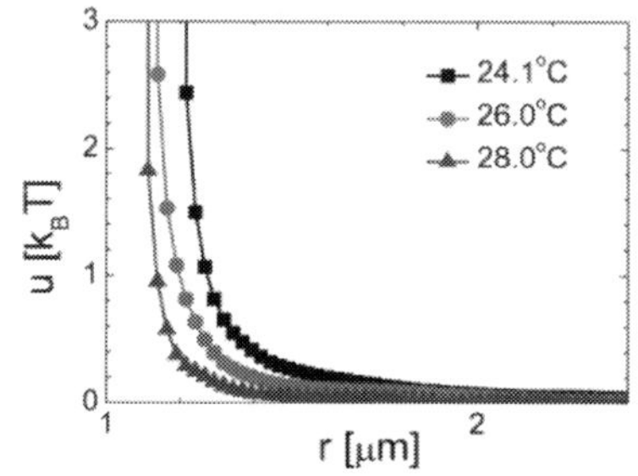

Figure 1. The pair potentials, $u(r)$, of NIPA spheres at 24.1°C, 26.0°C and 28.0°C.

Hard spheres confined between two walls exhibit a cascade of phases as the wall separation increases [15]: $1\triangle - 2\square - 2\triangle - 3\square - 3\triangle - 4\square$ and so on. Here $n\triangle$ denotes an n-layer triangular lattice (i.e. the (111) plane of a face-centered cubic (FCC) lattice), $n\square$ denotes an n-layer square lattice (i.e. the (100) plane of an FCC lattice). We observed such a cascade of phases in thin films of soft NIPA spheres. We found that films of more than 8 layers only have triangular lattices in the xy plane.

Colloidal suspensions were sucked into thin glass cells by a capillary force. Such a capillary flow can anneal crystals of $>$ 30 layers into mm- or cm-sized large FCC crystalline domains with the (111) plane along the xy direction. For crystals of $<$8 layers, however, the domain size was only about 10 to 100 μm, see Fig. 2(a). We further annealed the sample by increasing the temperature to nearly the melting point so that the polycrystal would become soft. Then we manually oscillate the microscope objective in the z direction to slightly deform the coverslip and induce a periodic water flow in the sample. Such a flow can effectively anneal an area of 0.1 mm^2. Then we moved the stage in the xy plane to anneal the neighboring areas and achieved crystalline domains of $>$1 mm^2 with $> 10^6$ particles per layer, see the subarea in Fig. 2(b). Such a large domain size is important for the melting study because long wavelength fluctuations and finite-size effects are important in low dimensions.

The sample temperature was increased in steps of 0.2°C. At each temperature, 5 minutes of video were recorded by bright-field microscopy, 2 to 4 minutes of video were recorded for each layer by confocal microscopy, and several confocal scans of the 3D static structure were taken along the z direction. Video rates were 30 frames/sec for bright-field microscopy and 7.5 frames/sec for confocal microscopy. The particle positions in each frame were obtained using standard 2D and 3D image analysis algorithms [16]. More than 100 layers of (111) planes can be seen through even in the bright field [9]. This unique advantage of microgel particles is due

Journal of Physics: Conference Series **319** (2011) 012010 doi:10.1088/1742-6596/319/1/012010

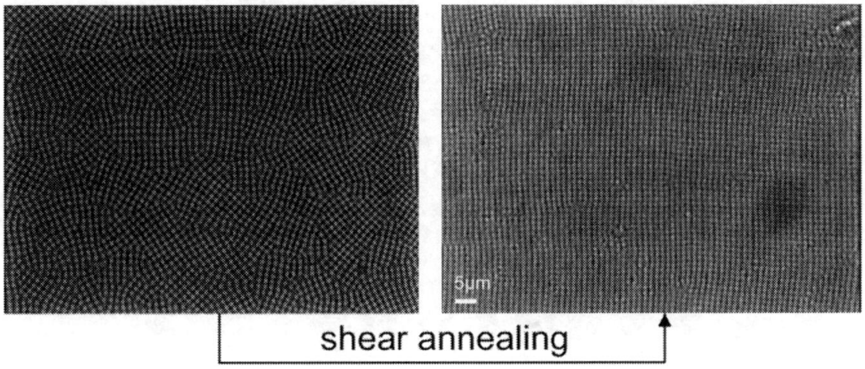

Figure 2. (a) The freshly made 3□ polycrystal consists of small domains. (b) The subarea of a ~1 mm^2 crystalline domain annealed by the water flow induced by the mechanical vibration.

to the fact that the refractive index of NIPA spheres is close to that of water and the diffraction of a triangular lattice in the xy plane matches the image of a neighboring (111) plane in an FCC crystal quite well.

3. Results of the experiment

3.1. Thick films (> 4 layers)

3D NIPA colloidal polycrystals melt only from grain boundaries [9]. Particles near smaller defects such as dislocations and partial dislocations have stronger motions than the bulk crystalline particles [9], but they never melt into small liquid domains. We observed that melting from dislocations became possible when the film thickness was between 5 to about 12 layers, see the coexistence of two liquid "lakes" melted from dislocations and one liquid strip melted from a grain boundary in Fig. 3. At the melting point, all the grain boundaries melt into liquid strips, and some dislocations melt into cylindrical liquid lakes with uniform cross-sections along the z direction. If the dislocation does not penetrate through the film, then neither does the lake, see Fig. 4.

The dislocation melting is often suppressed by a neighboring liquid strip or lake within about 20 lattice constants. A large crystalline domain can contain multiple lakes (see Fig. 3). Dislocations melt more easily in thinner films and in larger crystalline domains under higher heating rates. For example, 12-layer films do not melt from dislocations unless the heating rate is very high ($\sim 0.1°C/sec$), whereas 5- to 7-layer films with domains of more than 100 μm in size always melt from dislocations. This indicates that the difference in energy between the grain-boundary melting and the dislocation melting is comparable to the thermal energy.

The liquid strips and lakes have interesting interactions and dynamics. Unlike a dislocation which mainly glides along its Burgers vector at a relatively high speed, a lake randomly diffuses at a much lower speed independent of the direction of its Burgers vector. There is either an attraction or repulsion between two neighboring lakes, depending on which of the following two effects is dominant: 1) the surface tension at the liquid-crystal interfaces induces an effective attraction especially at a distance of less than 10 lattice constants; 2) the Burgers vectors of the two neighboring lakes usually differ by $\pi/3$, and such an acute angle between the effective "dipoles" will induce repulsion. We did not observe neighboring lakes with opposite Burgers vectors because their corresponding dislocations can be easily annihilated in the crystal phase before they melt into lakes. Lakes slowly diffuse and irreversibly merge into liquid strips. Large crystalline domains and large liquid strips coexist in the final equilibrium.

The 3D melting behaviors were observed in 300-layer samples. The crystal-liquid coexistence

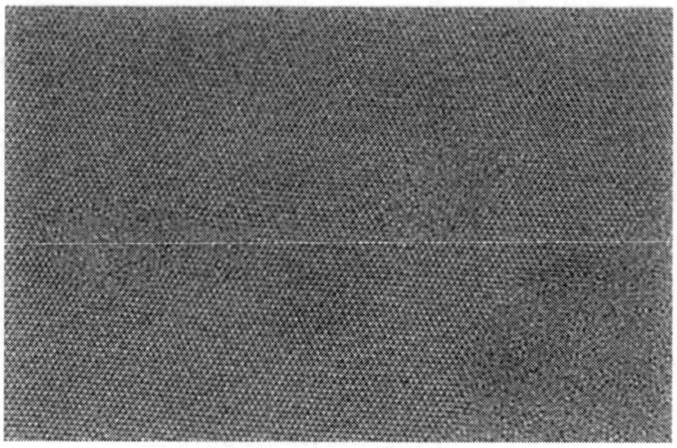

Figure 3. A 7-layer microgel thin film in the crystal-liquid coexistence phase. Two liquid "lakes" melted from dislocations are embedded in one large crystalline domain. The lower right corner is a liquid strip melted from a grain boundary.

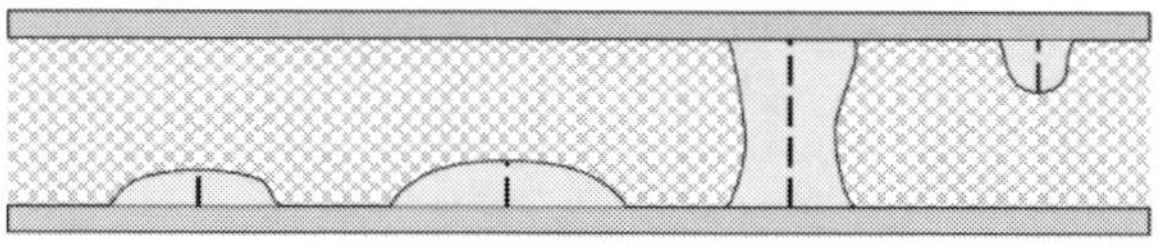

Figure 4. Partially melted 5- to 12-layer crystalline films confined between two glass walls. Dashed lines: dislocations. Yellow shaded area: liquid phase. Lakes do not penetrate through the film if the dislocations do not.

regime is about 6% in volume fraction [6], which is close to the 5% coexistence regime (from 49.5% to 54.5%) in the well-known 3D hard sphere phase diagram [17]. This 1% difference is likely due to the fact that our NIPA spheres are soft and the volume fraction cannot be rigorously defined. Interestingly, the crystal-liquid coexistence regime decreases with the film thickness and vanishes at the critical thickness of 4 layers, see Fig. 5.

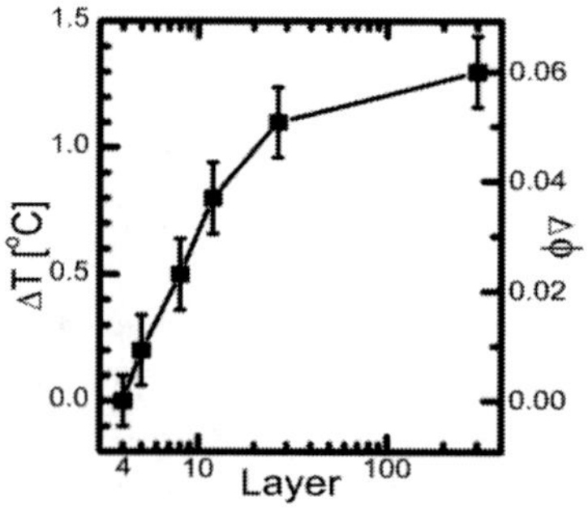

Figure 5. The liquid-solid coexistence regime (expressed as the temperature interval ΔT or volume fraction interval $\Delta\phi$) decreases with the film thickness and vanishes at 4 layers.

So where does the melting start if there are no grain boundaries? We carefully annealed thick films of > 30 layers and made cm-sized single crystals. These single crystals melted from glass walls and formed droplets, see Fig. 6. The droplets had different sizes and contact angles with the glass walls because the dislocations had different lengths, see Fig. 7(a). As the temperature was increased, the droplet became larger and then the whole surface layer melted as shown in Fig. 7(b). We never observed cylindrical lakes which can penetrate through the z direction as depicted in Fig. 4. Dislocations cannot melt if they are not connected to the walls. Melting from other defects is completely suppressed by such interfacial melting in films of > 30 layers.

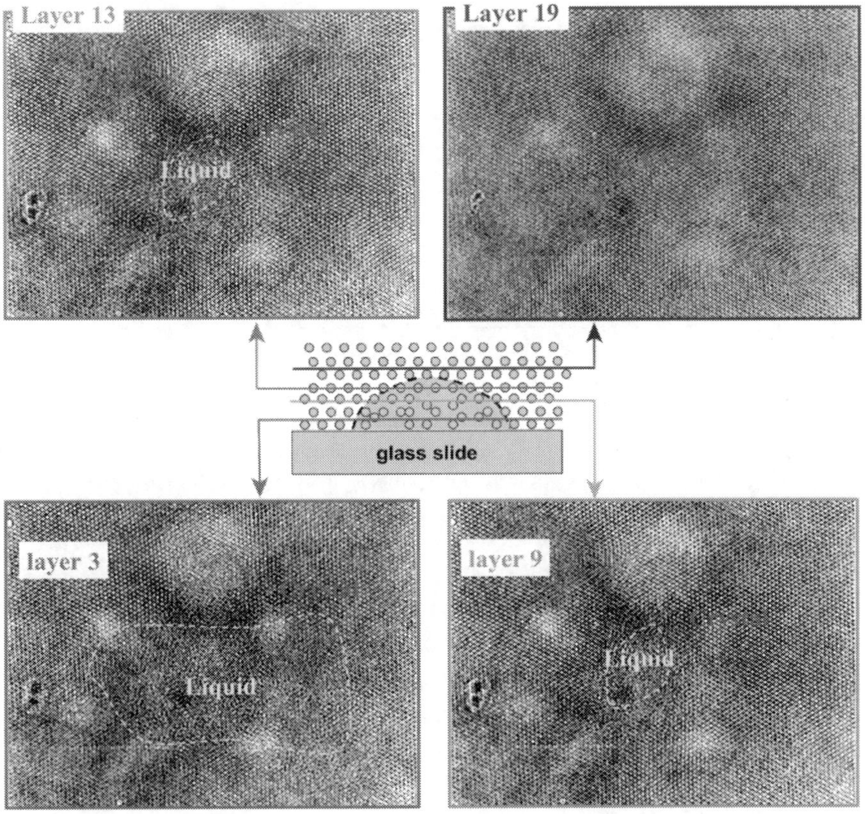

Figure 6. The single crystal melted from film-wall interfaces and formed liquid droplets. The cross-sections of a liquid droplet at the 3rd, 9th, 13th and 19th layers are shown.

3.2. Thin films (2-4 layers)

Thin films ($\leq$ 4 layers) do not have any preferential sites for melting as thick films do. They melt homogenously from everywhere in the xy plane, see the particle trajectories in Fig. 8. Increasing the temperature by just 0.2°C caused the whole crystalline structure to break down from the inside of domains. Disordered particles form small unstable defect clusters, but they never nucleate into large stable liquid domains such as lakes and strips. Since the melting is homogenous, we chose (60 μm)^2 ($\sim$2300 particles) representative areas in the centers of large domains and searched for the hexatic phase. In the KTHNY theory [1, 2], 2D crystals melt in two steps, from the crystal to the hexatic phase and then from the hexatic to the liquid phase. The intermediate hexatic phase has short-range translational and quasi-long-range orientational orders. The 2D crystal phase has quasi-long-range translational and long-range orientational orders. The liquid phase has short-range translational and orientational orders. Orders are characterized by the correlation functions of the order parameters as shown in Table 1. The correlation functions are defined in Table 2 and have three types of behaviors: approaching a

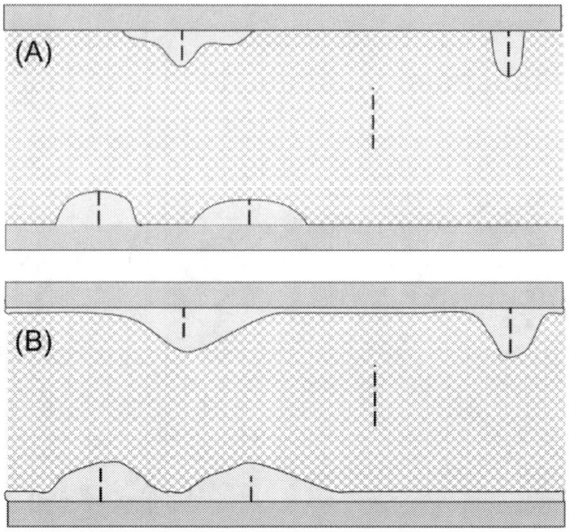

Figure 7. Surface melting of 3D single crystals with some dislocations. Dashed lines: dislocations. Yellow shaded area: liquid phase. (A) Melting starts from the intersection between dislocations and wall surfaces, forming lakes around the dislocations. No lake can penetrate through the film. Dislocations not connected to the walls cannot melt. (B) Further increasing the temperature results in melting from wall surfaces.

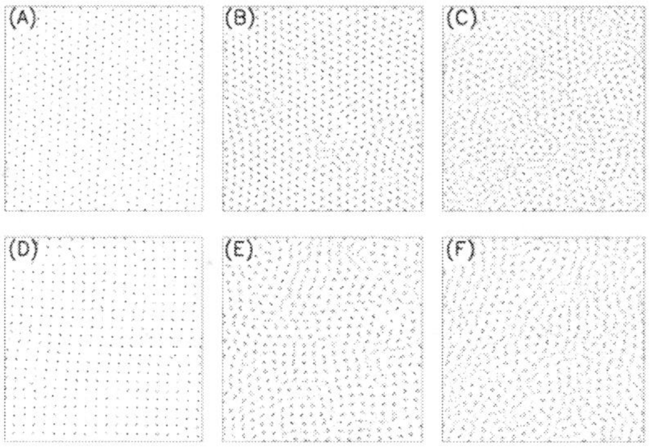

Figure 8. Typical 10-second trajectories of the homogenous melting in 2△ (a-c) and 3□ (d-f) films. Each figure is a subarea of a surface layer. (a) Crystal at 26.6°C; (b) At the melting point of 28.6°C; (c) Liquid phase at 28.8°C; (d) Crystal at 26.0°C; (e) At the melting point of 27.9°C; (f) Liquid phase at 28.1°C.

	2D crystal	Hexatic	Liquid
$g_6(r)$	nonzero constant at $r \to \infty$	$\sim r^{-\eta_{6r}}$, $0 < \eta_{6r} \leq 1/4$	$\sim e^{-r/\xi_{6r}}$
$g_6(t)$	nonzero constant at $t \to \infty$	$\sim r^{-\eta_{6t}}$, $0 < \eta_{6t} = 2\eta_{6r} \leq 1/4$	$\sim e^{-t/\tau}$
$g_T(r)$	$\sim r^{-\eta_T}$, $1/4 < \eta_T \leq 1/3$	$\sim e^{-r/\xi_T}$	$\sim e^{-r/\xi_T}$

Table 1. In the KTHNY theory, the 2D crystal, hexatic and liquid phases are characterized by different exponential or power-law correlations.

finite constant (long-range order); power-law decay (quasi-long-range order); and exponential decay (short-range order).

We have observed the hexatic phase in the melting of 2D microgel crystals in ref. [18]. For

	Translational order	Orientational order
Local order parameter	$\psi_{Tj}(\mathbf{r}) = e^{i\mathbf{G}\cdot\mathbf{r}_j}$	6-fold: $\psi_{6j}(\mathbf{r}) = (\sum_{k=1}^{nn} e^{i6\theta_{jk}})/nn$ 4-fold: $\psi_{4j}(\mathbf{r}) = (\sum_{k=1}^{nn_4} e^{i4\theta_{jk}})/nn_4$
Global order parameter	$\psi_T = \lvert\langle\psi_{Tj}\rangle\rvert$	$\psi_6 = \lvert\langle\psi_{6j}\rangle\rvert$
Susceptibility	$\chi_T = \lim_{A\to\infty} A(\langle\lvert\psi_T^2\rvert\rangle - \langle\lvert\psi_T\rvert\rangle^2)$	$\chi_6 = \lim_{A\to\infty} A(\langle\lvert\psi_6^2\rvert\rangle - \langle\lvert\psi_6\rvert\rangle^2)$
Spatial correlation	$g_T(r) = \langle\psi_{Ti}^*(r_i)\psi_{Tj}(r_j)\rangle$	$g_6(r = \lvert\mathbf{r}_i - \mathbf{r}_j\rvert) = \langle\psi_{6i}^*(r_i)\psi_{6j}(r_j)\rangle$
Time autocorrelation	$g_T(t) = \langle\psi_{Ti}^*(t_0)\psi_{Ti}(t_0+t)\rangle$	$g_6(t) = \langle\psi_{6i}^*(t_0)\psi_{6i}(t_0+t)\rangle$

Table 2. Translational and orientational order parameters, susceptibilities and correlations. $\mathbf{G}$ is a primary reciprocal lattice vector. nn is the number of nearest neighbors of particle j at position $\mathbf{r}_j = (x_j, y_j)$. nn_4 is the number of nearest neighbors within a distance of $1.2a$ where the lattice constant a is measured from the radial distribution function $g(r)$. θ_{jk} is the angle of the bond between particle j and its neighbor k. $\langle\ \rangle$ is the ensemble average over all particles. A is the system area. t is the time. The subscripts in order parameters ψ, susceptibility χ and correlations $g(r,t)$ are 6 for triangular lattices and 4 for square lattices.

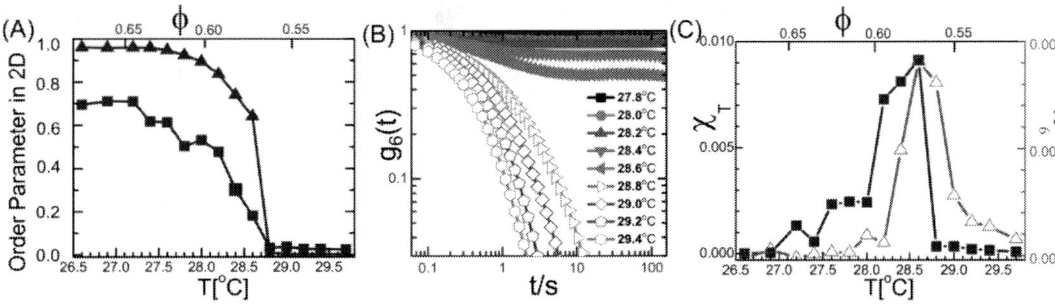

Figure 9. 2△-crystal melting measured from one of the two surface layers. (a) 2D translational and orientational order parameters, ψ_T, ψ_6 as a function of sample temperature T and volume fraction ϕ. The jumps at 28.7°C indicate the melting transition. (b) Orientational correlation functions $g_6(t)$ in time. Open symbols: liquid phase. Solid symbols: solid phase. (c) Translational susceptibility χ_T and orientational susceptibility χ_6. The peaks show the melting temperature at 28.7°C.

multilayer thin films, we measured the structures layer by layer because the particles were well stratified into separate layers even in the dense liquid phase near the melting point. From the particles' trajectories, we calculated their global order parameters, correlations and susceptibilities as defined in Table 2. The results for 2△ film melting are shown in Fig. 9. The correlations in Fig. 9(b) clearly show a crystal phase at $T > 28.7$°C and a liquid phase at $T < 28.7$°C. The hexatic phase characterized by the algebraic decay was not observed. The one-step phase transition without a hexatic phase can be further confirmed from the observations that ψ_T and ψ_6 jump at the same point (Fig. 9(a)) and that χ_T and χ_6 peak at the same point (Fig. 9(c)). Bulk layers and surface layers have similar correlations and susceptibilities, indicating that they melt at the same temperature. These results in 2△ melting were also observed in 3△, 3□, and 4□ film melting.

4. Summary and Discussion

Thick, thin and monolayer films exhibit different melting behaviors. Thick films (> 4 layers) melt heterogeneously with a crystal-liquid coexistence regime which decreases with the film thickness and vanishes at a critical thickness of 4 layers [6]. Moreover, 5- to 12-layer films exhibit a novel heterogeneous melting from dislocations and form lakes embedded in crystalline domains. Single crystals (>-30 layers) melt from film-wall interfaces and suppress other types of heterogeneous melting. Thin films (2-4 layers) melt homogenously in one step without a middle tetratic or hexatic phase. Monolayers melt homogenously in two steps with an intermediate hexatic phase [18]. Thinner films are softer and the extra long-wavelength fluctuations in low dimensions [2] may break the large crystalline domains from the inside and result in homogenous melting. These melting behaviors are robust for NIPA spheres of different sizes, but they are sensitive to particle interactions. Our preliminary experiments show that if particles have attractions, the critical thickness may be less than 4 layers or even exhibit crystal-liquid coexistence in monolayers.

The microgel colloid is a versatile system for melting and freezing studies. The surface effects can be avoided by focusing a beam of light into a single crystal to achieve local superheating and melting. By tuning the beam of light, we have been able to control the size and shape of a liquid domain embedded in a single crystal or of a crystalline island embedded in liquid phase. This setup can provide a more controllable way to explore the superheating, melting and freezing in 3D.

5. Acknowledgement

We thank Ahmed Alsayed and Arjun G. Yodh for the NIPA microgel spheres. The present study was financially supported by the RGC grant 601208 and the William Mong Institute of Nano Science and Technology at the Hong Kong University of Science and Technology.

References

[1] Nelson D R 2002 *Defects and Geometry in Condensed Matter Physics* (Cambridge University Press, UK)
[2] Strandburg K J 1988 Rev. Mod. Phys. **60**, 161
[3] Mak C H 2006 Phys. Rev. E **73**, 065104(R)
[4] Zhu D M and Dash J G 1986 Phys. Rev. Lett. **57**, 2959
[5] Pettersen M S, Lysek M J and Goodstein D L 1989 Phys. Rev. B **40**, 4938.
[6] Peng Y, Wang Z R, Alsayed A M, Yodh A G and Han Y 2011 Phys. Rev. E **83**, 011404
[7] Peng Y, Wang Z R, Alsayed A M, Yodh A G, and Han Y 2010 Phys. Rev. Lett. **104**, 205703
[8] Dash J G 1999 Rev. Mod. Phys. **71**, 1737
[9] Alsayed A M, Islam M F, Zhang J, Collings P J and Yodh A G 2005 Science **309**, 1207
[10] Chui S T 1982 Phys. Rev. Lett. **48**, 933
[11] Glaser M A and Clark N A 1993 Adv. Chem. Phys. **83**, 543
[12] Pelton R 2000 Adv. Colloid Interface Sci. **85**, 1
[13] Behrens S H and Grier D G 2001 Phys. Rev. E **64**, 050401
[14] Han Y and Grier D G 2003 Phys. Rev. Lett. **91**, 038302
[15] Schmidt M and Löwen H 1996 Phys. Rev. Lett. **76**, 4552
[16] Crocker J C and Grier D G 1996 J. Colloid Interface Sci. 298
[17] Pusey P N and van Megen W 1986 Nature **320**, 340
[18] Han Y, Ha N Y, Alsayed A M and Yodh A G 2008 Phys. Rev. E **77**, 041406

Continuum Models and Discrete Systems Symposia (CMDS-12) IOP Publishing
Journal of Physics: Conference Series **319** (2011) 012011 doi:10.1088/1742-6596/319/1/012011

Rheology and dynamical heterogeneity in frictionless beads at jamming density

Takahiro Hatano

Earthquake Research Institute, University of Tokyo, Tokyo 113-0032, Japan

E-mail: `hatano@eri.u-tokyo.ac.jp`

Abstract. We investigate the rheological properties of an assembly of inelastic (but frictionless) particles close to the jamming density using numerical simulation, in which uniform steady states with a constant shear rate $\dot{\gamma}$ is realized. The system behaves as a power-law fluid and the relevant exponents are estimated; e.g., the shear stress is proportional to $\dot{\gamma}^{1/\delta_S}$, where $\delta_S = 0.64(2)$. It is also found that the relaxation time τ and the correlation length ξ of the velocity increase obeying power laws: $\tau \sim \dot{\gamma}^{-\beta}$ and $\xi \sim \dot{\gamma}^{-\alpha}$, where $\beta = 0.27(3)$ and $\alpha = 0.23(3)$.

1. Introduction

The rheological properties of suspensions are important in various contexts ranging from industries to earth science. A certain class of suspensions may be modeled as repulsive beads in a medium. For systems of a very low Reynolds number, long-range hydrodynamic interactions may be ignored and thus model suspensions may be further simplified as an assembly of repulsive particles at zero temperature and no gravity. Such repulsive sphere models are also used as models for granular matter or a certain class of foams [1]. In such simple models, the important parameter that determines the rheology is the density. In particular, it is well known that the viscosity seems to diverge at a certain density. Furthermore, the repulsive particles are inevitably in contact above a certain density and the system acquires rigidity without crystallization [2, 3, 4], which is now referred to as the jamming transition.

It has been observed in many systems that the length and the time scales are observed to increase as the density approaches the critical jamming density [5, 6, 7, 8, 9, 10, 11, 12]. Importantly, however, such long wavelength fluctuations are observable only in a dynamic quantity such as the displacement of grains during a certain time lag [10]. This spatial heterogeneity is due to the cooperative nature of the particle rearrangement. Quite interestingly, a similar heterogeneity is also observed in supercooled liquids [13, 14], in which the correlation length and time increases as the temperature decreases or the density increases. Although the relation between these heterogeneities in very different systems is not clear, they may lead to an overarching concept in general amorphous systems.

It is recognized that the cooperative motion of particles also plays an essential role in the rheological properties [15, 16, 17]. However, to this date, we still do not understand the nature of the cooperative motion and how it affects dense granular rheology. In this paper, by molecular dynamics simulation, we investigate the nature of velocity correlation in dense granular flow at the jamming density. In particular, we define the correlation length and time to find that they grow as the shear rate decreases. We also show that they dominate the dynamic properties of dense granular flow, particularly diffusion and rheology.

2. Model

Here we consider a monodisperse particulate system, where the particle diamater and the mass is denoted by d and M, respectively. The volume fraction is fixed to be 0.639, which is the jamming density estimated for three dimensional systems within the precision of ± 0.001 [4]. The dimensions of the system is $L \times L \times L$, where we set $L = 40.32$ (80000 particles) unless otherwise indicated. We set the flow direction along the x axis and the velocity gradient along the y axis, and adopt Lees-Edwards boundary conditions [18].

The radius and the position of particle i are denoted by R_i and $\mathbf{r}_i$, respectively. Then using $\mathbf{r}_{ij} := \mathbf{r}_i - \mathbf{r}_j$, $\mathbf{n}_{ij} := \mathbf{r}_{ij}/|\mathbf{r}_{ij}|$, and $h_{ij} := (R_i + R_j) - |\mathbf{r}_{ij}|$, the repulsive force acting between particles i and j is written as $\mathbf{F}_{ij} := [kh_{ij} - \zeta \mathbf{n}_{ij} \cdot \dot{\mathbf{r}}_{ij}]\, \mathbf{n}_{ij}$. If $R_i + R_j < |\mathbf{r}_{ij}|$, particles i and j are not in contact so that the force vanishes. The coefficient of restitution, e, is given via the relation

$$e = \exp\left[-\frac{\pi}{\sqrt{4\bar{M}k/\zeta - 1}} \right], \tag{1}$$

where $\bar{M}$ is the reduced mass so that $\bar{M} = M/2$ in a monodisperse system. We choose $2Mk/\zeta = 1$, which corresponds to the vanishing coefficient of restitution. (We intended to model soft and viscous particles like tapioca). Throughout this study, we adopt the units in which $d = 1$, $M = 1$, and $k = 1$. (thus $\zeta = 2$.)

A constant shear rate $\dot{\gamma}$ is applied to the system through the Lees-Edwards boundary conditions [18]. Note that under these boundary conditions the system volume is constant. Thus, the important parameters here are the shear rate $\dot{\gamma}$ and the packing fraction denoted by ϕ. A steady state of uniform shear rate $\dot{\gamma}$ can be realized starting from a class of special initial conditions. Here we investigate such uniform steady states. Accordingly, it is convenient to define in the following discussions the velocity fluctuation with respect to steady shear flow. The velocity fluctuation is defined by $\mathbf{v}_i := \mathbf{r}_i - \dot{\gamma}\mathbf{n}_y \cdot \mathbf{r}_i$, where $\mathbf{n}_y$ is the unit vector along the y axis.

3. Velocity autocorrelation

First we investigate the cooperative nature of the velocity fluctuation $\mathbf{v}_i(t)$ by defining the velocity autocorrelation function of a tracer particle i.

$$C_V(t) = \frac{\langle \mathbf{v}_i(t) \cdot \mathbf{v}_i(0) \rangle}{\langle \mathbf{v}_i^2 \rangle}. \tag{2}$$

This correlation function can be calculated at several shear rates ranging $\dot{\gamma} = 10^{-1}$ to 10^{-6}. At each shear rate the correlation function exhibits a characteristic relaxation time denoted by τ. We estimate this time scale by rescaling the time with $\tau(\dot{\gamma})$ so as to $C_V(t/\tau)$ at each shear rate collapses. Such collapse is shown in Figure 1 (a). Recalling that the system is marginally jammed (i.e., the mean free path is zero), it may be remarked that the velocity correlation time is several times larger than $\sqrt{k/M} = 1$. Furthermore, this characteristic time increases obeying a power law as the shear rate decreases;

$$\tau \propto \dot{\gamma}^{-\beta}, \tag{3}$$

where $\beta = 0.27(3)$. See the inset of Figure 1 (a). It is also noteworthy that the relaxation time is still much smaller than $\dot{\gamma}^{-1}$, which is the structural relaxation time due to shear

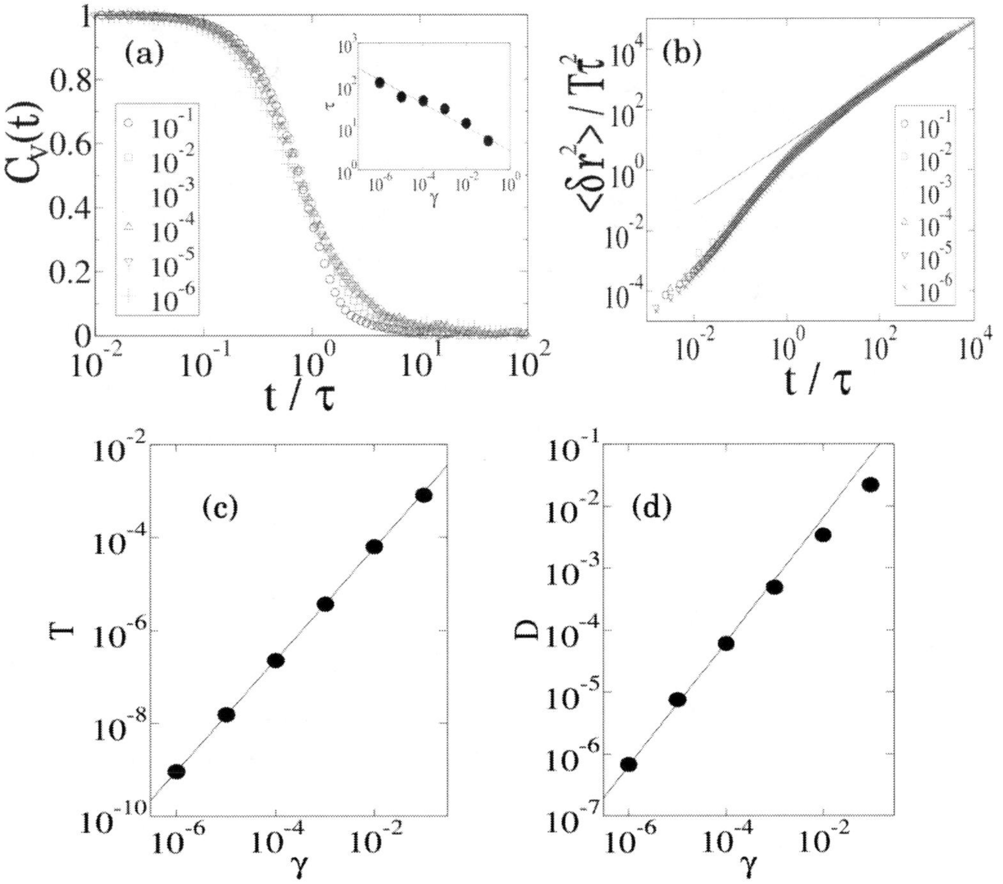

Figure 1. (a) Collapse of the velocity autocorrelation function $C_V(t)$ at various shear rates by rescaling the time with τ. The legends show the shear rate. The inset shows τ as a function of the shear rate, where the solid line is proportional to $\dot{\gamma}^{-0.27}$. (b) Collapse of the diffusion profiles at various shear rate. Here the mean square displacement $\langle \delta \mathbf{r}_i^2(t) \rangle$ is normalized by the kinetic temperature $T \equiv \langle \mathbf{v}_i^2 \rangle$ and τ^2. This means that the diffusion coefficient is scaled with $T\tau$. The solid line is proportional to t. (c) The shear-rate dependence of the kinetic temperature. The solid line is proportional to $\dot{\gamma}^{1.2}$. (d) The shear-rate dependence of the diffusion coefficient. The solid line is proportional to $\dot{\gamma}^{1.0}$.

flow. We can thus regard τ as the intermediate time scale at sufficiently low shear rates; i.e., $\sqrt{m/k} \ll \tau \ll \dot{\gamma}^{-1}$.

The velocity correlation time involves the nature of diffusion. The diffusion profile is defined by $\langle \delta \mathbf{r}_i^2(t) \rangle = | \int_0^t ds \mathbf{v}_i(s)|^2$. We compute this quantity at several shear rates and find that they collapse using the velocity correlation time τ and the kinetic temperature defined by $T := \langle \mathbf{v}_i^2 \rangle$. Namely,

$$\langle \delta \mathbf{r}_i^2(\tau) \rangle = T\tau^2 f(t/\tau), \tag{4}$$

where $f(\cdot)$ is a universal function. The scaling collapse is shown in the inset of Figure 1 (b). One can immediately conclude from this scaling that the diffusion coefficient D is scaled with $T\tau$. As shown in Figure 1 (c), the shear-rate dependence of the kinetic temperature is expressed as $T \propto \dot{\gamma}^{y_T}$, where $T = 1.20(3)$. Thus, $D \sim T\tau$ is expected to be proportional to the shear rate. We estimate D from the diffusion profile at each shear rate and find that D is proportional to $\dot{\gamma}$ at sufficiently low shear rates ($\dot{\gamma} \leq 10^{-3}$). This is shown in Figure 1 (d), where the solid line is

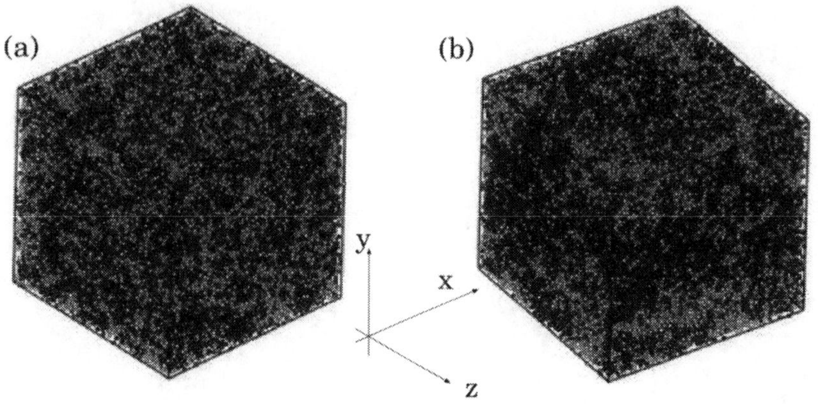

Figure 2. Heterogeneity of the velocity fluctuation defined by equation (5). Grains with larger velocity ($m_i < e^{-1}$) are brightly shown. (a) A snapshot at $\dot{\gamma} = 10^{-2}$. (b) A snapshot at $\dot{\gamma} = 10^{-5}$, where grains with large velocity form a lager cluster.

proportional to $\dot{\gamma}$. Note that the scaling of the diffusion coefficient is apparently anomalous at higher shear rates; it appears that $D \propto \dot{\gamma}^{y_D}$, where $y_D \simeq 0.8$. We remark that this anomalous scaling with almost the same exponent at higher shear rates has been observed in some similar systems such as a model foam under shear [20] and an experiment on colloidal system [21].

One may notice in Figure 1 (b) that diffusion occurs after sufficiently long time (i.e., $f(x) \propto x$ for $x \gg 1$), whereas superdiffusion occurs in the shorter time scale ($t \leq \tau$). It is important to notice that subdiffusion is not observed in contrast to other granular systems without flow [7, 8, 9, 10]. This indicates that the in-cage motion is not relevant to the present system and the structural relation takes place even in a shorter time scale $t \sim \tau$. This is because the applied shear breaks the cages. This enables us to detect the dynamical heterogeneity using a two-point correlation function instead of a four-point correlation function that is typically utilized to observe dynamical heterogeneities. This is discussed in the next section.

Another important quantity regarding the diffusion profile is the distance over which grains move cooperatively. This length scale is defined by $\sqrt{\langle \delta \mathbf{r}_i^2(\tau) \rangle}$ and denoted by l_c. As is apparent from Eq. (4), l_c is scaled with $\sqrt{T}\tau$. We confirm that this length scale is scaled with $\dot{\gamma}^x$ ($x \simeq 0.3$) and thus vanishes in the zero shear rate limit. This makes a quite contrast to supercooled liquids, in which l_c is almost constant and as large as the cage size [22]. Because l_c vanishes in the $\dot{\gamma} \to 0$ limit, it is expected that superdiffusion does not occur (i.e., diffusion is observed at all time scales) at sufficiently slow shear rates [19].

4. Spatial heterogeneity of velocity

The fluctuating velocity field $\mathbf{v}_i(t)$ exhibits spatial heterogeneity, which is enhanced at the slower shear rates. To quantify the spatial heterogeneity of $\mathbf{v}_i(t)$ clearly, it is indeed convenient to define the following quantity rather than the velocity itself.

$$m_i(t) \equiv \exp\left[-\frac{\mathbf{p}_i^2(t)}{\langle \mathbf{p}_i^2(t) \rangle_i}\right], \tag{5}$$

where $\langle \cdot \rangle_i$ denotes the average over all grains. The snapshot of $m_i(t)$ is shown in Figure 2, where the growing correlation length at lower shear rate is apparent. Then we define the correlation

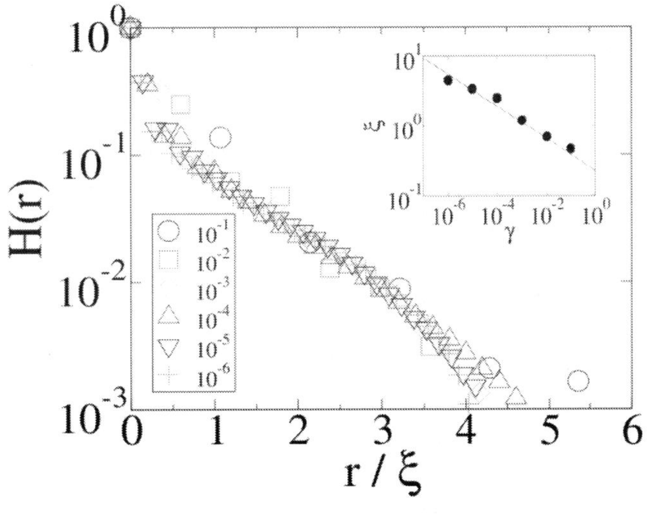

Figure 3. The correlation functions defined by Eqs. (5) and (6). The legends indicate the shear rate. Note that the length scale is normalized by ξ so that the correlation functions collapse. (Inset) The correlation length ξ as a function of $\dot{\gamma}$. The solid line is proportional to $\dot{\gamma}^{-0.23}$.

function as

$$h(r;t) \equiv \frac{\sum_{i \geq j} \delta m_i(t) \delta m_j(t) \delta(r - |\mathbf{x}_i(t) - \mathbf{x}_j(t)|)}{\sum_{i \geq j} \delta(r - |\mathbf{x}_i(t) - \mathbf{x}_j(t)|)}, \tag{6}$$

where $\delta m_i(t) \equiv m_i(t) - \langle m_i(t) \rangle_i$. We remark that $h(r;t)$ does not significantly fluctuate in time in the present system (containing 80000 grains) so that we drop the argument t in $h(r;t)$ to write it as $H(r)$. The correlation function $H(r)$ at various shear rates is shown in Figure 3. This clearly shows that the correlation length ξ increases as

$$\xi \propto \dot{\gamma}^{-\alpha}, \tag{7}$$

where the exponent is estimated as $\alpha = 0.23(3)$. The power-law divergence of the correlation length with respect to the shear rate is consistent with the fact that the critical point is located at $\dot{\gamma} = 0$ [23, 3]. Note that equations (5) and (6) respresent a two-point correlation. This makes a quite contrast to vibrated [10] or air-fluidized granular media [7, 8, 9] and other thermal glass formers such as supercooled liquids [14], where the spatial heterogeneity can be detected only in a four-point correlation. As discussed in the previous section, this may be one of the essential features of an athermal system under flow, in which the cage is washed away by the flow and therefore the kinetic arrest is absent.

Taking Eqs. (3) and (7) into account, the time constant of the cooperative motion is scaled with $\tau \propto \xi^z$, where $z = 1.3(1)$. Interestingly, this is comparable to an experiment on a two dimensional air-fluidized system [9].

We remark that the correlation length that diverges as $\xi \propto S^{-0.5}$ is also found in a simulation on a two-dimensional model foam [11]. This indeed leads to $\xi \sim \dot{\gamma}^{-0.21}$ as they obtain $S \propto \dot{\gamma}^{0.42}$ at the critical density. The exponent for the correlation length is coincident with that in the present study, although the models and the definition of correlation length are slightly different. This coincidence indicate that the exponent for the correlation length is insensitive to the dimensionality and the details of the model (i.e., viscosity of the medium and the nature of the dissipative force.)

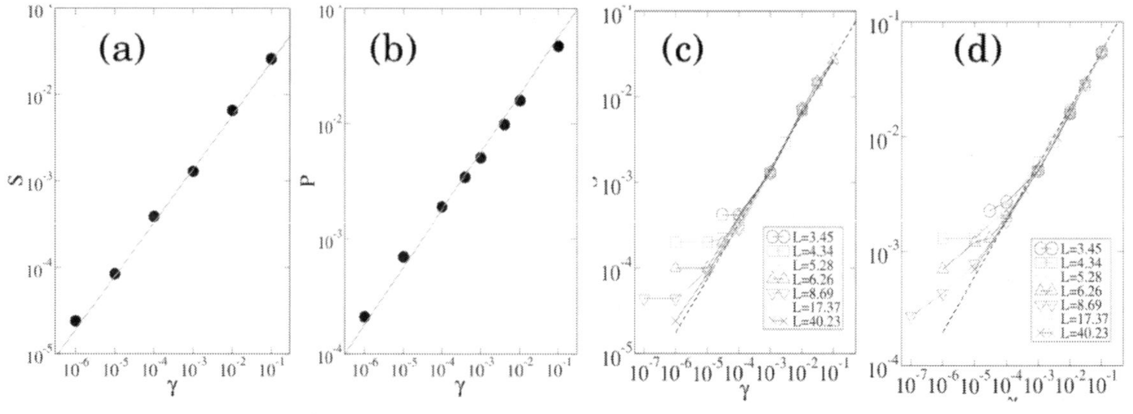

Figure 4. (a) The shear-rate dependence of the shear stress. The solid line is proportional to $\dot{\gamma}^{0.64}$. (b) The shear-rate dependence of the pressure. The solid line is proportional to $\dot{\gamma}^{0.5}$. (c) System-size dependence of the shear stress. The dashed line is proportional to $\dot{\gamma}^{0.64}$. (d) System-size dependence of the pressure. The dashed line is proportional to $\dot{\gamma}^{0.5}$.

We also remark that the correlation length is also found in an experiment on an inclined plane flow [25], in which the correlation length in terms of a four-point correlation increases as $V^{-0.25}$. This also supports Eq. (7) because the velocity profile is exponential so that $\dot{\gamma} \propto V$. The important implication of this experiment is the insensitivity of the exponent for the growing correlation length 0.25 on the nature of the interparticle force; i.e., whether it is frictional or frictionless.

5. Rheology

5.1. Granular matter as a power-law fluid

Next we discuss the rheology. As shown in Figure 4, the shear stress and the pressure exhibit power-law dependences on the shear rate: $S \propto \dot{\gamma}^{1/\delta_S}$ and $P \propto \dot{\gamma}^{1/\delta_P}$, where $1/\delta_S = 0.64 \pm 0.02$ and $1/\delta_P = 0.5 \pm 0.02$. Note that these exponents do not contradict those previously conjectured for a granular matter at the critical density [26, 27]. These exponents differ from 2.0, i.e., Bagnold's scaling [24], which holds only in unjammed regime where the stiffness of the particles does not affect the mean free time. The nontrivial power-law rheology arises where the mean free time (or the mean free path) vanishes. We remark that similar power-law behaviors of the shear stress have been observed in other amorphous particulate systems such as foam models [11, 28, 29], although the observed exponent is different from the present value. In foam models, the exponent $1/\delta_S$ has been estimated as 0.28 [29] or 1/2 [28], which are significantly lower than that in the present model. It has been shown that the rheological exponents are sensitive to the details of the force models [4, 26] and the nature of the medium [11, 26]. We wish to remark that the foam models [11, 28, 29] do not posses inertia and this may make the difference in the rheological exponents. However, notwithstanding a plausible theory on the rheological exponents [28], there is yet no consensus regarding their precise values.

5.2. Finite-size effect

Generally, the true critical behaviors cannot be observed in a finite system, as the growing correlation length eventually exceeds the system size. To detect such a finite size effect, we investigate the system size dependence of rheology in the following way. The control parameters here are the shear rate $\dot{\gamma}$ and the system size L so that the shear stress S and the pressure

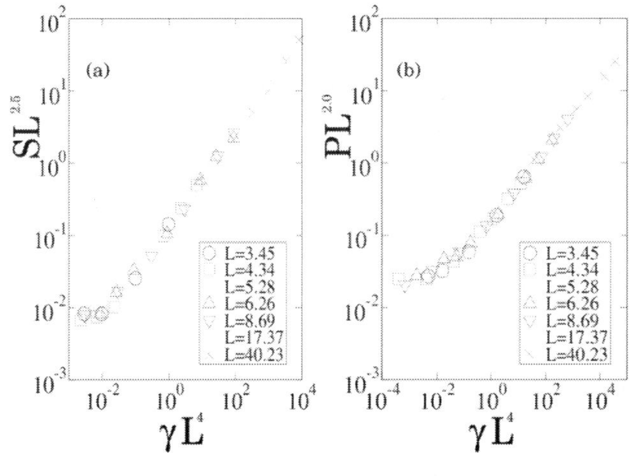

Figure 5. Collapse of the rheological data shown in Figure. 4 by finite-size scaling: (a) equation (8), and (b) equation (9).

P are formally expressed as $S = S(\dot{\gamma}, L^{-1})$ and $P = P(\dot{\gamma}, L^{-1})$. These relations obtained in our simulation are shown in Figure 4. Note that, at lower shear rates, the shear stress and the pressure are larger than expected from the power law behaviors. This tendency is remarkable for smaller systems; i.e., the crossover occurs at higher shear rate as the system becomes smaller. This system-size dependent rheology is described by finite size scaling, which are of the same form as those in conventional critical phenomena [30].

$$S(\dot{\gamma}, L) = L^{-1/y_S} f_S(\dot{\gamma} L^{1/y_{\dot{\gamma}}}), \quad (8)$$
$$P(\dot{\gamma}, L) = L^{-1/y_P} f_P(\dot{\gamma} L^{1/y_{\dot{\gamma}}}), \quad (9)$$

where f_S and f_P are scaling functions. Figure 5 shows the collapse of the rheological data using Eqs. (8) and (9), where the exponents are estimated as $1/y_S = 2.5 \pm 0.2$, $1/y_P = 2.0 \pm 0.2$, and $1/y_{\dot{\gamma}} = 4.0 \pm 0.5$. We can also see that the finite size effect comes into play where $\dot{\gamma} L^{1/y_{\dot{\gamma}}} \leq 0.1$; i.e., $L \leq \dot{\gamma}^{-y_{\dot{\gamma}}}$. This means that the correlation length increases as $\xi \sim \dot{\gamma}^{-y_{\dot{\gamma}}}$, because the finite size effect is generally due to the correlation length that exceeds the system size. Thus, $y_{\dot{\gamma}}$ must be identical to α in equation (7), which is estimated using a spatial correlation function, equations (5) and (6). This is indeed valid within the numerical precision. Therefore we can conclude that the finite-size effect in rheology is due to the fact that the correlation length of the dynamical heterogeneity exceeds the system size.

References

[1] Durian D J 1997 *Phys. Rev. E* **55** 1739
[2] Aharonov E and Sparks D 1999 *Phys. Rev. E* **60** 6890
[3] O'Hern C S, Langer S A, Liu A J, and Nagel S R 2002 *Phys. Rev. Lett.* **88** 075507
[4] O'Hern C S, Silbert L E, Liu A J, and Nagel S R 2003 *Phys. Rev. E* **68** 011306
[5] Drocco J A, Hastings M B, Olson Reichhardt C J, and Reichhardt C 2005 *Phys. Rev. Lett.* **95** 088001
[6] Dauchot O, Marty G, and Biroli G 2005 *Phys. Rev. Lett.* **95** 265701
[7] Abate A R and Durian D J 2006 *Phys. Rev. E* **74** 031308
[8] Abate A R and Durian D J 2007 *Phys. Rev. E* **76** 021306
[9] Keys A S, Abate A R, Glotzer S C, and Durian D J 2007 *Nature Phys.* **3** 260
[10] Lechenault F, Dauchot O, Biroli G, and Bouchaud J.-P. 2008 *Europhys. Lett.* **83** 46003
[11] Olsson P and Teitel S 2007 *Phys. Rev. Lett.* **99** 178001

[12] Hatano T 2009 *Phys. Rev. E* **79** 050301(R)
[13] Kob W et al. 1997 *Phys. Rev. Lett.* **79** 2827
[14] Yamamoto R and Onuki A 1998 *Phys. Rev. E* **58** 3515
[15] Jenkins J T 2006 *Phys. Fluids* **18** 103307
[16] Mitarai N and Nakanishi H 2007 *Phys. Rev. E* **75** 031305
[17] Bizon C, Shattuck M D, Swift J B, and Swinney H L 1999 *Phys. Rev. E* **60** 4340
[18] Allen M P and Tildesley D J 1989, *Computer Simulation of Liquids*, (Clarendon, Oxford)
[19] Heussinger C, Berthier L, and Barrat J.-L 2010 *EPL* **90** 20005
[20] Olsson P 2009 arXiv:0910.2623v1
[21] Besseling R, Weeks E R, Schofield A B, and Poon W C K 2007 *Phys. Rev. Lett.* **99** 028301
[22] Kob W and Andersen H C, *Phys. Rev. E* **51** 4626
[23] Liu A J and Nagel S 1998 *Nature* **396** 21
[24] See, for example, Silbert L E, Ertas D, Grest G S, Halsey T C, Levine D, and Plimpton S J 2001 *Phys. Rev. E* **64** 051302; Mitarai N and Nakanishi H 2005 *Phys. Rev. Lett.* **94** 128001 and references therein.
[25] Katsuragi H, Abate A, and Durian D J 2010 *Soft Matter* **6** 3023
[26] Hatano T 2008 *J. Phys. Soc. Jpn.* **77** 123002
[27] Hatano T, Otsuki M, and Sasa S 2007 *J. Phys. Soc. Jpn.* **76** 023001
[28] Tighe B et al. 2010 *Phys. Rev. Lett.* **105** 088303
[29] Olsson P and Teitel S 2011 *Phys. Rev. E* **83** 030302
[30] In general, finite-size scaling reads $S(\Phi, \dot{\gamma}, L) = L^{-1/y_S} f_S(\Phi L^{1/y_\Phi}, \dot{\gamma} L^{1/y_{\dot{\gamma}}})$. where Φ is the distance from the critical density. Thus, Φ must be zero (or $\Phi L^{1/y_\Phi} \simeq 0$) if one wishes to estimate $y_{\dot{\gamma}}$.

Journal of Physics: Conference Series **319** (2011) 012012 doi:10.1088/1742-6596/319/1/012012

Thermo-microstretch elastic bodies and plane waves

E Inan[1] **and A Kiris**[2]

[1] Faculty of Arts and Sciences, Işık University, 34980, Şile, İstanbul, Turkey
[2] Faculty of Science and Letters, İstanbul Technical Univ., 34469, Maslak, İstanbul, Turkey

E-mail: inan@isikun.edu.tr, kiris@itu.edu.tr

Abstract. In the present work, vibration problems of rectangular plates are considered for the determination of upper bounds to the unknown microstretch material properties. The frequencies are obtained by extending the Ritz method to this case. The analysis shows that some additional frequencies characterizing the microstretch effects appear among the classical frequencies. Furthermore, by the increasing values of the microstretch constants, the additional frequencies disappear and only the classical frequencies remain in the spectrum. Considering this phenomenon, an optimization problem is established for the identification of the upper bounds of microstretch elastic constants. In the second part of the work, thermal effects are considered and several theories are discussed. Finally, propagation of the plane waves is investigated.

1. Introduction

It is well known that the linear theory of elasticity is unable to explain the behaviour of the materials having complex microstructure such as polymers, porous media and micro damaged materials, etc. Modelling such materials with Eringen's [1] microstretch theory which assumes that every particles of the material may do both microrotation and volumetric microelongation in addition to the bulk deformation is more convenient. Success of this theory depends on the correctness of the choice of system parameters. Dependence of the dynamic response of the medium to material properties allows us to establish a vibration analysis [2] for the determination of such material constants similar to the classical cases [3–11]. This analysis is based on 3-D microstretch theory and extended Ritz method. Following Zhou et al. [12], the triplicate Chebyshev polynomial series are used to describe plate deflections. The wave propagation problems in micropolar and microstretch media are discussed in [13, 14], and it is shown that two and three new non-classical waves appear in micropolar and microstretch cases, respectively. Parallel to these results, we found some additional frequencies due to the microstretch character, among the classical frequencies [2]. As it is expected, these additional frequencies disappear when the microstretch material constants are taken zero. Besides, we observed that these additional frequencies are more sensitive to the variation of micro elastic constants than the classical frequencies. Therefore, the values of additional frequencies rapidly increase by the increase of micro constants and then considerable amount of them move out among classical frequencies. During the variation of microstretch constants, the values of classical frequencies remain same up to some critical values, but they also begin to deviate after these critical values. This phenomenon tells us that the microstructure becomes more dominant and starts to affect the macro properties. For instance, considering this model as representing a damaged body, we may conclude that the development and the growth of micro cracking start to affect the macro properties of the body. So, we may define these threshold values as the upper bounds for the microstretch material parameters. That means, the material is not a microstretch body anymore. Beyond these threshold values of the micro elastic constants, the material loses its microstretch character. So, this is the key point of this study that will be used to construct an optimization problem to determine the upper bounds of the microstretch properties. To find more specific values for the micro constants, we need to know some experimental data for the frequencies of the material. But, it needs more sophisticated equipments and experimental techniques that are not available at the moment. Thus, we may use experimental results due to classical cases which give us only the upper bounds for micro constants.

Another important problem of damage mechanics is to include thermal effects, and thermo-microstretch theory can be used. In this case, some additional microthermal constants will occur. To determine upper bounds for the new constants, same method may not work properly due to the special form of the thermal analysis and some modifications should be done in the method. Then, it is left to a further study. Thus, we give here an analysis of the comparison of several thermal theories and then investigate the plane wave propagation in a thermo-microstretch material.

2. Fundamental equations and wave propagation in microstretch medium

The constitutive equations for a linear homogeneous, isotropic, microstretch elastic solid may be given as follows [15]

$$\begin{aligned}
&t_{kl} = \lambda\,\varepsilon_{mm}\,\delta_{kl} + (\mu+\kappa)\,\varepsilon_{kl} + \mu\,\varepsilon_{lk} + \lambda_0\,\theta\,\delta_{kl},\\
&m_{kl} = \alpha\,\gamma_{mm}\,\delta_{kl} + \beta\,\gamma_{kl} + \gamma\,\gamma_{lk},\\
&m_k = a_0\,\theta_{,k},\\
&s-t = \lambda_1\,\theta + \lambda_0\,\varepsilon_{kk}.
\end{aligned} \tag{1}$$

Here, t_{kl}, m_{kl} are the stress and couple stress tensors, m_k is the microstretch vector and $s = s_{kk}$, $t = t_{kk}$. Strain tensors of a microstretch medium are

$$\varepsilon_{kl} = u_{l,k} + e_{lkm}\phi_m, \quad \gamma_{kl} = \phi_{k,l}, \quad \gamma_k = \theta_{,k}. \tag{2}$$

Here, λ, μ are Lamé constant and shear modulus, κ, α, β, γ are micropolar constants, λ_0, λ_1 and a_0 are microstretch constants, ρ is the mass density, j is micro-inertia, and $\boldsymbol{u}$, $\boldsymbol{\phi}$ and θ are displacement and microrotation vectors and microstretch scalar, respectively.

Accordingly, the equations of motion in a linear homogeneous and isotropic microstretch elastic solid are given as

$$\begin{aligned}
&(c_1^2+c_3^2)\nabla\nabla\cdot\boldsymbol{u} - (c_2^2+c_3^2)\nabla\times\nabla\times\boldsymbol{u} + c_3^2\nabla\times\boldsymbol{\phi} + \bar{\lambda}_0\boldsymbol{\nabla}\theta = \ddot{\boldsymbol{u}}\\
&(c_4^2+c_5^2)\nabla\nabla\cdot\boldsymbol{\phi} - c_4^2\nabla\times\nabla\times\boldsymbol{\phi} + \omega_0^2\nabla\times\boldsymbol{u} - 2\omega_0^2\boldsymbol{\phi} = \ddot{\boldsymbol{\phi}}\\
&c_6^2\Delta\theta - c_7^2\theta - c_8^2\nabla\cdot\boldsymbol{u} = \ddot{\theta}
\end{aligned} \tag{3}$$

where

$$\begin{aligned}
&c_1^2 = \tfrac{(\lambda+2\mu)}{\rho}, \quad c_2^2 = \tfrac{\mu}{\rho}, \quad c_3^2 = \tfrac{\kappa}{\rho}, \quad c_4^2 = \tfrac{\gamma}{\rho j}, \quad c_5^2 = \tfrac{(\alpha+\beta)}{\rho j},\\
&c_6^2 = \tfrac{2a_0}{\rho j}, \quad c_7^2 = \tfrac{2\lambda_1}{3\rho j}, \quad c_8^2 = \tfrac{2\lambda_0}{3\rho j}, \quad \omega_0^2 = \tfrac{c_3^2}{j}, \quad \bar{\lambda}_0 = \tfrac{\lambda_0}{\rho}.
\end{aligned} \tag{4}$$

The displacement and microrotation vectors can be decomposed to their solenoidal and rotational parts by using Helmholtz decomposition as

$$\begin{aligned}
&\boldsymbol{u} = \nabla\bar{u} + \nabla\times\boldsymbol{U}, \quad \nabla\cdot\boldsymbol{U} = 0,\\
&\boldsymbol{\phi} = \nabla\bar{\boldsymbol{\phi}} + \nabla\times\boldsymbol{\Phi}, \quad \nabla\cdot\boldsymbol{\Phi} = 0.
\end{aligned} \tag{5}$$

Now substituting these potentials into equations of motion, we arrive at the following equations:

$$
\begin{aligned}
&(c_1^2+c_3^2)\Delta\bar{u}+\bar{\lambda}_0\theta=\ddot{\bar{u}},\\
&c_6^2\Delta\theta-c_7^2\theta-c_8^2\Delta\bar{u}=\ddot{\theta},\\
&(c_2^2+c_3^2)\Delta\boldsymbol{U}+c_3^2\nabla\times\boldsymbol{\Phi}=\ddot{\boldsymbol{U}},\\
&c_4^2\Delta\boldsymbol{\Phi}-2\omega_0^2\boldsymbol{\Phi}+\omega_0^2\nabla\times\boldsymbol{U}=\ddot{\boldsymbol{\Phi}},\\
&(c_4^2+c_5^2)\Delta\bar{\phi}-2\omega_0^2\bar{\phi}=\ddot{\bar{\phi}}.
\end{aligned}
\tag{6}
$$

3. 3D vibration analysis of the microstretch plate

We consider a homogeneous isotropic rectangular plate with length a, width b and thickness h [2]. A Cartesian coordinate system (x_1, x_2, x_3) is located at the mid-point of the plate. The maximum energy functional of a plate is

$$\Pi = V_{\max} - T_{\max} \tag{7}$$

where $V_{\max}$ and $T_{\max}$ are linear elastic strain energy and kinetic energy, respectively. Assuming harmonic-time dependence, displacement, microrotation and microstretch components of the microstretch plate undergoing free vibration may be written in terms of amplitude functions as follows

$$\{\boldsymbol{u}(x_1,x_2,x_3,t),\boldsymbol{\phi}(x_1,x_2,x_3,t),\theta(x_1,x_2,x_3,t)\}=\{\boldsymbol{U}(x_1,x_2,x_3),\boldsymbol{\Phi}(x_1,x_2,x_3),\Theta(x_1,x_2,x_3)\}\,e^{i\omega t} \tag{8}$$

Here, ω denotes the natural frequency. Now introducing the following non dimensional parameters;

$$\xi=\frac{2x_1}{a},\quad \eta=\frac{2x_2}{b},\quad \zeta=\frac{2x_3}{h}, \tag{9}$$

the elastic strain energy $V_{\max}$ and the kinetic energy $T_{\max}$ take the following forms:

$$
\begin{aligned}
V=\tfrac{h}{4\alpha_1}\int_{-1}^{1}\int_{-1}^{1}\int_{-1}^{1}&\left[\lambda\,\bar{\Lambda}_1^2+(2\mu+\kappa)\bar{\Lambda}_2+2\mu\bar{\Lambda}_3+(\mu+\kappa)\bar{\Lambda}_4+\alpha\bar{\Lambda}_5^2+(\beta+\gamma)\bar{\Lambda}_6\right.\\
&\left.+2\beta\bar{\Lambda}_7+\gamma\bar{\Lambda}_8+a_0\bar{\Lambda}_9\right]d\xi\,d\eta\,d\zeta\\
+\tfrac{b\,h}{8}\int_{-1}^{1}\int_{-1}^{1}\int_{-1}^{1}&\left[\kappa\,\bar{\bar{\Lambda}}_3+2\,\lambda_0\,\Theta\,\bar{\Lambda}_1\right]d\xi\,d\eta\,d\zeta+\tfrac{a\,b\,h}{16}\int_{-1}^{1}\int_{-1}^{1}\int_{-1}^{1}\left[\kappa\bar{\bar{\bar{\Lambda}}}_3+\lambda_1\Theta^2\right]d\xi\,d\eta\,d\zeta
\end{aligned}
\tag{10}
$$

and

$$T=\frac{\rho}{16}a\,b\,h\,\omega^2\int_{-1}^{1}\int_{-1}^{1}\int_{-1}^{1}\left\{\left[U_1^2+U_2^2+U_3^2\right]+j\left[\Phi_1^2+\Phi_2^2+\Phi_3^2\right]+3\,j\,\Theta^2\right\}d\xi\,d\eta\,d\zeta \tag{11}$$

where

$$
\begin{aligned}
&\bar{\Lambda}_1=\bar{\varepsilon}_{\xi\xi}+\bar{\varepsilon}_{\eta\eta}+\bar{\varepsilon}_{\zeta\zeta},\quad \bar{\Lambda}_2=\bar{\varepsilon}_{\xi\xi}^2+\bar{\varepsilon}_{\eta\eta}^2+\bar{\varepsilon}_{\zeta\zeta}^2,\\
&\bar{\Lambda}_3={}_1\bar{\varepsilon}_{\xi\eta}\,{}_1\bar{\varepsilon}_{\eta\xi}+{}_1\bar{\varepsilon}_{\xi\zeta}\,{}_1\bar{\varepsilon}_{\zeta\xi}+{}_1\bar{\varepsilon}_{\eta\zeta}\,{}_1\bar{\varepsilon}_{\zeta\eta},\\
&\bar{\bar{\Lambda}}_3=\left({}_1\bar{\varepsilon}_{\xi\eta}-{}_1\bar{\varepsilon}_{\eta\xi}\right)\,{}_2\bar{\varepsilon}_{\xi\eta}+\left({}_1\bar{\varepsilon}_{\xi\zeta}-{}_1\bar{\varepsilon}_{\zeta\xi}\right)\,{}_2\bar{\varepsilon}_{\xi\zeta}+\left({}_1\bar{\varepsilon}_{\eta\zeta}-{}_1\bar{\varepsilon}_{\zeta\eta}\right)\,{}_2\bar{\varepsilon}_{\eta\zeta},\\
&\bar{\bar{\bar{\Lambda}}}_3={}_2\bar{\varepsilon}_{\xi\eta}^2+{}_2\bar{\varepsilon}_{\xi\zeta}^2+{}_2\bar{\varepsilon}_{\eta\zeta}^2,\quad \bar{\Lambda}_4={}_1\bar{\varepsilon}_{\xi\eta}^2+{}_1\bar{\varepsilon}_{\eta\xi}^2+{}_1\bar{\varepsilon}_{\xi\zeta}^2+{}_1\bar{\varepsilon}_{\zeta\xi}^2+{}_1\bar{\varepsilon}_{\eta\zeta}^2+{}_1\bar{\varepsilon}_{\zeta\eta}^2,\\
&\bar{\Lambda}_5=\bar{\gamma}_{\xi\xi}+\bar{\gamma}_{\eta\eta}+\bar{\gamma}_{\zeta\zeta},\quad \bar{\Lambda}_6=\bar{\gamma}_{\xi\xi}^2+\bar{\gamma}_{\eta\eta}^2+\bar{\gamma}_{\zeta\zeta}^2,\\
&\bar{\Lambda}_7=\bar{\gamma}_{\xi\eta}\,\bar{\gamma}_{\eta\xi}+\bar{\gamma}_{\xi\zeta}\,\bar{\gamma}_{\zeta\xi}+\bar{\gamma}_{\eta\zeta}\,\bar{\gamma}_{\zeta\eta},\quad \bar{\Lambda}_8=\bar{\gamma}_{\xi\eta}^2+\bar{\gamma}_{\eta\xi}^2+\bar{\gamma}_{\xi\zeta}^2+\bar{\gamma}_{\zeta\xi}^2+\bar{\gamma}_{\eta\zeta}^2+\bar{\gamma}_{\zeta\eta}^2,\\
&\bar{\Lambda}_9=\bar{\Theta}_{,\xi}^2+\bar{\Theta}_{,\eta}^2+\bar{\Theta}_{,\zeta}^2
\end{aligned}
\tag{12}
$$

and

$$\begin{aligned}
&\bar{\varepsilon}_{\xi\xi} = \frac{\partial U_1}{\partial \xi}, \quad \bar{\varepsilon}_{\eta\eta} = \alpha_1 \frac{\partial U_2}{\partial \eta}, \quad \bar{\varepsilon}_{\zeta\zeta} = \frac{\alpha_1}{\alpha_2}\frac{\partial U_3}{\partial \zeta}, \quad {}_1\bar{\varepsilon}_{\xi\eta} = \frac{\partial U_2}{\partial \xi}, \quad {}_1\bar{\varepsilon}_{\eta\xi} = \alpha_1 \frac{\partial U_1}{\partial \eta}, \quad {}_1\bar{\varepsilon}_{\xi\zeta} = \frac{\partial U_3}{\partial \xi},\\
&{}_1\bar{\varepsilon}_{\zeta\xi} = \frac{\alpha_1}{\alpha_2}\frac{\partial U_1}{\partial \zeta}, \quad {}_1\bar{\varepsilon}_{\eta\zeta} = \alpha_1 \frac{\partial U_3}{\partial \eta}, \quad {}_1\bar{\varepsilon}_{\zeta\eta} = \frac{\alpha_1}{\alpha_2}\frac{\partial U_2}{\partial \zeta}, \quad {}_2\bar{\varepsilon}_{ij} = e_{jik}\Phi_k, \quad (i, j, k = \xi,\, \eta,\, \zeta),\\
&\bar{\gamma}_{\xi\xi} = \frac{\partial \Phi_1}{\partial \xi}, \quad \bar{\gamma}_{\eta\eta} = \alpha_1 \frac{\partial \Phi_2}{\partial \eta}, \quad \bar{\gamma}_{\zeta\zeta} = \frac{\alpha_1}{\alpha_2}\frac{\partial \Phi_3}{\partial \zeta}, \quad \bar{\gamma}_{\xi\eta} = \alpha_1 \frac{\partial \Phi_1}{\partial \eta}, \quad \bar{\gamma}_{\eta\xi} = \frac{\partial \Phi_2}{\partial \xi}, \quad \bar{\gamma}_{\xi\zeta} = \frac{\alpha_1}{\alpha_2}\frac{\partial \Phi_1}{\partial \zeta},\\
&\bar{\gamma}_{\zeta\xi} = \frac{\partial \Phi_3}{\partial \xi}, \quad \bar{\gamma}_{\eta\zeta} = \frac{\alpha_1}{\alpha_2}\frac{\partial \Phi_2}{\partial \zeta}, \quad \bar{\gamma}_{\zeta\eta} = \alpha_1 \frac{\partial \Phi_3}{\partial \eta}, \quad \bar{\Theta}_{,\xi} = \frac{\partial \Theta}{\partial \xi}, \quad \bar{\Theta}_{,\eta} = \alpha_1 \frac{\partial \Theta}{\partial \eta}, \quad \bar{\Theta}_{,\zeta} = \frac{\alpha_1}{\alpha_2}\frac{\partial \Theta}{\partial \zeta},\\
&\alpha_1 = \frac{a}{b}, \quad \alpha_2 = \frac{h}{b}.
\end{aligned} \tag{13}$$

Here, following [2, 12], each amplitude functions in (8) are expressed in terms of triplicate series of Chebyshev polynomials multiplied by proper admissible boundary functions, i.e.,

$$\begin{aligned}
&U_1(\xi,\eta,\zeta) = F_{u_1}(\xi,\eta)\sum_{i,j,k=1}^{\infty} A_{ijk} P_i(\xi)P_j(\eta)P_k(\zeta), \qquad \Phi_1(\xi,\eta,\zeta) = F_{\Phi_1}(\xi,\eta)\sum_{\hat{i},\hat{j},\hat{k}=1}^{\infty} \hat{A}_{\hat{i}\hat{j}\hat{k}} P_{\hat{i}}(\xi)P_{\hat{j}}(\eta)P_{\hat{k}}(\zeta),\\
&U_2(\xi,\eta,\zeta) = F_{u_2}(\xi,\eta)\sum_{l,m,n=1}^{\infty} B_{lmn} P_l(\xi)P_m(\eta)P_n(\zeta), \quad \Phi_2(\xi,\eta,\zeta) = F_{\Phi_2}(\xi,\eta)\sum_{\hat{l},\hat{m},\hat{n}=1}^{\infty} \hat{B}_{\hat{l}\hat{m}\hat{n}} P_{\hat{l}}(\xi)P_{\hat{m}}(\eta)P_{\hat{n}}(\zeta),\\
&U_3(\xi,\eta,\zeta) = F_{u_3}(\xi,\eta)\sum_{p,q,r=1}^{\infty} C_{pqr} P_p(\xi)P_q(\eta)P_r(\zeta), \qquad \Phi_3(\xi,\eta,\zeta) = F_{\Phi_3}(\xi,\eta)\sum_{\hat{p},\hat{q},\hat{r}=1}^{\infty} \hat{C}_{\hat{p}\hat{q}\hat{r}} P_{\hat{p}}(\xi)P_{\hat{q}}(\eta)P_{\hat{r}}(\zeta),\\
&\Theta(\xi,\eta,\zeta) = F_{\Theta}(\xi,\eta)\sum_{\hat{\hat{i}},\hat{\hat{j}},\hat{\hat{k}}=1}^{\infty} \hat{\hat{A}}_{\hat{\hat{i}}\hat{\hat{j}}\hat{\hat{k}}} P_{\hat{\hat{i}}}(\xi)P_{\hat{\hat{j}}}(\eta)P_{\hat{\hat{k}}}(\zeta).
\end{aligned} \tag{14}$$

One dimensional i^{th} Chebyshev polynomial is given as

$$P_i(\chi) = cos\left[(i-1)\, arccos(\chi)\right]. \tag{15}$$

Finally, substituting the series (14) into the energy functional (7) and minimizing this functional with respect to the coefficients of the Chebyshev polynomials, i.e.,

$$\frac{\partial \Pi}{\partial A_{ijk}} = 0, \quad \frac{\partial \Pi}{\partial B_{lmn}} = 0, \quad \frac{\partial \Pi}{\partial C_{pqr}} = 0, \quad \frac{\partial \Pi}{\partial \hat{A}_{\hat{i}\hat{j}\hat{k}}} = 0, \quad \frac{\partial \Pi}{\partial \hat{B}_{\hat{l}\hat{m}\hat{n}}} = 0, \quad \frac{\partial \Pi}{\partial \hat{C}_{\hat{p}\hat{q}\hat{r}}} = 0, \quad \frac{\partial \Pi}{\partial \hat{\hat{A}}_{\hat{\hat{i}}\hat{\hat{j}}\hat{\hat{k}}}} = 0, \tag{16}$$

leads to the following eigenvalue problem,

$$\left(\boldsymbol{K} - \Omega^2 \boldsymbol{M}\right) \boldsymbol{Z} = \boldsymbol{0} \tag{17}$$

with

$$\Omega = \omega\, a \sqrt{\rho}. \tag{18}$$

Here, the column vector $\boldsymbol{Z}$ may be written with its sub-column vectors

$$\boldsymbol{Z} = \{\boldsymbol{A}, \boldsymbol{B}, \boldsymbol{C}, \hat{\boldsymbol{A}}, \hat{\boldsymbol{B}}, \hat{\boldsymbol{C}}\hat{\hat{\boldsymbol{A}}}\}. \tag{19}$$

and each sub-column vector is in the following form [2]

$$\boldsymbol{A} = \{A_{111}, \ldots, A_{11K}, \ldots, A_{1k1}, \ldots, A_{1kK}, \ldots, A_{I11}, \ldots, A_{IJK}\}. \tag{20}$$

4. Analysis of the spectrum of frequencies and the construction of the optimization problem

Here, we consider a plate made of Gauthier material [16] with the properties $\nu = 0.4$, $E = 5.29\,GPa$, $j = 1.96 \times 10^{-7} m^2$, $\alpha_1 = 1$, $\alpha_2 = 0.1$. And the plate is considered as simply supported. The number of terms used in equation (14) is taken as $8 \times 8 \times 4$ [2]. In addition

Journal of Physics: Conference Series **319** (2011) 012012 doi:10.1088/1742-6596/319/1/012012

to the classical constants; the Poisson ratio, ν, and the modulus of elasticity, E, the number of the unknown material coefficients of a linear isotropic microstretch material is seven, namely $\{\kappa,\ \alpha,\ \beta,\ \gamma,\ a_0,\ \lambda_0,\ \lambda_1\}$.

As it is explained above, the existence of the additional frequencies due to the microstructure of the material makes the spectrum different than the one observed in the classical theory. Therefore, an objective function similar to the one used in the classical theory [3–11] could not be sufficient to obtain microstretch elastic properties. Now, to construct a well posed optimization problem, we use two objective functions that should be minimized; the first one is the difference between experimental frequencies and classical frequencies stored in a set (A^{clas}) and the second is the number of the additional frequencies stored in another set (B^{micro}).

The optimization problem may be constructed by superposing two objective functions

$$\underset{subject\ to\ \boldsymbol{X}}{Minimize} \quad \tau_1\left(\sum_{i=1}^{I}(f_i^e - f_i(\boldsymbol{X}))^2\right) + \tau_2\left(\ length(B^{micro}(\boldsymbol{X}))\right). \tag{21}$$

Here, the vector $\boldsymbol{X}$ represents the unknown parameters, i.e.

$$\boldsymbol{X} = \{\kappa,\ \alpha,\ \beta,\ \gamma,\ a_0,\ \lambda_0,\ \lambda_1\}. \tag{22}$$

And, I is the number of the experimental and classical frequencies under consideration, f_i^e is i^{th} experimental frequency, $f_i(\boldsymbol{X})$ is the ith calculated classical frequency corresponding to f_i^e. $B^{micro}(\boldsymbol{X})$ is the set of the additional frequencies due to the microstructure of the medium, τ_1 and τ_2 are the weight functions of the first and second objective functions, respectively and

$$\tau_1 + \tau_2 = 1. \tag{23}$$

The direct search algorithm (DSA) and micro genetic algorithm (mGA) are used to solve inverse optimization problem. The results for Gauthier material [16] is given in the following table [2].

Table 1. Microstretch constants $\{\kappa, \alpha, \beta, \gamma, a_0, \lambda_0, \lambda_1\}$ for Gauthier material [16] obtained from DSA.

$\kappa = 1.3234 \times 10^{-4}\ GPa,\quad \alpha = 8.3255 \times 10^{-2} kN,\quad \beta = 0.10282 kN,\quad \gamma = 3.3349 kN,$ $a_0 = 15.947 kN,\quad \lambda_0 = 0.57702 kPa,\quad \lambda_1 = 34.650 kPa$	
Obtained Freq. (Hz) with above microstretch parameters: 148.534, 353.269, 353.269, 465.074, 465.074, 540.921, 657.715, 658.414, 658.414, 824.744, 825.204	Error 0.0000734863

5. Thermo-microstretch theories

Different theories may be used for thermal problems, such as; static thermoelasticity, uncoupled and coupled quasi-static thermoelasticity, uncoupled and coupled dynamic thermoelasicity etc. Additional approaches may also be given.

As it is well known, the classical theory of heat conduction in solids uses the hypothesis that heat flux is proportional to the gradient of the temperature distribution. As a result of this hypothesis, the corresponding equation appears as a parabolic partial differential equation which results that a thermal disturbance in the body instantaneously affects all points of the body. This infinite speed is contrary with the physical realities. To remove this paradox in the classical theory, generalized thermoelasticity was developed [17]. Generalized theory introduces a short time parameter to the theory to establish a steady state heat conduction when a temperature gradient is suddenly produced in the solid. This is called thermal relaxation time. Thermoelasticity concept has led to a wide range of extension of the classical theory of thermoelasticity. Some theories may be given as [18]:

(i) Coupled thermoelasticity [19].
(ii) Thermoelasticity where the Fourier law of heat conduction is modified by taking into consideration a single relaxation time (Lord and Shulman model) [17].
(iii) Thermoelasticity where the constitutive relations for the stress tensor and the entropy are generalized by introducing two different relaxation times (Green and Lindsay model) [20].
(iv) Thermoelasticity without energy dissipation, proposed by Green and Naghdi [21] where, the Fourier law is replaced by a heat flux rate-temperature gradient relaxation.

These generalizations are the most well-known ones. Their corresponding theories in microstretch theory may be given as follows

1-Couple thermo-microstretch theory (Eringen [1]). The additional terms may be seen in the following equations:

$$\begin{aligned} t_{kl} &= \lambda u_{r,r}\delta_{kl} + \mu(u_{k,l}+u_{l,k}) + \kappa(u_{l,k}-\varepsilon_{klm}\phi_r) + \lambda_0\theta\delta_{kl} - \beta_0 T\delta_{kl} \\ a_0\theta_{,kk} &+ \tfrac{1}{3}\beta_1 T - \tfrac{1}{3}\lambda_1\theta - \tfrac{1}{3}\lambda_0 u_{k,k} + \rho(l - \tfrac{3}{2}j\ddot{\theta}) = 0 \\ \rho\gamma_0\dot{T} &+ \beta_0 T_0\dot{u}_{k,k} + \beta_1 T_0\dot{\theta} - \kappa\nabla^2 T - ph = 0. \end{aligned} \tag{24}$$

And the first field equation may be given in the vector form as:

$$(\lambda+2\mu+\kappa)\nabla\nabla\cdot\boldsymbol{u} - (\mu+\kappa)\nabla\times\nabla\times\boldsymbol{u} + \kappa\nabla\times\boldsymbol{\phi} + \lambda_0\nabla\theta - \beta_0\nabla T + \rho(\boldsymbol{f}-\ddot{\mathbf{u}}) = \mathbf{0} \tag{25}$$

2- Corresponding equations to the second theory with one relaxation time are

$$\begin{aligned} t_{kl} &= \lambda u_{r,r}\delta_{kl} + \mu(u_{k,l}+u_{l,k}) + \kappa(u_{l,k}-\varepsilon_{klm}\phi_r) + \lambda_0\theta\delta_{kl} - \beta_0 T\delta_{kl}, \\ a_0\theta_{,kk} &+ \tfrac{1}{3}\nu_1 T - \tfrac{1}{3}\lambda_1\theta - \tfrac{1}{3}\lambda_0 u_{k,k} + \rho(l - \tfrac{3}{2}j\ddot{\theta}) = 0, \\ \rho\gamma_0(\dot{T}+\tau_0\ddot{T}) &+ \nu T_0(\dot{u}_{k,k}+\tau_0\ddot{u}_{k,k}) + \nu_1 T_0(\dot{\theta}-\tau_0\ddot{\theta}) + \kappa\nabla^2 T - ph = 0. \end{aligned} \tag{26}$$

Here

$$\nu = (3\lambda+2\mu+\kappa)\alpha_{t_1}, \quad \nu_1 = (3\lambda+2\mu+\kappa)\alpha_{t_2}, \tag{27}$$

and α_{t_1}, α_{t_2} are the additional coefficients of linear thermal expansion. Then, the corresponding field equation may be written as

$$(\lambda+2\mu+\kappa)\nabla\nabla\cdot\boldsymbol{u} - (\mu+\kappa)\nabla\times\nabla\times\boldsymbol{u} + \kappa\nabla\times\boldsymbol{\phi} + \lambda_0\nabla\theta - \nu\nabla T + \rho(\boldsymbol{f}-\ddot{\mathbf{u}}) = \mathbf{0} \tag{28}$$

3- Equations of the coupled theory with two relaxation times are

$$\begin{aligned} t_{kl} &= \lambda u_{r,r}\delta_{kl} + \mu(u_{k,l}+u_{l,k}) + \kappa(u_{l,k}-\varepsilon_{klm}\phi_r) + \lambda_0\theta\delta_{kl} - \nu(T+t_1\dot{T})\delta_{kl} \\ a_0\theta_{,kk} &+ \tfrac{1}{3}\nu_1(T+t_1\dot{T}) - \tfrac{1}{3}\lambda_1\theta - \tfrac{1}{3}\lambda_0 u_{k,k} + \rho(l - \tfrac{3}{2}j\ddot{\theta}) = 0 \\ \rho\gamma_0(\dot{T}+\tau_0\ddot{T}) &+ \nu T_0\dot{u}_{k,k} + \beta_1 T_0(\dot{\theta}-\tau_0\ddot{\theta}) + \kappa\nabla^2 T - ph = 0, \end{aligned} \tag{29}$$

and the same field equation for this case becomes,

$$(\lambda+2\mu+\kappa)\nabla\nabla\cdot\boldsymbol{u} - (\mu+\kappa)\nabla\times\nabla\times\boldsymbol{u} + \kappa\nabla\times\boldsymbol{\phi} + \lambda_0\nabla\theta - \nu(1+t_1\frac{\partial}{\partial t})\nabla T + \rho(\boldsymbol{f}-\ddot{\mathbf{u}}) = \mathbf{0}. \tag{30}$$

As it can be followed from the above expressions, the structure of the equations does not allow us to apply the method used in [2]. With some modifications, second and third theories may be considered, which contain two new unknowns as $\nu(\alpha_{t_1})$, $\nu_1(\alpha_{t_2})$. This part of the work is left to further studies.

6. Propagation of plane waves

To investigate the waves propagating along x_1 direction, we consider the solution of equations (3) with the inclusion of thermal terms as seen in Eq.(24). Here, we take $f_i = l_{ij} = 0$ and $u_i = u_i(x_1, t)$, $\phi_i = \phi_i(x_1, t)$, $\theta = \theta(x_1, t)$, $T = T(x_1, t)$. Then, we find

$$\begin{array}{ll}
1- & (c_1^2 + c_3^2)u_{1,11} + \bar{\lambda}\theta_{,1} - \beta_0^{*2}T = \ddot{u}_1 \\
2- & (c_2^2 + c_3^2)u_{2,11} - c_3^2\phi_{3,1} = \ddot{u}_2 \\
3- & (c_2^2 + c_3^2)u_{3,11} - c_3^2\phi_{2,1} = \ddot{u}_3 \\
4- & (c_4^2 + c_5^2)\phi_{1,11} - 2\omega_0^2\phi_1 = \ddot{\phi}_1 \\
5- & c_4^2\phi_{2,11} - 2\omega_0^2\phi_2 - \omega_0^2 u_{3,1} = \ddot{\phi}_2 \\
6- & c_4^2\phi_{3,11} - 2\omega_0^2\phi_3 - \omega_0^2 u_{2,1} = \ddot{\phi}_3 \\
7- & c_4^2\theta_{,11} - c_7^2\theta - c_8^2 u_{1,1} + \beta_1^{*2}T = \ddot{\theta} \\
8- & \bar{\gamma}\dot{T} + \beta_0^{*2}\dot{u}_{k,k} + \bar{\beta}_1^2\dot{\theta} - c_3^2\nabla^2 T - h = 0.
\end{array} \tag{31}$$

The additional coefficients are

$$\beta_0^{*2} = \frac{\beta_0 T_0}{\rho}, \quad \bar{\beta}_1^2 = \frac{\beta_1 T_0}{\rho}, \quad c_3^2 = \frac{\kappa}{\rho}. \tag{32}$$

Here, we have eight equations for eight unknowns, $u_1, u_2, u_3, \phi_1, \phi_2, \phi_3, \theta$ and T. As it can be followed from above equations, the first, seventh and the last equations are coupled for the unknowns, u_1, θ and T.

Since the direction of propagation is x_1, In the first equation, u_1 represents the longitudinal wave. So, we may call it *longitudinal displacement wave*. The other terms, θ and T may be considered as representing *micro-dilatational* and *thermal* waves respectively.

Second and sixth equations of (31) are also coupled equations for transverse displacement for u_2 and transverse micro-shear ϕ_3. Then, they may be regarded as the equations for representing the propagation of *transverse displacement* and *transverse micro-shear* waves. Third and fifth equations of (31) represent another such pair of coupled equations, specifically showing a coupling between transverse displacement u_3 and transverse micro-shear ϕ_2. And they may be naturally called as the equations for representing the propagation of another couple of *transverse displacement* and *transverse micro-shear* waves in the medium.

To find the dispersion relations for plane harmonic waves, we consider the following forms

$$\begin{array}{ll}
u_i = U_i \exp\left[i(\xi x - \omega t)\right], & \phi_i = \Phi_i \exp\left[i(\xi x - \omega\, t)\right] \\
\theta = \Theta \exp\left[i(\xi x - \omega t)\right], & T = \Gamma \exp\left[i(\xi x - \omega\, t)\right]
\end{array} \tag{33}$$

where U_i, Φ_i, Θ and Γ are unknown constants, ξ is the wave number, ω is the angular frequency.

a- First, we will consider the uncoupled fourth equation of (31) for ϕ_1. It represents the longitudinal transverse micro shear. To find the dispersion relation for ϕ_1, we use definitions (33) and set the coefficient ϕ_1 to zero, which gives

$$c^2 = \left(\frac{\omega}{\xi}\right)^2 = (c_4^2 + c_5^2)\frac{\omega^2}{\omega^2 - \omega^{*2}}. \tag{34}$$

Here c is the phase velocity and

$$\omega^{*2} = \omega_0^2 = \frac{2\kappa}{\rho j}. \tag{35}$$

Now we may write

$$\begin{array}{ll}
\omega > \omega^* & c\ is\ real, \\
\omega = \omega^* & c\ is\ infinite, \\
\omega < \omega^* & c\ is\ imaginary.
\end{array} \tag{36}$$

Here ω^* is a cut off frequency which depends on the material characteristic κ and it is positive. Without any micro crack or void in the medium, we arrive the limit case of $\omega^* = 0$ and find $c_1 \Rightarrow \mu^*/\rho_0$ which seems reasonable. As a result, we may conclude that we have here a dispersive wave which propagate only at a frequency higher than ω^*.

b- As an example for coupled equations, we may consider second and fifth equations of (31). These are given for transverse displacement u_2 and transverse micro-shear ϕ_3. They represent the propagation of *transverse displacement* and *transverse micro-shear* waves. Following the same procedure, we find the dispersion equation as,

$$c^4 - c^2\left[c_2^2 + 2c_3^2 + c_4^2 + 2c^{*2}\right] + (c_2^2 + c_3^2)(c_2^2 + 2c^{*2}) = 0. \tag{37}$$

The investigation of this equation is similar to the one given in [13]. Thus, we shall not give the details here.

c- Last group of coupled equations are given for u_1, θ and T representing *longitudinal displacement* wave, *longitudinal micro-dilatational* and *thermal* waves respectively. Substituting above definitions for u_1, θ and T into the corresponding equations, we obtain

$$\begin{array}{l} \left[\omega^2 - \left(c_1^2 + c_3^2\right)\xi^2\right]U + i\xi\bar{\lambda}\Theta + \bar{\beta}_0\,(i\xi)\,\Gamma = 0 \\ -c_8^2 i\xi U + \left[3\omega^2 - \left(\xi^2 c_6^2 + c_7^2\right)\right]\Theta + \omega_1^2\Gamma = 0 \\ \beta_0^{*2}\omega\xi U - \beta_1^{*2} i\omega\Theta - \left(c_3^2\xi^2 - \gamma\omega i\right)\Gamma = 0. \end{array} \tag{38}$$

To obtain nonzero solutions for the unknowns $\boldsymbol{U}$, Θ, Γ in the above system of equations, the determinant of the coefficient matrix must be zero. This gives the dispersion relation for the propagation of longitudinal displacement wave, longitudinal micro-dilatational and thermal waves.

Since it is rather complicated, we would like to consider a more simple problem just to see the effects of the thermal field. Considering

$$\beta_1 << \lambda_0 < \beta_0, \tag{39}$$

we may neglect the small terms in the first step in the corresponding equations and take

$$T = \left(\frac{\varepsilon}{\xi}\right)U. \tag{40}$$

Here

$$\varepsilon = \frac{\beta_0 T_0}{\kappa}. \tag{41}$$

Now, substituting this approximate form of T in first and eight equations of (31), we obtain the following dispersion relation,

$$\omega^4 - [\alpha + \varepsilon]\,\omega^2 - \beta + \gamma\varepsilon = 0. \tag{42}$$

Here

$$\alpha = \left(\frac{1}{3}\right)c_6^2\xi^2 + \left(\frac{1}{3}\right)c_7^2 + \left(c_1^2 + c_3^2\right), \quad \beta = \bar{\lambda}_0 c_8^2\xi^2, \quad \gamma = \left(\frac{1}{3}\right)\left(c_6^2\xi^2 + c_7^2\right). \tag{43}$$

It is important to see here whether ω^2 is real. For this condition the radical must be nonzero, which gives,

$$\begin{array}{l} \frac{1}{9}\left(c_6^2\xi^2 + c_7^2\right)^2 + \left(c_1^2 + c_3^2\right)^2 + \frac{2}{3}\left(c_6^2\xi^2 + c_7^2\right)\left(c_1^2 + c_3^2\right) + \\ 2\left[\frac{1}{3}\left(c_6^2\xi^2 + c_7^2\right) + \left(c_1^2 + c_3^2\right)\right]\varepsilon + \varepsilon^2 + 4\beta - \frac{4}{3}\left(c_6^2\xi^2 + c_7^2\right)\varepsilon > 0. \end{array} \tag{44}$$

To satisfy this condition for all values of ξ, we must have

$$\begin{aligned} &1.\quad c_6^4 > 0, \\ &2.\quad \tfrac{1}{3}c_7^2\left(c_1^2 + c_3^2\right) + 6\lambda_0 \tfrac{c_8^2}{c_6^2} > \varepsilon, \\ &3.\ \left[\tfrac{1}{3}c_7^2 + \left(c_1^2 + c_3^2\right)\right]^2 + 2\left(c_1^2 + c_3^2\right)\varepsilon + \varepsilon^2 > 0. \end{aligned} \tag{45}$$

Here, the first and third conditions are obviously satisfied. A condition similar to the second one is also obtained by Eringen for the microstretch bodies by comparing the dispersion relation with the corresponding one in lattice dynamics [1].

7. Conclusions

The upper bounds of microstretch elastic constants of a microstretch material are found by constructing an optimization problem with two objective functions. So, these results may be used in damaged bodies modelled by microstretch theory. An example with the results obtsined here is given in a recent work of the authors and found the displacement field of a damaged hollow cylinder [22].

In the second part of this work, a discussion is given for the investigation of thermal problems of microstructured materials with different approaches. Later, the propagation of plane waves for thermo-microstretch material is studied. Determination of the microthermal constants is left to further studies.

References

[1] Eringen A C 1990 *Int. J. Eng. Sci.* **28** 1291-1301

[2] Kiris A and Inan E 2008 *Int. J. Eng. Sci.* **46** 585-597

[3] Deobald L R and Gibson R F 1988 *J. Sound and Vibration* **124** 269-283

[4] Sol H and Oomens C 1997 *Material Identification Using Mixed Numerical Experimental Methods* (Boston: Kluwer Academic Publishers)

[5] Ayorinde E O and Yu L 2005 *J. Sound and Vibration* **283** 243-262

[6] Pedersen P and Frederiksen P S 1992 *Measurement* **10** 113-118

[7] Frederiksen P S 1995 *Identification of elastic constants including transverse shear moduli of thick orthotropic plates*, Report N: **500** Technical University of Denmark

[8] Mota Soares C M , Freitas M M, Araújo A L and Pedersen P 1993 *Composite Structures* **25** 277-285

[9] Bledzki A K, Kessler A, Rikards R and Chate A 1999 *Composites Sci. Tech.* **59** 2015-2024

[10] De Wilde W P 1991 *Numerical identification of composites* ed A Vautrin and H Sol (London/New York: Elsevier Applied Science) pp 1-15

[11] Liu G R and Han X 2003 *Computational inverse techniques in nondestructive evaluation* (Boca Raton: CRC Press)

[12] Zhou D, Cheung Y K, Au F T K and Lo S H 2002 *Int. J. Solids and Struct.* **39** 6339-6353

[13] Parfitt V R and Eringen A C 1969 *J. Acoustical Soc. Am.* **45** 1258-1272

[14] Singh B and Kumar R 1998 *Int. J. Eng. Sci.* **36** 891-912

[15] Eringen A C 1999 *Microcontinuum Field Theories: Foundations and Solids* (New York: Springer Verlag)

[16] Gauthier R D and Jahsman W E 1975 *J. Appl. Mech.* **42** 369374

[17] Lord H W and Shulman Y 1967 *J. Mech. and Phys. Solids* **15** 299-309

[18] Al-Qahtani H and Datta S K 2004 *J. Appl. Phys.* **96** 3645-3658

[19] Oden J T 1969 *Nuclear Engineering and Design* **10** 465-475

[20] Green A E and Lindsay K A 1972 *J. Elasticity* **2** 1-7

[21] Green A E and Naghdi P M 1992 *J. Thermal Stresses* **15** 253-264

[22] Inan E and Kiris A (will appear) *Vibration Problems ICOVP2011* ed B Marvalova (Springer Proceedings in Physics Series)

Journal of Physics: Conference Series **319** (2011) 012013 doi:10.1088/1742-6596/319/1/012013

Multi-scale random sets: from morphology to effective properties and to fracture statistics

Dominique Jeulin

Centre de Morphologie Mathématique, Mathématiques et Systèmes, Mines ParisTech 35 rue Saint-Honoré, F77300 Fontainebleau, France

E-mail: `dominique.jeulin@mines-paristech.fr`

Abstract. Complex microstructures in materials often involve multi-scale heterogeneous textures, modelled by random sets derived from Mathematical Morphology. Starting from 2D or 3D images, a complete morphological characterization by image analysis is performed, and used for the identification of a model of random structure. From morphological models, simulations of realistic microstructures are introduced in a numerical solver to compute appropriate fields (electric, elastic stress or strain, ...) and to estimate the effective properties by numerical homogenization, accounting for scale dependent statistical fluctuations of the fields. Our approach is illustrated by various examples of multi-scale models: Boolean random sets based on Cox point processes and various random grains (spheres, cylinders), showing a very low percolation threshold, and therefore a high conductivity or high elastic moduli for a low volume fraction of a second phase. Multiscale Cox point processes are also a source of instructive models of fracture statistics, such as multiscale weakest link models.

1. Introduction

In the context of many materials, such as nanocomposites, it is common to encounter a non-homogeneous dispersion of a charge in a matrix [1]. For instance arrangement of aggregates (like carbon black) appear at different scales. This multiscale arrangement has implications on the prediction of the effective properties of such composites (like the conductivity, the dielectric permittivity or the elastic moduli), from the properties of the two components (charge and matrix), and their spatial distribution. Similarly, in the context of fracture statistics, crack initiation can occur on random defects with a mulstiscale distribution.

In this paper, we introduce a general methodology for modelling these situations [2]. It is based on the theory of random sets [3, 4, 5, 6]. The morphology is summarized and simulated by multi-scale random models accounting for the heterogeneous distribution of aggregates. Identification of the model is made from image analysis.

In the next sections, we develop the following points:

- Principle of random structure modeling
- Models of multi scale random sets, simulation of microstructures, and percolation
- Prediction of effective properties by numerical simulations
- Multi-scale and multi-criteria Fracture Statistics models

2. Principle of random structure modeling

The heterogeneity of materials can be handled through a probabilistic approach, which enables us to generate models and simulation of the microstructures.

When considering two-phase materials (for instance a set of particles A embedded in a matrix A^c), we use a model of random closed set (RACS) A [3, 4, 5, 6], fully characterized from a probabilistic point of view by its Choquet capacity $T(K)$ defined on the compact sets K, from (1) below, where P denotes a probability:

$$T(K) = P(K \cap A \neq \emptyset) = 1 - Q(K) = 1 - P(K \subset A^c). \tag{1}$$

In practice, $T(K)$ can be estimated by area fraction measurements on 2D images, or from volume fraction estimation on 3D images (from true microstructures, or from simulations), after a morphological dilation of the set A by the set K [3, 4, 5, 6], noted $A \oplus K$, or calculated for a given theoretical model. Equation (1) is used for the identification of a model (estimation of its parameters, and test of its validity). Particular cases of morphological properties deduced from (1) are the volume fraction V_v, the covariance (a useful tool to detect the presence of scales or anisotropies), the distribution of distances of a point in A^c to the boundary of A. The access to 3D images of microstructures by means of X-ray microtomography [7] makes it possible to use 3D compact sets K (like balls $B(r)$ with various radii r) to characterize the random set.

3. One scale models

The archetype of random structure model is a random point process. It represents the most simple kind of random structure, like small defects isolated in a matrix. As a particular case of RACS, a random point process is characterized by its Choquet capacity $T(K)$. For a locally finite point process, use can be made of the probability generating function $G_K(s)$ of the random variable $N(K)$ (random number of points of the process contained in K). We have

$$G_K(s) = E\{s^n\} = \sum_0^\infty P_n(K) s^n,$$

where $P_n(K) = P\{N(K) = n\}$, P being a probability. From this definition, we have $1 - T(K) = G_K(0) = P\{N(K) = 0\}$. For example, a typical model is the Poisson point process. It is the prototype of random point process without any order, since the numbers of points of the process contained in any pair of sets without intersection are independent random variables. For the homogeneous Poisson point process with intensity θ (i.e. average number of points per unit volume in 3D), we have, $\mu(K)$ being the Lebesgue measure (volume in 3D) of the compact set K:

$$P_n(K) = \frac{(\theta\mu(K))^n}{n!} \exp(-\theta\mu(K)) \text{ and } G_K(s) = \exp(\theta\mu(K)(s-1))$$

Starting from a point process, more general models, called grain models, can be generated, like the Boolean model, multi-scale models from intersection of random sets, or the multi-scale Cox Boolean model.

3.1. Boolean model

Some materials (like porous media or composite materials) can be simulated by means of a basic random set model, namely by distributions of overlapping objects. A random set model, the Boolean model, was proposed by G. Matheron [3, 4] to reproduce this situation. First, we consider a location of centres of grains by means of a Poisson point process with intensity θ. In this condition, a volume V contains a random number of centres N following a Poisson distribution with average value θV. Then a random set is obtained by the union of random grains (any compact set). For this model, the Choquet capacity is given by (2), where $\overline{V}(A' \oplus \check{K})$ is the average volume of the random primary grain A', dilated by K:

$$T(K) = 1 - \exp -\theta\overline{V}(A' \oplus \check{K}) \tag{2}$$

As a particular case, the volume fraction V_V of the grains, is given by (3):

$$q = 1 - V_V = 1 - \exp -\theta\overline{V}(A'). \tag{3}$$

Using for K a pair of points, 3 points, or a ball with radius r gives access to the covariance, the third order moment, and to the spherical contact distribution (namely the distribution of the distance of a random point in A^c to the boundary of A). For a given population of random grains, these morphological functions are theoretically available from (2). In the special case of spherical grains, the distribution function of the spheres radii can be estimated from the covariance [8], and therefore the model can be identified from the two points statistics, which is a one-dimensional information. For more general situations, the population of grains requires higher order moments for the identification of the model.

3.2. Percolation of the Boolean model

For materials made of components with a high contrast of properties, like for instance carbon black in a polymeric matrix, there is a strong effect on the macroscopic properties when a given phase percolates through the structure, inducing connected paths in the samples of the medium. Analytical estimations of the percolation of Boolean models of cylinders are available with the excluded volume model [9]. More recent analytical estimates of the percolation threshold of isotropic Boolean models with convex grains are based on the zeroes of the connectivity number [8, 10, 11]. They give an estimate of the percolation threshold of the grains p_c^1 and of A^c . These two percolation thresholds are different, as a result of the fact that the two sets A and A^c are non-symmetrical. The percolation threshold can also be estimated on simulations of the microstructure. It appears that a Boolean model with anisotropic primary grains (for instance sphero-cylinders) shows a much lower percolation threshold (0.01145 for an aspect ratio $l/r = 100$ [9]) than for isotropic grains (0.2895 for spheres [12, 13]. This can explain the expected outstanding mechanical, electrical or chemical properties of composites containing carbon nanotubes, mainly due to their shape, giving a low percolation threshold. The percolation of the complementary set of a Boolean model of spheres obtained analytically from the connectivity number and by simulations are given by 0.05698 and 0.0540 ± 0.005 respectively [14].

4. Models of multi-scale random sets, simulation of microstructures, and percolation

4.1. Combination of independent random sets to generate multi-scale random media

Starting from the basic models, more complex structures, such as superposition of scales, or fluctuations of the local volume fraction p of one phase can be generated in a simple way by a combination of various random sets, considering independent realizations. A convenient construction of multi-scale models makes use of the union or intersection of independent random sets A_i with different scales. In the case of intersections, $A = \cap_i A_i$, and it is easy to show that we have:

$$P(K) = P\{K \subset (\cap_i A_i)\} = \Pi_i P\{K \subset A_i\}. \tag{4}$$

The result (4) is exact without any approximation, whatever the independent random sets A_i and their scales. For instance the overall volume fraction is the product of the volume fractions of the A_i. Similarly the binary covariance $C(h) = P\{x \in A, x+h \in A\}$, and more generally the n points probabilities are obtained as a product of the corresponding individual n points probabilities. A lower bound of the corresponding percolation threshold p_c can be estimated by the products of percolation thresholds p_c^i when the scales are widely separated: $p_c \simeq \Pi_i p_c^i$. This model was used to simulate the random distribution of carbon black in composites by means of the intersection of three Boolean models of spheres at different scales, reproducing the carbon black particles, aggregates, and zones of exclusion in the matrix [15]. The identification of the model is made from the measurement of the overall binary covariance on $C(h)$ images obtained by transmission electron microscope on thin sections of the material.

4.2. The multi-scale Cox Boolean model

In another way to account for a non homogeneous distribution of random grains, it is possible to replace the Poisson point process by a Cox point process [16]. Consider a positive random function (RF) giving a non homogenous intensity $\theta(x)$. For any realization of this RF, a Poisson point process with intensity $\theta(x)$ is generated. For any realization, the number of points in a domain D follows a Poisson distribution with average $\theta(D) = \int_D \theta(dx)$:

$$P_n(D) = P\{N(D) = n\} = \frac{\theta(D)^n}{n!}\exp(-\theta(D)).$$

Furthermore, for each realization, the number of points falling in any family of domains without any intersection are independent random variables. This is not anymore the case when considering the ensemble of realizations of the random intensity.

If $\varphi_K(\lambda)$ is the Laplace transform of the positive random variable $E_{A'}\{\theta(A' \oplus \check{K})\}$, where $E_{A'}$ states for the mathematical expectation with respect to the random set A', and E_θ the mathematical expectation with respect to the random function $\theta(x)$, we have:

$$T(K) = 1 - E_\theta\{\exp(-E_{A'}\{\theta(\check{A}' \oplus K)\})\} = 1 - \varphi_K(1). \tag{5}$$

An interesting particular case is obtained by means of a constant intensity θ inside a first random set A. For instance A is a Boolean model of spheres with a large radius R. We then keep the points of a Poisson point process contained in A, as germs for centers of spheres with a smaller radius r [12]. The random measure becomes $\theta(dx) = \theta 1_A(x)dx$, where $1_A(x)$ is the indicator function of the set A ($1_A(x) = 1$ if $x \in A$ and $1_A(x) = 0$ if $x \in A^c$). We have

$$T(K) = 1 - \Phi_K(\theta), \tag{6}$$

where $\Phi_K(\lambda)$ is the Laplace transform of the random variable $E_{A'}\{V((\check{A}' \oplus K) \cap A)\}$ obtained on the realizations of the random set A, after averaging over the realizations of the primary grain A'. As a particular case for a deterministic grain A' the Choquet capacity of the Boolean Cox model is deduced from the distribution of the change of support of the set A over the compact set $\check{A}' \oplus K$ (or probability law of $V((\check{A}' \oplus K) \cap A)$). It is usually difficult to access to this law for any random set A, but it can be easily estimated from simulations. We can therefore estimate the theoretical covariance (or higher order moments) of the model.

In the case of a large separation of scales between the two sets A and A' we can obtain an approximation of the $T(K)$ and then of $P(K)$ when assuming that $V(\check{A}' \oplus K) \to 0$. Therefore, the covariance and the three points probability are asymptotically equal to the corresponding theoretical values (4) given for the intersection of independent random sets. This result is valid for any n points probability, giving an approximation of the corresponding moments for this Cox Boolean model. It can be obtained as follows, to first order in $V((\check{A}' \oplus K)$, writing $P_1(K) = P\{K \subset A\}$, $Q_1(K) = P\{K \subset A^c\}$:

$$\Phi_K(\lambda) \simeq P_1(\check{A}' \oplus K)\exp\left(-\lambda V(\check{A}' \oplus K)\right) + Q_1(\check{A}' \oplus K))$$

and therefore, noting A_2 the standard Boolean model made of the union of grains A' located on Poisson points with intensity θ, $P_2(K) = P\{K \subset A_2\}$, $Q_2(K) = P\{K \subset A_2^c\}$

$$\begin{aligned} Q(K) &\simeq P_1(\check{A}' \oplus K)\exp\left(-\theta V(\check{A}' \oplus K)\right) + 1 - P_1(\check{A}' \oplus K) \\ P(K) &\simeq 1 - Q(K) \simeq P_1(\check{A}' \oplus K)(1 - Q_2(K)) \simeq P_1(\check{A}' \oplus K)P_2(K) \\ &\simeq P_1(K)P_2(K) \end{aligned}$$

For thin sections in electron microscopy, we can notice that the available experimental information is a projection of the set A through the thickness e, and therefore we can estimate $T(K \oplus e)$, from which the model can be identified, after replacing the primary grain A' by $A' \oplus e$.

For carbon black nanocomposites, it is usual to model the distribution of carbon black particles by means of a three scale model [15, 12, 1]: spherical carbon particles (with a possible distribution of radii $f_1(r)$) are located on Poisson points inside inclusion zones (Boolean model of spheres with a distribution of radii $f_2(r)$) and outside of exclusion zones (Boolean model of spheres with a distribution of radii $f_3(r)$). The identification of the parameters of the model is made by means of an iterative optimization process, minimizing the difference between probabilistic properties of simulations and of images of the material: in [1], the multi-scale model is identified from measurements of the covariance, the third order moment, and of the area fraction after 2D closings on transmission electron images.

4.3. Percolation of the Cox Boolean model

As expected, the generation of aggregates produces random media with a lower percolation threshold: for a large separation of scales in a two scales model, the lower bound of the overall percolation threshold p_c is given by the product $p_c^1 p_c^2$ of the corresponding thresholds, as in the case of the intersection of independent random sets. Models involving iteration of scales generate microstructure with a very low percolating threshold, and therefore with improved macroscopic properties when the percolating component presents the higher property (e.g. conductivity or elastic moduli). It shows that a typical homogeneous distribution of grains in space (like for the standard Boolean model) does not produce a medium with optimal properties. The percolation threshold of various multi-scale models was estimated in various situations [12, 17, 14] . In [14], we get $p_c = 0.0849$ for a scale factor equal to 30 in a two scale Cox Boolean model of spheres, close to $0.2897^2 = 0.0839$. In [17] the percolation threshold of sphero-cylinders (aspect ratio $l/r = 100$) with centres located in a primary Boolean model of spheres (with $V_v = 0.32$) percolates for $p_c = 0.0056$. Again, considering the case of aggregates of carbon nanotubes, high performances are expected from such a microstructure for a very low volume fraction of charge, due to the presence of clusters.

5. Prediction of effective properties by numerical simulations

An efficient way to solve the problem of homogenization of physical properties, for instance to predict the dielectric permittivity or the effective mechanical properties of heterogeneous media, makes use of numerical solutions of the corresponding partial differential equations (PDE) solutions, before estimating the effective properties by spatial averaging of the solution. This requires the input of 3D images (real images, as obtained by confocal microscopy, or by X-ray microtomography) or simulations, so-called digital materials. In a second step a computational code (Finite Elements, Fast Fourier Transform, PDE numerical solver) is implemented.

We use a method derived from [18, 19], to estimate the equivalent macroscopic dielectric constant ε^* from the structure of a material and the properties of its constituents [20]. The numerical solution is obtained by means of the Green function of a homogeneous reference medium for the corresponding PDE for the electrical field or for the elastic field problem. For this, we determine the polarization field $P(x)$ inside the material knowing the local dielectric permittivity $\varepsilon(x)$, by application of a constant electric field E_0, and of periodic boundary conditions. An iterative calculation in the Fourier space is used to estimate $P(x)$, giving ε^* by means of a spatial average $<>$. We have

$$\varepsilon^* = \frac{< D >}{< E >} = \frac{< (\varepsilon(x) - \varepsilon_0)^{-1}\varepsilon(x)P(x) >}{< (\varepsilon(x) - \varepsilon_0)^{-1}P(x) >} \tag{7}$$

ε_0 being the dielectric permittivity of a reference medium used for the iterative solution, and D the electric displacement ($D(x) = \varepsilon(x)E(x)$).

Using the FFT approach is very versatile and does not require any meshing of the microstructure, in contrast with other numerical methods such as Finite Elements. In elasticity, stress and strain field maps are obtained on the scale of microstructure, to get a detailed study of the effect of the microstructure on the local fields. With the Morph-Hom code developed by F. Willot [21, 22] images of large sizes (up to 1500^3) can be handled for any contrast of properties, and therefore the electrostatic or elastic behavior of porous media or of rigid media could be successfully studied by numerical techniques. This was applied to the Boolean model of spheres with any volume fractions [21] and to two-scales and three-scales Cox Boolean models of spheres [22]. The reinforcing effect of the iteration of scales with a large separation is clearly observed in the case of a very high conductivity or of a rigid phase, as a result of a lower percolation threshold. In the case of porous media, the reverse effect is observed, as would appear in the case of a damage at the interface between clusters of rigid particles and the matrix. A similar approach was applied to a microtomographic image of a mortar material [23]. A detailed study of the local enhancement of the stress field by the microstructure was made by means of image analysis of the maps obtained by FFT. This is a promising approach to study the sensitivity of microstructural elements on damage.

Finally, when working on images of a material or on realizations of a random medium, a natural question arises [24, 25]: what is the representativeness of the effective property estimated on a bounded domain of a microstructure? In other words, what is the size of a so-called "Representative Volume Element" RVE. We address this problem by means of a probabilistic approach giving size-dependent intervals of confidence, initially developed in the framework of the homogenization of the elastic moduli of random media [26], and based on the size effect of the variance of the effective properties of simulations of random media.

6. Multi-scale and multi-criteria Fracture Statistics models

Fracture Statistics models are necessary in order to predict the probability of fracture according to the microstructure, the size of specimens and the applied stress field. These models are of major importance for engineering purpose.

In a medium with heterogeneities at different scales, several populations of defects may exist in various sub-volumes V_i inside a domain with volume V (with $V = \sum_{i=1}^{i=r} V_i$).

We propose for this type of situation theoretical models based on the weakest link model. After a recall of the weakest link model, connected to the Poisson points process, we introduce multi-criteria and multi-scale weakest link models. This is illustrated by some continuum models.

6.1. Recall of the weakest link model

The standard weakest link model is based on the assumption that fracture in a brittle material is initiated on the most critical defect, that controls the full fracture process. For this model, it means that when there is at least one point x in a specimen where the applied principal stress component $\sigma(x)$ is larger than the local critical stress $\sigma_c(x)$, the specimen is broken. Usually it is assumed that the occurrence or absence of critical defects (generating fracture) of any volume elements generate a set of independent events. After a decomposition of the volume V into links v_i and assuming that there is a fracture of the volume V when a single link v_i is broken, a classical computation for independent events gives:

$$P\{\text{Non fracture}\} = \prod_i P\{\text{Non fracture of } v_i\}$$

For $v_i \longrightarrow 0$

$$P\{\text{Non fracture of } V\} = \exp(-\int_V \Phi(\sigma(x))dx) = \exp(-V\Phi(\sigma_{eq}))$$

where the equivalent stress is defined from

$$\Phi(\sigma_{eq}) = \frac{1}{V}\int_V \Phi((\sigma(x))dx$$

This assumption is equivalent to a distribution of point defects in a matrix with $\sigma_c = \infty$, according to a Poisson point process in space, with intensity $\Phi(\sigma)$, where $\Phi(\sigma)$ is the average number per unit volume of defects with a critical stress σ_c lower than σ. For the Weibull model, the function $\Phi(\sigma)$ is power law in σ.

For a homogeneous applied stress field $\sigma(x) = \sigma$,

$$P\{\text{Non fracture of } V\} = \exp(-V\Phi(\sigma))$$

6.2. Multi-criteria and multi-scale weakest link model

We consider now r populations of defects and a decomposition of the specimen V in various sub-volumes V_i : $V = \cup_{i=1}^{i=r} V_i$, with $V = \sum_{i=1}^{i=r} V_i$. Every sub-volume V_i contains a sub-class of Poisson defects with intensity $\Phi_i(\sigma)$. In what follows, no specific assumption on the type of $\Phi_i(\sigma)$ is made (it should increase monotonously with the loading [27, 24]), and therefore possible mixing of various kinds of point defects are considered, like Weibull models for power laws, with different moduli. The sub-volumes generate a family of random variables summing to V, with a multivariate Laplace transform $\phi(\lambda_1, \lambda_2, ..., \lambda_r)$:

$$\phi(\lambda_1, \lambda_2, ..., \lambda_r) = E\{\exp(-\sum_{i=1}^{i=r} \lambda_i V_i)\}$$

.

Applying the weakest link assumption to each sub-volume, the fracture probability of volume V is given by:

- For fixed volumes V_i, in the case of a homogeneous applied stress:

$$1 - P\{\text{fracture}\} = \exp(-\sum_{i=1}^{i=r} V_i\Phi_i(\sigma)) \tag{8}$$

- Taking the mathematical expectation of expression (8) with respect to V_i, we obtain the fracture probability:

$$1 - P\{\text{fracture}\} = \phi(\Phi_1(\sigma), \Phi_2(\sigma), ..., \Phi_r(\sigma)) \tag{9}$$

For a non homogeneous applied stress field $\sigma(x)$, equation (8) becomes, when the V_i are given:

$$1 - P\{\text{fracture}\} = \exp(-\sum_{i=1}^{i=r} V_i\Phi_i(\sigma_{eq}^i(V_i))) \tag{10}$$

noting

$$V_i\Phi_i(\sigma^i_{eq}(V_i)) = \int_{V_i} \Phi_i(\sigma(x))dx \tag{11}$$

Taking the mathematical expectation with respect to the random variables $V_i\Phi_i(\sigma^i_{eq}(V_i))$, the fracture probability is expressed by means of the Laplace transform of these variables, noted:

$$\phi_H(\lambda_1, \lambda_2, ..., \lambda_r) = E\{\exp(-\sum_{i=1}^{i=r} \lambda_i V_i \Phi_i(\sigma^i_{eq}(V_i)))\}$$

$$1 - P\{\text{fracture}\} = \phi_H(1, 1, ..., 1) \tag{12}$$

For very large samples, compared to the microstructure, modelled by a stationary multi-component random set, with components A_i, every $\Phi_i(\sigma^i_{eq}(V_i))$ can be replaced by

$$\Phi_i(\sigma^i_{eq}) = \frac{1}{V}\int_V \Phi_i(\sigma(x))dx,$$

after substitution of the average in a random V_i by the overall average in the total volume V. With this approximation, the fracture probability is given by (9), replacing each $\Phi_i(\sigma)$ by $\Phi_i(\sigma^i_{eq})$:

$$1 - P\{\text{fracture}\} = \phi(\Phi_1(\sigma^1_{eq}), \Phi_2(\sigma^2_{eq}), ..., \Phi_r(\sigma^r_{eq})) \tag{13}$$

6.3. Continuous model with two modes of fracture

We consider a random set A (and its complementary set A^c), with $p = P\{x \in A\}$. We assume that the random set A contains Poisson point defects with intensity $\Phi_1(\sigma)$, and A^c contains Poisson point defects with intensity $\Phi_2(\sigma)$. The probability of fracture can be deduced from the bivariate distribution of the random volumes V_1 and V_2 (with $V_1 + V_2 = V$) of A and A^c in the domain with volume V.

This bivariate distribution is not known in general, but can be assumed to be deduced from the Beta distribution. Let X be the random variable giving the proportion of A in a sample. The pdf $f(x)$ of the random variable X $(0 \leq X \leq 1)$ is assumed to follow a Beta distribution, with parameters α and β as for a Poisson sequence generated on a line in a 1D process:

$$f(x) = \frac{\Gamma(\alpha+\beta)}{\Gamma(\alpha)\Gamma(\beta)} x^{\alpha-1}(1-x)^{\beta-1} \tag{14}$$

The average and the variance of this distribution are given by:

$$\begin{aligned} E(X) &= \frac{\alpha}{\alpha+\beta} \\ Var(X) &= \frac{\alpha\beta}{(\alpha+\beta)^2(\alpha+\beta+1)} \end{aligned}$$

For a random set A with volume fraction p and with centred covariance $\overline{W}_2(h)$, we have

$$\begin{aligned} E(X) &= p \\ Var(X) &= \frac{1}{V^2}\int_V\int_V \overline{W}_2(x-y)\, dxdy \end{aligned} \tag{15}$$

For large volumes, when $V \gg A_n$ (A_n being the integral range deduced from the integral of the covariance), the variance becomes

$$Var(X) = p(1-p)\frac{A_n}{V} \tag{16}$$

The parameters α and β are expressed as a function of p and of $Var(X)$ or of A_n for large specimens, that are microstructural properties. We have:

$$\begin{aligned}\frac{\alpha}{\alpha+\beta} &= p\\ \frac{\alpha\beta}{(\alpha+\beta)^2(\alpha+\beta+1)} &= \frac{\alpha}{\alpha+\beta}\frac{\beta}{\alpha+\beta}\frac{A_n}{V}\end{aligned}$$

and

$$\begin{aligned}\alpha &= p(\frac{V}{A_n}-1)\\ \beta &= (1-p)(\frac{V}{A_n}-1)\end{aligned}$$

The Laplace transform of the bivariate distribution of variables XV and $(1-X)V$ is defined by:

$$\begin{aligned}\phi(\lambda_1,\lambda_2) &= E\{\exp(-\lambda_1 VX - \lambda_2 V(1-X))\} \\ &= \exp(-\lambda_2 V)E\{\exp(-VX(\lambda_1-\lambda_2))\}\\ &= \phi((\lambda_1-\lambda_2)V)\exp(-\lambda_2 V)\end{aligned} \tag{17}$$

where $\phi(\lambda)$ is the Laplace transform of the Beta distribution (14). We have:

$$\phi(\lambda) = \frac{\Gamma(\alpha+\beta)}{\Gamma(\alpha)}\sum_{k=0}^{k=\infty}\frac{(-1)^k\lambda^k}{k!}\frac{\Gamma(\alpha+k)}{\Gamma(\alpha+\beta+k)} \tag{18}$$

and

$$\begin{aligned}\frac{\Gamma(\alpha+\beta)}{\Gamma(\alpha)}\frac{\Gamma(\alpha+k)}{\Gamma(\alpha+\beta+k)} &= \frac{(\alpha+k-1)...(\alpha+1)a}{(\alpha+\beta+k-1)....(\alpha+\beta)}\\ &= \frac{(p(\frac{V}{A_n}-1)+k-1)...p(\frac{V}{A_n}-1)}{(\frac{V}{A_n}-1+k-1)...(\frac{V}{A_n}-1)}\end{aligned} \tag{19}$$

The probability of fracture is obtained by replacing (19) in (17, 18). Replacing the term V by $V(\check{A}'\oplus K)$ in 19, we obtain an estimation of the Choquet capacity of a Cox Boolean random set from equation (6), when the random variable $V((\check{A}'\oplus K)\cap A)$ introduced in section 4.2 follows a Beta distribution. This can occur for sets $\check{A}'\oplus K$ that cannot be considered as negligible with respect to A, while a large separation of scales was considered previously.

For very large samples ($V\to\infty$), the term in (19) becomes p^k and $\phi(\lambda)=\exp(-p\lambda)$. Then,

$$\phi(\lambda_1,\lambda_2) = \exp(-\lambda_2 V)\exp -pV(\lambda_1-\lambda_2) = \phi(\lambda_1,\lambda_2) = \exp -V(p\lambda_1+(1-p)\lambda_2)$$

and the fracture probability is given by:

$$1-P\{\text{rupture}\} = \exp -V(p\Phi_1(\sigma)+(1-p)\Phi_2(\sigma))$$

6.4. Multi-scale weakest link model

We can specialize the previous results to a single population of defects $\Phi(\sigma) = \Phi_1(\sigma)$, using $\Phi_2(\sigma) = 0$ in the previous developments. This corresponds to a multi-scale model, where the defects are located in a random set A. We have, using the Laplace transform of the Beta distribution $\phi(\lambda)$ and equation (17):

$$1 - P\{\text{fracture}\} = \phi(\Phi(\sigma)V)$$

For very large samples ($V \to \infty$),the term in (19) becomes p^k and $\phi(\lambda) = \exp(-p\lambda)$. Then $1 - P\{\text{fracture}\} = \exp -V(p\Phi(\sigma))$. Therefore, very large scale effects of this model are similar to the standard model, but different scale effects are obtained at a very small scale and at intermediary scales.

As in [27], this approach can be generalized to fracture controlled by a critical density of defects. The fracture probability is then worked out from the probability $P\{N \text{ defects} = k\}$ available for the multi-scale models, from its generator function $G_V(s) = E\{s^N\}$:

$$G_V(s) = \phi((s-1)\Phi(\sigma)V) \tag{20}$$

The probability to find k critical defects in a domain with volume V is deduced from (20) by use of the appropriate Laplace transform ϕ (for instance the Laplace transform of the Beta distribution).

6.5. Continuous model with r modes of fracture

The previous model can be generalized to the case of r modes of fracture, the Beta distribution being replaced by a multivariate Dirichlet distribution with parameters α_1, α_2,...., α_r. We have

$$f(x_1, x_2, ..., x_r) = \frac{\Gamma(\alpha_0)}{\Pi_{i=1}^{i=r}\Gamma(\alpha_i)}\Pi_{i=1}^{i=r} x_i^{\alpha_i - 1}$$

with $\alpha_0 = \sum_{i=1}^{i=r} \alpha_i$ and $x_r = 1 - \sum_{i=1}^{i=r-1} x_i$.

Each random variable X_i follows a Beta distribution with parameters α_i and $\alpha_0 - \alpha_i$, with average $E(\ X_i) = \frac{\alpha_i}{\alpha_0}$ and variance $Var(X_i) = \frac{\alpha_i(\alpha_0-\alpha_i)}{(\alpha_0)^2(\alpha_0+1)}$. For illustration we consider a random mosaic built on a random tesselation of space, by affecting independently to each class (or cell) the mode of fracture i with the probability p_i. On a large scale for a random mosaic, $Var(X_i) = p_i(1-p_i)\frac{A_n}{V}$ (the Dirichlet distribution implies that the integral range A_n does not depend on the mode i).

For this model we have $\alpha_0 = 1 - \frac{V}{A_n}$ and $\alpha_i = p_i(1 - \frac{V}{A_n})$.

The expansion of the multivariate Laplace transform $\phi(\lambda_1, \lambda_2, ..., \lambda_r)$ depends on the moments $\mu_{k_1....k_r}$:

$$\mu_{k_1....k_r} = E\left\{\Pi_{j=1}^{j=r} X_j^{kj}\right\} = \frac{\Gamma(\sum_{i=1}^{i=r}\alpha_i)}{\Pi_{i=1}^{i=r}\Gamma(\alpha_i)}\Pi_{i=1}^{i=r}\frac{\Gamma(\alpha_i + k_i)}{\Gamma(\sum_{i=1}^{i=r}(\alpha_i + k_i))}$$

The Laplace transform of the Dirichlet distribution is obtained from:

$$\phi(\lambda_1, \lambda_2, ..., \lambda_r) = \sum_{n=0}^{n=\infty} \frac{(-1)^n}{n!}\left(\sum_{k_1+...+k_r=n} \lambda_1^{k_1}...\lambda_r^{k_r}\mu_{k_1....k_r}\right) \tag{21}$$

The fracture probability is given by replacing in (21) the λ_i by $V\Phi_i(\sigma)$. In practice, the expansion can be stopped after a finite rank n.

7. Conclusion

Multi-scale models of random media provide a wide variety of morphologies to simulate complex microstructures for application to real materials. These models are able to capture phenomena like the observation of very low percolation thresholds, explaining the enhancement of some effective properties for a low volume fraction of additions in a matrix. Our approach is based on measurements obtained by image analysis, to test and select appropriate models, and to estimate their parameters. Combined to predictive models, by analytical means, or more generally by numerical simulations, the multi-scale models give access to the optimization of microstructures with respect to the required properties. They are also the source of new fracture statistics behavior in multi-scale distributions of defects, that would be helpful for the explanation of the observed behavior of very heterogeneous materials [28]. This approach can be used in many domains of application, like materials, nanocomposites, porous media, or biological media.

References

[1] Jean A, Jeulin D, Forest S, Cantournet S and N'Guyen F 2011 *J. Microsc.* **241** (3) 243
[2] Jeulin D 2009 *Thermec 2009 (*Materials Science Forum, Vol. 638 - 642) ed T Chandra N Wanderka W Reimers and M Ionescu, pp 81-6
[3] Matheron G 1967 *Eléments pour une théorie des milieux poreux* (Paris: Masson)
[4] Matheron G 1975 *Random sets and Integral Geometry* (New York: Wiley)
[5] Serra J 1982 *Image analysis and Mathematical Morphology* (London: Academic Press)
[6] Jeulin D 2000 *Statistics and Computing* **10** 121
[7] Parra-Denis E, Barat C, Jeulin D and Ducottet C 2008 *Materials Characterization* **59** 338
[8] Bretheau T and Jeulin D 1989 *Revue Phys. Appl.* **24** 861-9
[9] Balberg I, Anderson C H, Alexander S and Wagner N 1984 *Phys. Rev.* B **30** N° 7 3933
[10] Mecke K and Stoyan D 2005, in *Space, Structure and Randomness* (Lecture Notes in Statistics vol 183) ed Bilodeau M, Meyer F and Schmitt M (Berlin: Springer Verlag) p 151
[11] Jeulin D 2010, in *Mathematical morphology: from theory to applications* ed Talbot H and Najman L (New York: ISTE /J Wiley) Chapter 19
[12] Jeulin D and Moreaud M 2005 *Proc. 9th European Congress on Stereology and Image Analysis*, (Zakopane, Poland) **1** 341
[13] Rintoul M D and Torquato S 1997 *J. Phys. A: Math. Gen.* **30** L585-L92
[14] Jeulin D and Moreaud M 2006 *Proc. Matériaux 2006* (Dijon)
[15] Savary L, Jeulin D and Thorel A 1999 *Acta Stereol.* **18**, N3 297
[16] Jeulin D 1996 Modeling heterogeneous materials by random structures, Invited lecture, European Workshop on Application of Statistics and Probabilities in Wood Mechanics, Bordeaux (22-23 March 1996) *N-06/96/MM Paris School of Mines Publication*
[17] Jeulin D and Moreaud M 2006 *Proc. S4G, 6th Int. Conf. Stereology, Spatial Statistics and Stochastic Geometry* (Prague, 26-29 June 2006) Union Czech Mathematicians and Physicists, ed R Lechnerova I Saxl and V Benes, p 269
[18] Eyre D J and Milton G W 1999 *Eur. Phys. J. Appl. Phys.* **6** 41
[19] Moulinec H and Suquet P 1998 *Comp. Meth. in Applied Mech. and Eng.* **157** 69
[20] Paciornik S, Gomes O F M, Delarue A, Schamm S, Jeulin D and Thorel A 2003 *E PJ Applied Physics* **21** 17
[21] Willot F and Jeulin D 2009 *International Journal of Engineering Sciences* **47** 313-24
[22] Willot F and Jeulin D 2011 to appear in *International Journal of Multiscale Computational Engineering*
[23] Escoda J, Willot F, Jeulin D, Sanajuja J, Toulemonde C 2011 *Cement and Concrete Research* **41** 542
[24] Jeulin D 2001, in *Mechanics of Random and Multiscale Microstructures*, ed D Jeulin and M Ostoja-Starzewski (Berlin: Springer Verlag) p 33
[25] Jeulin D 2005, in *Space, Structure and Randomness* (Lecture Notes in Statistics vol 183) ed M Bilodeau M, Meyer F and Schmitt M (Berlin:Springer Verlag) p 183
[26] Kanit T, Forest S, Galliet I, Mounoury V and Jeulin D 2003 *Int. J. Sol. Str.* **40** 3647
[27] Jeulin D 1991 *Thèse de Doctorat d'Etat ès Sciences Physiques*, Université of Caen
[28] Andrieu A, Pineau A, Besson J, Bouaziz O and Ryckelynck D 2011 submitted to *Engng Fract Mech*

Journal of Physics: Conference Series **319** (2011) 012014 doi:10.1088/1742-6596/319/1/012014

Control of crack pattern using memory effect of paste

Akio Nakahara, Yuu Shinohara and Yousuke Matsuo

Laboratory of Physics, College of Science and Technology, Nihon University, Funabashi 274-8501, Japan

E-mail: nakahara@phys.ge.cst.nihon-u.ac.jp

Abstract. A densely packed colloidal suspension, called as a paste, remembers the direction of external mechanical fields, such as flow and vibration. When the pastes are dried, memories in pastes are visualized as macroscopically anisotropic crack patterns, such as lamellar, radial, ring and spiral. Here, we experimentally investigate how pastes remember such experiences by using paste with different size distribution of colloidal particles. We find that a paste with smaller particles have a better memory, in the sense it remembers external mechanical fields at smaller solid volume fraction, which implies that interparticle forces between colloidal particles play an important role in memory effects, causing a quantitative change in the phase diagram for the same material. This result supports the hypothesis that memories in pastes are maintained as microscopically anisotropic network structure of colloidal particles, connected via interparticle forces between colloidal particles, such as van der Waals interaction.

1. Introduction

When cracks are formed, they emerge suddenly and unavoidably, and it is often difficult to control the formation of these cracks [1, 2]. For example, if you drop a glass cup to the ground and the cup is broken, it is difficult to predict the precise shape of each fragments. In these situations, there appear statistical laws instead, and a mass distribution of fragments obeys a kind of scaling laws [3, 4]. Only few successes were reported where the morphology of 1-dimensional cracks is controlled in a case of quenched glass plate. A stability of a propagating straight crack is studied experimentally and a transition from straight to oscillatory and branched cracks is obtained by controlling cracking condition like a cooling rate [5, 6, 7, 8].

Recently, it is reported that the morphology of 2-dimensional crack pattern can also be controlled in the case of desiccation cracks [9]. Usually when a paste made of colloidal particles and water dries, isotropic and cellular desiccation crack patterns appear, with the characteristic spacing between cracks almost same as the depth of the paste [10, 11, 12, 13]. When a paste is vibrated or fluidized before drying, however, pastes remember the direction of flow and vibration, and memories in pastes are visualized as morphology of desiccation crack patterns [9, 14, 15, 16]. That is, the plasticity of paste enables itself to remember external mechanical fields [17, 18, 19, 20]. This result opens a new way to control crack formation, because, by imprinting memory into paste beforehand, the future desiccation crack patterns can be controlled as we want [15, 16, 21, 22]. We would like to know the mechanism of memory effects for better control of crack formation.

This paper is constructed as follow; In Sec. 2, we review the experimental findings of memory effects in paste and distinguish memory of flow from memory of vibration. In Sec. 3, we present morphological phase diagrams of desiccation crack patterns, and show experimental conditions in which pastes remember external mechanical fields, such as vibration and flow. Section 4 is the main part of this paper. We perform experiments using pastes with different size distribution of colloidal particles to investigate the effects of interparticle forces on the formation of memories in pastes. Finally, Sec. 5 is devoted to conclusions with further applications.

2. Memory effects of paste

As was introduced above, it was found that a densely packed colloidal suspension, called as a paste, has memories of external mechanical fields, such as flow and vibration. Here, the size of colloidal particles are distributed from sub micron to several ten micron. For example, a mixture of calcium carbonate ($CaCO_3$) and water remembers the direction of vibration, when it is vibrated in the case of a water-poor condition. When we vibrate a water-poor $CaCO_3$ paste horizontally in one direction for a short period, stop the vibration, and dry the paste at 25°C and 30% humidity, a lamellar desiccation crack pattern appears with the directions of crack propagation all perpendicular to the direction of the initial vibration [9, 14]. We imagine that, when the water-poor $CaCO_3$ paste was vibrated, a longitudinal density fluctuation was formed inside paste along the direction of the initial vibration, and that deformed structure remains due to the presence of the plasticity. Note that paste of $CaCO_3$ remembers only the direction of vibration and cannot remember flow direction.

On the other hand, it is reported that some pastes can remember not only the vibration but also a flow. Such pastes can remember the direction of flow when they are fluidized in a water-rich condition. Paste of magnesium carbonate hydroxide, carbon paste (mixture of activated charcoal with water), and clay paste such as kaolin are examples of pastes having memory of flow. When a water-rich paste is vibrated, it is easily fluidized, and the resultant desiccation crack pattern has a structure which is similar to the flow pattern that we observed at the initial

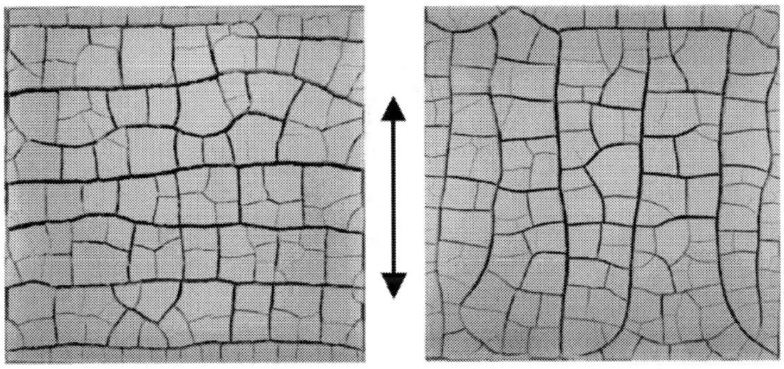

(a) Memory of vibration (b) Memory of flow

Figure 1. Transition in the direction of lamellar crack patterns as the value of the solid volume fraction at the initial vibration is decreased. The paste used in this figure is magnesium carbonate hydroxide (Kanto Chemical, Tokyo). The arrow indicates the direction of the initial vibration, where the amplitude r and the frequency f of the vibration are 15mm and 2Hz, respectively, i.e., the maximum strength of the vibration $4\pi^2 rf^2$ is 2.4m/s^2. We vibrate the container horizontally in one direction for 1 min., stop the vibration, and dry the paste until desiccation cracks are formed. The lengths of the sides of both square acrylic containers are 200mm, and the mass of powder in each container is set to be 100g so that the final spacing between cracks will be almost the same. (a) Memory of vibration. The value of the solid volume fraction ρ is 12.5%, so the paste is in a water-poor condition. Such a water-poor paste remembers a direction of the initial vibration, and the direction of lamellar cracks is perpendicular to the direction of the initial vibration. (b) Memory of flow. The value of the solid volume fraction ρ is 6.7%, so the paste is in a water-rich condition. Such a water-rich paste remembers a flow direction, and the direction of lamellar cracks is parallel to the direction of the flow induced by the initial vibration [15].

Table 1. Memory effects

	Memory of vibration	Memory of flow
calcium carbonate ($CaCO_3$)	YES	NO
magnesium carbonate hydroxide	YES	YES
activated charcoal (carbon)	YES	YES
clay (kaolin)	YES	YES

vibration. That is, when a paste has a memory of flow, the directions of desiccation cracks become parallel to the flow direction [15]. In Table 1, we summarize which kind of memory effects each paste has. Figure 1 shows that there exists a transition in the memory of paste from a memory of vibration to a memory of flow, which can be visualized as a transition in the direction of crack propagation.

To distinguish memory of flow from memory of vibration, we perform experiments by using containers with various shapes. As for the memory of vibration shown in Fig. 2(a), the directions of lamellar cracks are anywhere perpendicular to the direction of the initial vibration. On the other hand, as for the memory of flow shown in Fig. 2(b), the direction of lamellar cracks in square container is parallel to the direction of the initial vibration, while in parallelogrammatic containers, the direction of lamellar cracks is parallel to the direction of the flow along the oblique direction and not to the direction of the initial vibration. That is, we can distinguish memory of flow from memory of vibration.

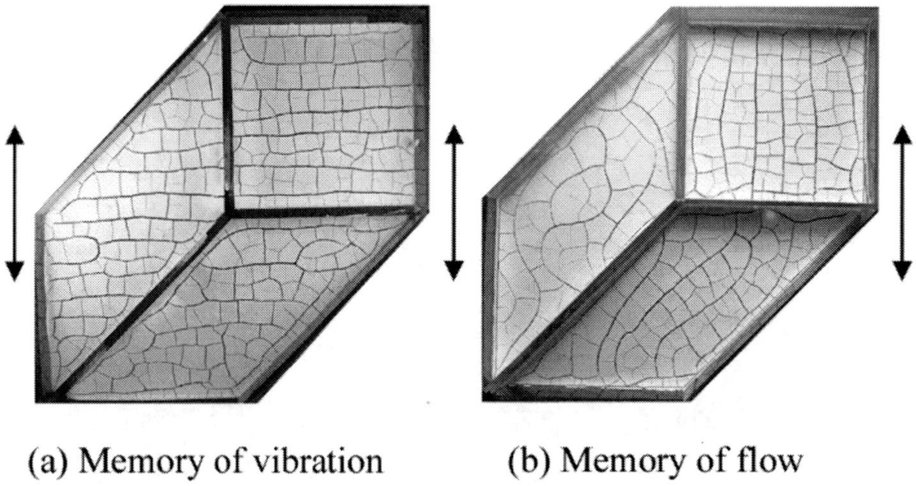

Figure 2. Comparison between two memory effects. The length of each side of the square container is 200mm and the lengths of the neighboring sides of the two parallelogrammatic containers are 200mm and $200\sqrt{2}$mm. The arrows indicate the direction of the initial vibration, where the amplitude r and the frequency f of the vibration are 15mm and 1Hz, respectively, i.e., the maximum strength of the vibration $4\pi^2 r f^2$ is 0.6m/s^2. (a) Memory of vibration. Here, we use a water-poor $CaCO_3$ paste with a solid volume fraction of $\rho = 44\%$. The direction of crack propagation is anywhere perpendicular to the direction of the initial vibration. As for the memory of vibration, the boundary condition for the movement of paste is the bottom of the container. (b) Memory of flow induced by the initial vibration [15]. Here, we use a water-rich paste of magnesium carbonate hydroxide with a solid volume fraction of $\rho = 7.7\%$. The flow patterns caused by the initial external vibration are memorized and are exhibited by the crack patterns. As for the memory of flow, the boundary condition for flow motion is the side walls of containers.

Journal of Physics: Conference Series **319** (2011) 012014 doi:10.1088/1742-6596/319/1/012014

3. Morphological phase diagram of desiccation crack patterns

To investigate the mechanism of memory effects, systematic experiments are performed to obtain a morphological phase diagram of desiccation crack patterns of pastes. The pastes we use here are magnesium carbonate hydroxide and activated charcoal (carbon) (Kanto Chemical, Tokyo). Both pastes have memories of both vibration and flow. First, we observe sizes and shapes of dry powder particles by Scanning Electron Microscope (SEM) (Fujitsu, Tokyo). Figure 3 shows that particles of magnesium carbonate hydroxide are much more smaller than those of activated charcoal (carbon). Figure 3 also shows that the shape of magnesium carbonate hydroxide is disk-like, while the shape of activated charcoal (carbon) is crashed rod-like.

To make morphological phase diagrams, the solid volume fraction, i.e., the volume fraction of powder in the mixture, plays an important role in the memory effect. When the solid volume fraction is small, colloidal particles are distributed sparsely, so that the mixture is regarded as a Newtonian viscous fluid. On the other hand, when the solid volume fraction exceeds a threshold value (Liquid-Limit), a network of colloidal particles is formed via van der Waals attractive interaction, and the rheology of the mixture changes drastically. This very concentrated suspension of colloidal particles is regarded as a paste. Also, a strength of the initial external vibration should take an appropriate value to imprint a memory into a paste. Here, the maximum strength of the vibration is given by $4\pi^2 r f^2[\mathrm{m/s^2}]$, where $r[\mathrm{m}]$ is the amplitude and $f[1/\mathrm{s}]$ is the frequency of the vibration.

Morphological phase diagrams of desiccation crack patterns are shown in Fig. 4, as a function of both the solid volume fraction and the strength of the initial vibration. Here, Fig. 4(a) shows a case where we use magnesium carbonate hydroxide, and Fig. 4(b) shows the one where we use activated charcoal (carbon). The size of the square container is 200mm with each side. The mass of colloidal particles in the mixture in each container is fixed as 100g, so that the final thickness of paste with different solid volume fractions is equalized when they dry and thus the characteristic sizes of final crack patterns are also equalized. We vibrate the container

(a) magnesium carbonate hydroxide (b) activated charcoal (carbon)

Figure 3. Images of dry particles of magnesium carbonate hydroxide and active charcoal (carbon) taken by Scanning Electron Microscope (SEM) . (a) magnesium carbonate hydroxide. The scale in the figure indicates that the distance between neighboring small white dots just above parameters is 0.20μm, so the horizontal size of the figure is 8.0μm. The shape of magnesium carbonate hydroxide is disk-like. (b) active charcoal (carbon). The scale in the figure indicates that the distance between neighboring small white dots just above parameters is 5.00μm, so the horizontal size of the figure is 200.0μm. The shape of activated charcoal (carbon) is crashed rod-like. From (a) and (b), we see that particles of magnesium carbonate hydroxide are much more smaller than those of activated charcoal (carbon).

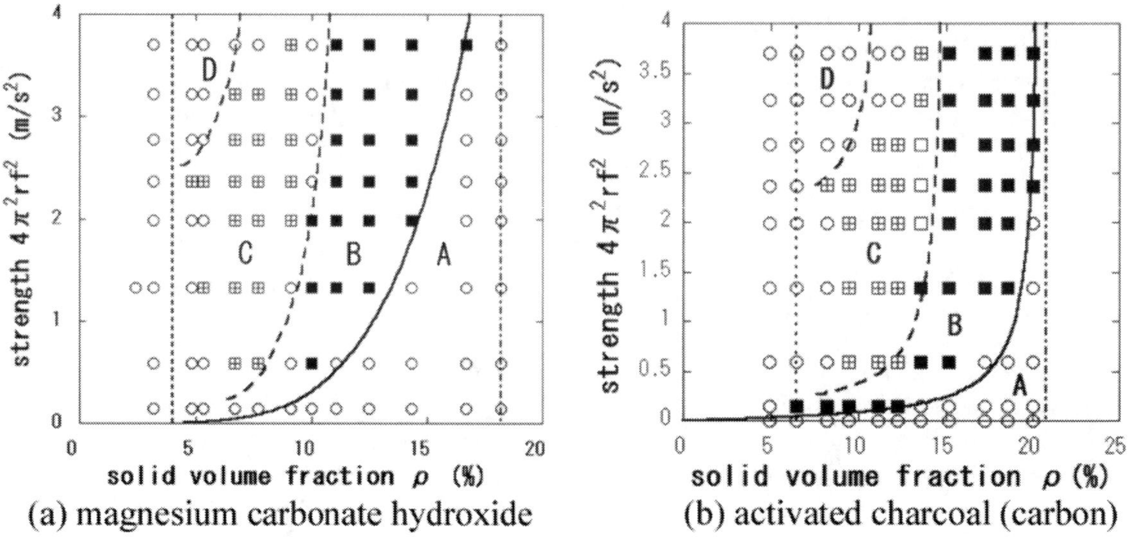

Figure 4. Morphological phase diagram of desiccation crack patterns, shown as a function of the solid volume fraction ρ and the maximum strength of the initial vibration $4\pi^2rf^2$. The length of each side of the square container is 200 mm, and mass of powder in each container is fixed as 100g. The vertical dotted lines represent the Liquid-Limit and the vertical dashed-and dotted lines represent Plastic-Limit. The solid lines represent the yield stress lines, on which the maximum value of the shear stress induced by the vibration becomes same as that of the yield stress of paste. The broken lines represent boundaries between regions. Open circles denote isotropic cellular crack patterns. Solid squares denote lamellar crack patterns, the direction of which is perpendicular to the direction of the initial vibration, and is the visualization of the memory of vibration. Open squares with a plus inside denote lamellar crack patterns, the direction of which is parallel to the direction of the flow induced by the initial vibration, and is the visualization of the memory of flow. (a) Paste made of magnesium carbonate hydroxide and water [15]. (b) Paste made of activated charcoal (carbon) and water. For both pastes, in regions A below yield stress line, pastes do not move at all, so they have no memory and only isotropic cellular crack patterns appear. Pastes remember vibration in regions B, and remember flow in regions C, but in regions D where pastes were in turbulent states we only get cellular cracks again. We see that both pastes have an ability to remember the directions of vibration and flow.

horizontally in one direction for 1 min., stop the vibration, and dry the paste until desiccation cracks are formed.

The paste of magnesium carbonate hydroxide, shown in Fig. 4(a), has memories of both flow and vibration. The region between two vertical lines, i.e., Liquid-Limit at 4% and Plastic-Limit at 18%, is divided by the solid and the dashed curves into four regions, A, B, C, and D. The solid curve corresponds to the yield stress line, on which the maximum strength of the initial vibration is equal to the strength of the yield stress of the paste. Isotropic cellular crack patterns appear in regions A and D. In region A, the maximum value of the strength of the initial vibration is smaller than that of the yield stress of paste, so paste did not experience any motion at the initial vibration. On the other hand, in region D, the paste were in a turbulent state at the initial vibration. We consider that the paste remembers the direction of the turbulent flow, but it is difficult to confirm the relation between the direction of turbulent flow and the morphology of random cellular cracks. In regions B just above the yield stress line, a water-poor paste is vibrated just like the east part of Japan was hit by an huge earthquake at 11 March 2011, and the paste remembers the direction of the vibration even after the vibration stops. In region C where a water-rich paste is fluidized in one direction like a lamellar flow (i.e., not in a turbulent state), the paste memorizes the direction of the flow induced by the initial vibration. We also get a similar result when we perform experiments using paste of activated charcoal (carbon), and its result is shown in Fig. 4(b).

On the other hand, the paste of calcium carbonate cannot remember flow direction. That is, even when water-rich $CaCO_3$ paste is fluidized just like in regions C in Fig. 4, $CaCO_3$ paste cannot remember flow direction, and only isotropic cellular crack patterns appear. Recent experiments[23] have clarified the reason why $CaCO_3$ paste cannot remember flow direction: it is due to the existence of long-range Coulombic repulsive interaction between charged colloidal particles in water. In fact, like pastes which remember flow direction, such as magnesium carbonate hydroxide and activated charcoal (carbon), colloidal particles are not charged in water. If the Coulombic repulsion is screened by adding sodium chloride into $CaCO_3$ paste, $CaCO_3$ paste gets an ability to remember flow direction.

4. Size dependence on memory effect

Then, why can pastes remember the external mechanical fields, such as vibration and flow? In this section, we perform experiments where we use pastes with different size distribution of colloidal particles. If the memory effect is just a result of jamming phenomena of inelastic particles at high solid volume fraction, we will see no change in memory effects as we keep the solid volume fraction constant and decrease the mean size of colloidal particles. Here, we speculate that the memory of paste is maintained in a network structure of colloidal particles, which are connected each other via interparticle forces between colloidal particles. When the mean size of colloidal particles decreases but the value of the solid volume fraction is still kept constant, the ratio of the surface of colloidal particles to their volume increases, and the effect of interparticle forces increases. We will check whether we can observe any changes in memory effects as we change the size distribution of colloidal particles in paste.

Here, we use Japan Industrial Standards (JIS) test powders of $CaCO_3$ particles with different size distribution. They are called test powders, because they are used to examine whether sizes of pores in filters fit to JIS. We use JIS Z 8901 test powder 1-16 ($CaCO_3$) and JIS Z 8901 test powder 1-17 ($CaCO_3$). First, we observe sizes and shapes of dry powder particles by Scanning Electron Microscope (SEM). From Fig. 5, we see that particles of JIS Z 8901 test powder 1-16

(a) JIS Z 8901 test powder 1-16 (CaCO3) (b) JIS Z 8901 test powder 1-17 (CaCO3)

Figure 5. Images of dry particles of calcium carbonate ($CaCO_3$) taken by Scanning Electron Microscope (SEM) . The scale in each figure indicates that the distance between neighboring small white dots just above parameters is 0.50μm, so the horizontal size of each figure is 20.0 μm. (a) Japan Industrial Standards (JIS) Z 8901 test powder 1-16 ($CaCO_3$). (b) JIS Z 8901 test powder 1-17 ($CaCO_3$). The shapes of both $CaCO_3$ test powder particles in (a) and (b) are similar to each other and are rough just like rocks and sand particles.

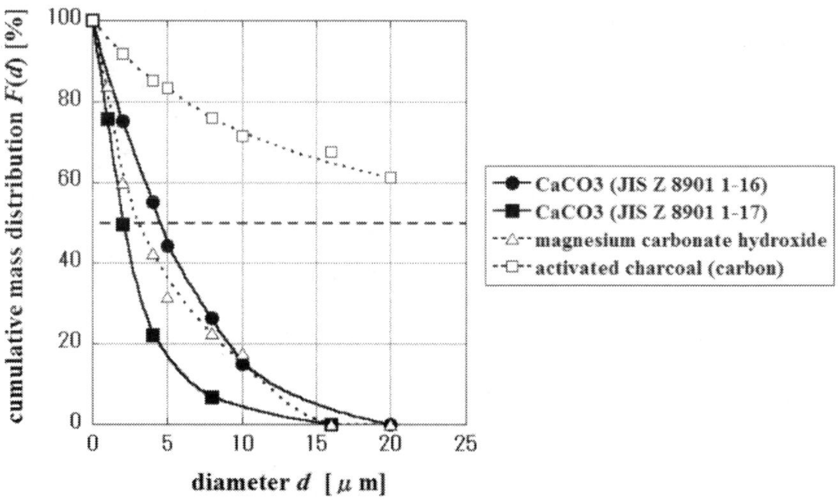

Figure 6. Cumulative mass distribution of colloidal particles, i.e., the fraction of the total mass of colloidal particles with diameters larger than $d[\mu m]$, expressed as $F(d)[\%]$, is represented as a function of d. Solid circles and its solid line denote $F(d)[\%]$ of JIS Z 8901 test powder 1-16 ($CaCO_3$), solid squares and its solid line denote $F(d)[\%]$ of JIS Z 8901 test powder 1-17 ($CaCO_3$), open triangles and its dotted line denote $F(d)[\%]$ of magnesium carbonate hydroxide, and open squares and its dotted line denote $F(d)[\%]$ of activated charcoal (carbon). The dashed line denotes the value of $F(d) = 50\%$, which corresponds to the characteristic size of colloidal particles. Except particles of activated charcoal which are much larger than other powder particles, the characteristic sizes of each colloidal particles can be estimated to be $4.5\mu m$ (JIS Z 8901 test powder 1-16 ($CaCO_3$)), $2.0\mu m$ (JIS Z 8901 test powder 1-17 ($CaCO_3$)), and $3.0\mu m$ (magnesium carbonate hydroxide), respectively.

($CaCO_3$) are larger than that of JIS Z 8901 test powder 1-17 ($CaCO_3$). Shapes of both $CaCO_3$ test powder particles in (a) and (b) are similar to each other and are rough just like rocks and sand particles. That is, we do not see any big differences in shapes between two JIS test powders.

Next, we measure the size distribution of colloidal particles in each paste using sedimentation method based on Stokesian approximation. Figure 6 shows cumulative mass distribution of colloidal particles, $F(d)[\%]$, i.e., the fraction of total mass of colloidal particles with diameters larger than d $[\mu m]$. Here, let us define a characteristic size of colloidal particles with a diameter where the value of $F(d) = 50\%$. From Fig. 6, we see that the characteristic size of JIS Z 8901 test powder 1-16 ($CaCO_3$) is $4.5\mu m$, while that of JIS Z 8901 test powder 1-17 ($CaCO_3$) is $2.0\mu m$. That is, the particles of JIS Z 8901 test powder 1-16 ($CaCO_3$) are almost two times larger than that of JIS Z 8901 test powder 1-17 ($CaCO_3$). Since the characteristic size of magnesium carbonate hydroxide is $3.0\mu m$ and is just between those of two $CaCO_3$ JIS test powders, the ability to remember flow direction is not determined solely by the value of the characteristic size of colloidal particles, as was confirmed in [23].

Morphological phase diagrams of desiccation crack patterns of $CaCO_3$ pastes are shown in Fig. 7, as a function of both the solid volume fraction and the maximum strength of the initial vibration. Here, Fig. 7(a) shows experimental results where we use JIS Z 8901 test powder 1-16 ($CaCO_3$), and Fig. 7(b) shows the one where we use JIS Z 8901 test powder 1-17 ($CaCO_3$). As was discussed above, $CaCO_3$ pastes cannot remember flow directions. The size of the square container is 200mm with each side, and the mass of $CaCO_3$ particles in each container is fixed as 360g in both (a) and (b).

Figure 7 shows that, by decreasing the characteristic size of colloidal particles, the values of Liquid-Limit and Plastic-Limit decrease, and the value of the yield stress of paste increases. This

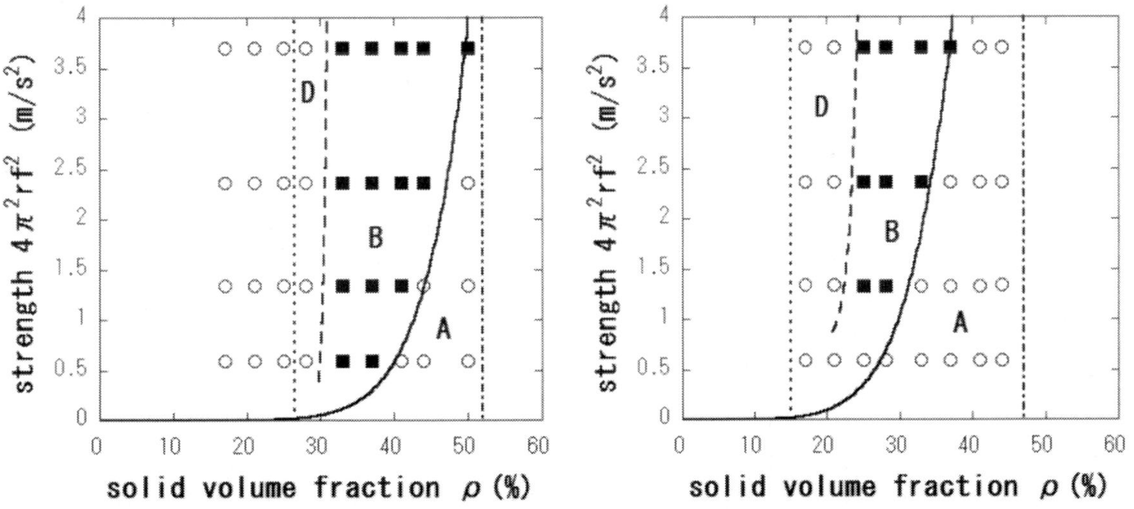

Figure 7. Morphological phase diagram of desiccation crack patterns, shown as a function of the solid volume fraction ρ and the maximum strength of the initial vibration $4\pi^2 rf^2$. The length of each side of the square container is 200 mm, and mass of powder in each container is fixed as 360g. The vertical dotted lines represent the Liquid-Limit and the vertical dashed-and dotted lines represent Plastic-Limit. The solid lines represent the yield stress lines, on which the maximum value of the shear stress induced by the vibration becomes same as that of the yield stress of paste. The broken lines represent boundaries between regions. Open circles denote isotropic cellular crack patterns. Solid squares denote lamellar crack patterns, the direction of which is perpendicular to the direction of the initial vibration, and is the visualization of the memory of vibration. (a) Paste made of JIS Z 8901 test powder 1-16 ($CaCO_3$). The characteristic size of colloidal particles is given by 4.5μm from Fig. 6. (b) Paste made of JIS Z 8901 test powder 1-17 ($CaCO_3$). The characteristic size of colloidal particles is given by 2.0μm from Fig. 6. For both pastes, in regions A below yield stress line, pastes do not move at all, so they have no memory and only isotropic cellular crack patterns appear. Pastes remember vibration in regions B, but in regions D where pastes were fluidized we only get cellular cracks again. That is, $CaCO_3$ pastes cannot remember flow directions. Comparing (a) and (b) we see that, by decreasing the characteristic size of colloidal particles, the values of Liquid-Limit and Plastic-Limit decrease, the value of the yield stress increases, the strength of viscosity and plasticity increase, and pastes come to remember vibration even at smaller solid volume fraction. Since decreasing the characteristic size of colloidal particles corresponds to increasing the ratio of the surface of colloidal particles to their volume, all these results indicate that interparticle forces play important roles in memory effects, as they connect colloidal particles to form intelligent network structure.

means that the strength of viscosity and plasticity increase, and the position of yield stress line in Fig. 7 shifts toward the smaller solid volume fraction. As for the ability to have memories, a paste with smaller particles have a better memory, in the sense it remembers the external mechanical fields at smaller solid volume fraction. Please note that, the yield stress lines in Fig. 7(a) and Fig. 7(b) coincide to the boundaries between regions A and B. Only when the value of shear stress exceeds the value of the yield stress of paste, the paste experiences a kind of movement like vibration and remembers in which direction it was vibrated. In both figures in Fig. 7, $CaCO_3$ pastes remember only the direction of the vibration. Even if we change the size distribution of colloidal particles, $CaCO_3$ pastes cannot get an ability to remember flow direction.

Since decreasing the characteristic size of colloidal particles corresponds to increasing the ratio of the surface of colloidal particles to their volume, we realize that interparticle forces between colloidal particles, such as van der Waals interaction, play important roles in memory effects. Of course, for memory effects to hold, the size of colloidal particles must not exceed several hundred micron. When we performed same experiments using carbon paste made of huge carbon particles with a characteristic size of particles about 300μm, we could not get desiccation crack patterns. That is, all these huge carbon particles are separated into individual particles as they dry, just like drying an assembly of sand particles.

We speculate that the origin of the memory effect is the deformation of network structure of colloidal particles, which are connected with each other via interparticle forces [24, 25, 26, 27]. When a water-poor paste is vibrated just like an earthquake, the longitudinal density fluctuation emerges inside the network structure. On the other hand, when a water-rich paste is fluidized, the dilute network structure is elongated along flow direction, unless particles are charged and the network is easily broken. These structural changes under external mechanical fields survive due to its plasticity, and are visualized as the morphology of desiccation crack patterns in situations where weaker points must be broken.

5. Conclusions

Pastes remember the directions of vibration and flow that they have suffered. We find that a paste with smaller particles have a better memory, in the sense it remembers external mechanical fields at smaller solid volume fraction. This result implies that interparticle forces between colloidal particles play an important role in memory effects, causing a quantitative change in the phase diagram for the same material. We speculate that memories in pastes are maintained as microscopically anisotropic network structure of colloidal particles, connected via interparticle forces between colloidal particles, such as van der Waals interaction. Memories in pastes can be visualized as crack patterns which appear when pastes dry. By using the memory effect of paste, we can control the morphology of desiccation crack patterns and make any geometrical crack patterns that streamlines can make. Recently there appear new challenges to control the formation of desiccation cracks by using electrical and magnetic fields [28, 29, 30, 31]. We would like to combine all these methods for the better control of crack formation.

Acknowledgments

We would like to acknowledge K. Hoshino and H. Nakayama for performing experiments with us and Ooshida Takeshi, M. Otsuki, S. Kitsunezaki, S. Goto, T. Matsumoto, T. Hatano, N. Ito, Ferenc Kun, T. Dutta and S. Tarafdar for valuable discussions. This work was supported by Grant-in-Aid for Scientific Research (KAKENHI) (B) 22340112 and (C) 21540388 and 23540452 of Japan Society for the Promotion of Science (JSPS). This project was also supported by JSPS and HAS under the Japan - Hungary Research Cooperative Program and by JSPS and DST under the Japan - India Science Cooperative Program.

References

[1] Herrmann H J and Roux S 1990, *Statistical Models for the Fracture in Disordered Media*, (Amsterdam: North-Holland)
[2] Lawn B 1993, *Fracture of Brittle Solids, 2nd ed.*, (Cambridge: Cambridge)
[3] Oddershede L, Dimon P, and Bohr J 1993, *Phys. Rev. Lett.* **71** 3107
[4] Herrmann H J, Wittel F K, and Kun F 2006, *Physica A* **371** 59
[5] Yuse A and Sano M 1993, *Nature* **362** 329
[6] Marder M 1994, *Phys. Rev. E* **49** R51
[7] Hayakawa Y 1994, *Phys. Rev. E* **49** R1804
[8] Sasa S -I, Sekimoto K, and Nakanishi H 1994, *Phys. Rev. E* **50** R1733
[9] Nakahara A and Matsuo Y 2005, *J. Phys. Soc. Jpn.* **74** 1362
[10] Groisman A and Kaplan E 1994, *Europhys. Lett.* **25** 415
[11] Allain C and Limat L 1995, *Phys. Rev. Lett.* **74** 2981
[12] Komatsu T S and Sasa S 1997, *Jpn. J. Appl. Phys.* **36** 391
[13] Kitsunezaki S 1999, *Phys. Rev. E* **60** 6449

[14] Nakahara A and Matsuo Y 2006, *J. Stat. Mech.: Theory Exp.* P07016
[15] Nakahara A and Matsuo Y 2006, *Phys. Rev. E* **74** 045102(R)
[16] *Physics Today* 2007, vol 60, no 9, p 116
[17] Ooshida Takeshi and Sekimoto K 2005, *Phys. Rev. Lett.* **95** 108301
[18] Otsuki M 2005, *Phys. Rev. E* **72** 046115
[19] Ooshida Takeshi 2008, *Phys. Rev. E* **77** 061501
[20] Ooshida Takeshi 2009, *J. Phys. Soc. Jpn.* **78** 104801
[21] Kitsunezaki S 2009, *J. Phys. Soc. Jpn.* **78** 064801
[22] Kitsunezaki S 2010, *J. Phys. Soc. Jpn.* **79** 124802
[23] Matsuo Y and Nakahara A, arXive:1101.0953v1 [cond-mat.soft].
[24] Tanaka H and Araki T 2000, *Phys. Rev. Lett.* **85** 1338
[25] Butter K, Bomans P H H, Frederik P M, Vroege G J and Philipse A P: *Nature Materials* **2** 88
[26] Butt H-J, Kappl M 2010, *Surface and Interfacial Forces*, (Weinheim: WILEY-VCH)
[27] Israelachvili J N 2011, *Intermolecular and Surface Forces, 3rd ed.*, (San Diego: Elsevier)
[28] Mal D, Sinha S, Middya T R, and Tarafdar S 2007, *Physica A* **384** 182
[29] Pauchard L, Elias F, Boltenhagen P, Cebers A and Bacri J C 2008, *Phys. Rev. E* **77** 021402
[30] Ngo A T, Ricardi J and Pileni M P 2008, *Nano Letters* **8** 2485
[31] Ngo A T, Ricardi J and Pileni M P 2008, *J. Phys. Chem. B* **112** 14409

Hysteresis and Power-Law Statistics during temperature induced martensitic transformation

Arya Paul[1], **Surajit Sengupta**[2], **Madan Rao**[3,4]
[1]S.N.Bose National Centre for Basic Sciences, JD-Block, Sector-III, Salt Lake, Kolkata 700098, India
[2]Indian Association for the Cultivation of Science, 2A & 2B Raja S.C.Mullick Rd, Jadavpur, Kolkata 700032, India
[3] National Center for Biological Sciences, GKVK, Bellary Road, Bangalore 560065, India
[4]Raman Research Institute, C.V. Raman Avenue, Sadashivanagar, Bangalore 560080, India.
E-mail: `aryapaul@gmail.com`

Abstract. We study hysteresis in temperature induced martensitic transformation using a 2D model solid exhibiting a square to rhombic structural transition. We find that upon quenching, the high temperature square phase, martensites are nucleated at sites having large non-affineness and ultimately invades the whole of the high temperature square phase. On heating the martensite, the high temperature square phase is restored. The transformation proceeds through avalanches. The amplitude and the time-duration of these avalanches exhibit power-law statistics both during heating and cooling of the system. The exponents corresponding to heating and cooling are different thereby indicating that the nucleation and dissolution of the product phase follows different transformation mechanism.

1. Introduction

As a direct consequence of breaking of translational and orientational symmetry, a solid, with a fixed composition, can exist in several forms with distinct crystallographic symmetries. Changing thermodynamic parameters like temperature, pressure, external stress, magnetic or electric fields [1, 2, 3, 4, 5], one can cause transitions between these structures. These transitions can be either continuous, appearing due to mechanical instabilities in the solid, or first order, when both the parent and the product phases are locally stable [6]. The transformation in this case proceeds via nucleation and growth [7]. Unlike most kinds of phase transitions, solids, however, rarely go to equilibrium [8] following a quench across a structural transition, often getting arrested in long lived metastable states called "microstructures" [9]. Most microstructures are crystalline over short length scales but lack long range order beyond some characteristic length set by the presence of defects like dislocations, grain boundaries etc. The final microstructure is determined by, among other things, depth of quench rate and transformation protocol [9]. For example, transformations at high temperatures are generally accompanied by large scale rearrangements of atomic positions where elasticity plays a relatively minor role in determining the final microstructure [8, 9, 10]. On the other hand, at low temperatures, the large scale structure is mainly determined by elasticity [8, 9]. This is especially true when the parent and product solids are crystallographically "*incompatible*" giving rise to interfacial strain across a parent-product interface [11]. How this strain is accommodated then determines, to a large

extent, the microstructures.
In this work, we look into the dynamics of microstructures during a temperature induced martensitic transformation in a model 2D solid exhibiting a square to rhombic structural transition. The system is gradually cooled from a high temperature square phase resulting in complex twinned microstructures. The product phase nucleates at sites determined by a large non-affineness. On heating the microstructure, we recover the initial square phase. The dissipation during the transition follows a power-law statistics both during heating and cooling with different exponents.
In section 2, we revisit elasto-plastic theory [13]. Starting from a basic Lagrangian and motivated by the molecular dynamics study of the model system [12], we derive dynamical equations for the relevant variables. Section 3 deals with the procedure undertaken to manifest temperature-induced martensites and the analysis of the results concerning microstructure evolution and the statistics of dissipation during the transition. Lastly in section 4, we discuss our results and outline possible future work within the framework of elasto-plastic theory.

2. The model

In this section we revisit our Elasto-plastic continuum theory [12, 13] for structural transitions in solids. Our theory allows us to determine the microstructure and the overall shape of a solid undergoing a square to a general oblique lattice. Unlike previous work in this subject [14, 15] we allow for the presence of non-affine, or plastic deformations [12, 13].

We assume that the components of the strain tensor $\mathbf{e}$ may be additively decomposed into elastic (affine) and plastic parts [17], *viz.* $e_{ij} = e^A_{ij} + e^P_{ij}$ where $i, j = x, y$. The plastic strains e^P_{ij} represents the total contribution from space and time dependent displacements which cannot be derived from an affine deformation. Such displacements may or may not be associated with individual microscopic defects, which, in the spirit of our continuum approach, need not be resolved into individual, microscopic defects. Non-affineness introduces multivaluedness in the particle displacement field $\mathbf{u} = (u_x(\mathbf{r}), u_y(\mathbf{r}))$ and cause the elastic parts of the strain to violate the St. Venant's compatibility condition [18], viz.

$$\nabla \times (\nabla \times \mathbf{e})^T = 0 \tag{1}$$

Note that this implies that the *total* strain tensor $\mathbf{e}$, nevertheless, does satisfy the compatibility condition.

The structural transition is driven by the non-linear response of the solid to one or more components of the strain - the order parameter (OP). For example, for the ($p4mm \to p2$) transition, the order parameter strains are $e_2 = e_{xx} - e_{yy}$ and $e_3 = e_{xy} + e_{yx}$ while the remaining volumetric strain $e^A_1 = e_{xx} + e_{yy}$ is a non-order parameter (NOP) strain [19]. We shall only consider the case when plasticity exists in the NOP sector.

The transition is described by the following free energy functional:

$$\begin{aligned} \mathcal{F} &= \frac{1}{2}\int \mathrm{dxdy}\ \Big[a_1(e^A_1 + e^p_1)^2 + a_2 e_2^2 + a_3 e_3^2 - \frac{1}{2} b_3 e_3^4 \\ &+ \frac{1}{3} d_3 e_3^6 + c_1(\nabla e^A_1)^2 + c_2(\nabla e_2)^2 + c_3(\nabla e_3)^2\Big]. \end{aligned} \tag{2}$$

In Eqn.2 we have made the simplifying assumption that the product oblique lattice is actually a rhombus so that the equilibrium value of $e_2 = 0$. This choice is motivated by MD simulation of our particular model solid[12] though we do not doubt that our theory may be easily extended to the general case. The three elastic constants a_1, a_2 and a_3 define the linear elasticity of the square (parent) phase. The connection with external control parameters such as temperature (T) is,

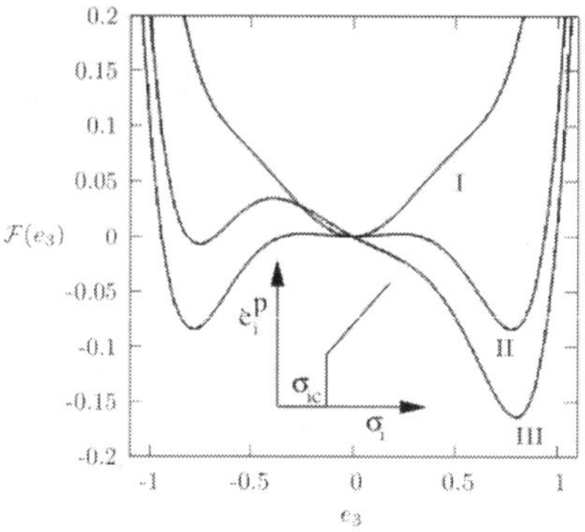

Figure 1. The homogeneous part of the free energy functional $\mathcal{F}(e_3)$ as a function of the OP strain e_3 for different values of the parameters a_3 and the external stress σ_3^{ext} viz. $a_3 = 0.5, \sigma_3^{ext} = 0$ (I), $a_3 = 0.1, \sigma_3^{ext} = 0$ (II) and $a_3 = 0.1, \sigma_3^{ext} = .1$ (III). The rest of the parameters $b_3 = -1$ and $d_3 = 1$ throughout. Inset shows a schematic representation of the typical dynamics of plastic strain $\dot{e}_i^P$ as a function of the local stress σ_i used in our calculations. Similar dynamics is assumed for volumetric (NOP, $i = 1$) as well as shear (OP, $i = 3$) plasticity. The threshold stress σ_{ic} is, of course, different in the two cases.

as usual, through the temperature dependence of these coefficients especially $a_3 \propto (T - T^*)/T^*$ where T^* is the limit of stability of the square solid. Reducing a_3 by cooling the solid stabilizes the rhombic phase (see Fig.1). The rest of the coefficients parametrize non-linearities and may be taken to be constants.

The Lagrangian is given by [15],

$$\mathcal{L} = \int \frac{\rho}{2} \left[(\dot{u}_x^2 + \dot{u}_y^2)\right] dxdy - \mathcal{F} \tag{3}$$

To obtain the equation of motion in the displacement fields, we need to solve the Euler Lagrange equation given by Eqn.4.

$$\frac{d}{dt}\frac{\partial \mathcal{L}}{\partial \dot{u}_i} - \frac{\partial \mathcal{L}}{\partial u_i} = -\frac{\partial \mathcal{R}}{\partial \dot{u}_i}. \tag{4}$$

The Rayleigh dissipation functional [16] is given by $\mathcal{R} = \frac{1}{2}\int \left[\gamma_1(\dot{e}_1^A)^2 + \gamma_2\dot{e}_2^2 + \gamma_3\dot{e}_3^2\right] dxdy$, where the coefficients γ_1, γ_2 and γ_3 are the corresponding viscosity coefficients of the system.
Since there is symmetry breaking in the space of the OP strains, the dynamics of e_2 and e_3 – the broken symmetry variables – is much slower than that of the affine NOP strain e_1^A. Therefore e_1^A reaches a steady state much faster compared to e_2 and e_3 and we can simplify by assuming that e_1^A is slaved, at all times, to the OP strains e_2 and e_3. Also note that, the St. Venant's compatibility condition (Eqn.1) may be used to eliminate e_2. Taking note of the above arguments and using Eqns.2, 3 and 4, we obtain the following equations of motion for the affine strains e_1, e_2 and e_3:

$$\nabla^2 e_1^A = 2\left(\frac{a_2 - a_3}{a_1 + a_2}\right)\frac{\partial^2 e_3}{\partial x \partial y} - \nabla^2 e_1^P, \tag{5}$$

$$\rho \ddot{e}_3 = \nabla^2\left(\frac{\delta \mathcal{F}}{\delta e_3} + \frac{\delta \mathcal{R}}{\delta \dot{e}_3}\right) + 2\frac{\partial^2}{\partial x \partial y}\left(\frac{\delta \mathcal{F}}{\delta e_1^A} + \frac{\delta \mathcal{R}}{\delta \dot{e}_1^A}\right). \tag{6}$$

Lastly, the dynamics of the plastic strain e_1^P needs to be specified. We assume that this is given by the following simple phenomenological equation of motion used by us (Fig.1 inset) is,

$$\begin{aligned} \dot{e}_1^P &= -\frac{1}{h_1}\,\sigma_1 + c_p\nabla^2 e_1^P && \text{if } |\sigma_1| > \sigma_{1c} \\ &= c_p\nabla^2 e_1^P && \text{otherwise} \end{aligned} \tag{7}$$

where $\sigma_1 = a_1 e_1$ is the volumetric stress, c_p is the diffusion coefficient in the plastic strain dynamics and h_1 is the viscosity. In reality, plastic strain production depends on the history of the system. Also defects once produced tends to reorganize themselves [23]. However, we have chosen not to incorporate any term(s) accounting for history dependence or defect reorganization in the plastic strain dynamics dynamics.

3. Temperature Induced Martensite

Martensitic transformations may be achieved by rapidly decreasing the temperature in the system (quenching) [20, 21]. In this work, we study the hysteresis behavior and microstructure evolution of martensites subjected to changes in temperature. Starting from an initial state

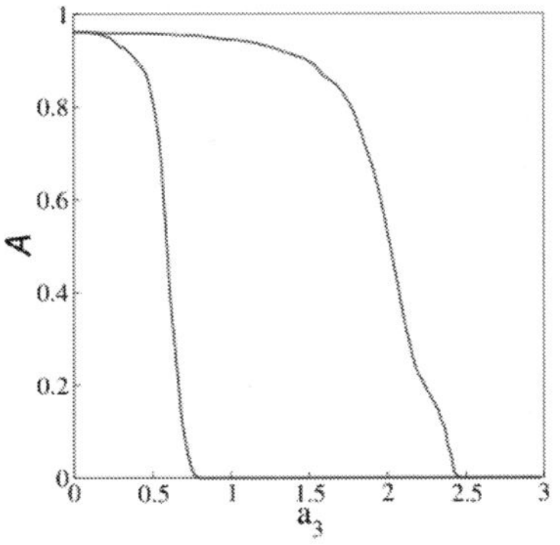

Figure 2. Plot of the area fraction A occupied by martensitic phase as a function of the degree of undercooling a_3.

with random e_1^p and a temperature, represented by the degree of undercooling a_3 and a_2, where the austenite (square) phase is the stable phase, the system is cooled from the austenite phase ($e_3 = 0$) to the martensite phase (twinned rhombic) by simultaneously decreasing a_3 and a_2 from $a_3 = a_2 = 3$ to $a_3 = a_2 = 0$. We let the system evolve according to Eqn.5 to Eqn.6. We choose a simple dynamics for the plastic strain as in Eqn.7. After each temperature drop, the system is allowed to relax for a time interval of 20 units of time. Once the temperature reaches 0, the system is then heated back to the austenite phase ($a_3 = a_2 = 3$) and a cycle of transformation is completed.

3.1. Hysteresis and Microstructure Evolution

Fig.2 shows the plot of the area fraction of the martensite phase embedded in the austenite phase as a function of the degree of undercooling a_3. During cooling, twinned martensites start appearing inside the matrix of the austenite phase when the temperature reaches below $a_3 = 0.8$. This is analogous to the martensitic start temperature. At this temperature, the austenite becomes energetically unstable compared to martensites. On further decreasing the temperature, martensites start appearing everywhere until it invades the whole of the austenite phase when the degree of undercooling, a_3 and a_2, reaches 0.2. This can be identified with the

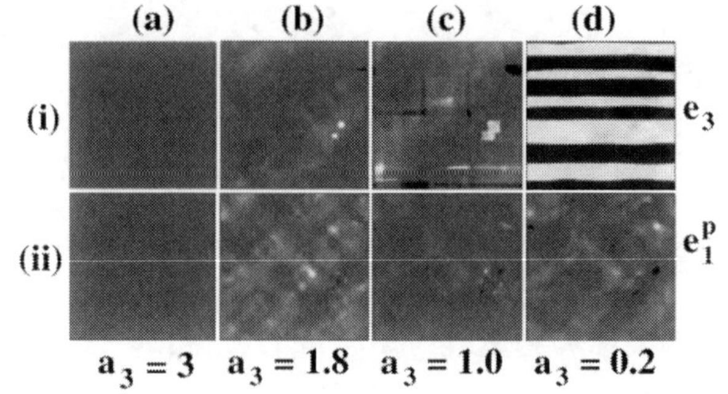

Figure 3. Plot of e_3 profile and e_1^p profile with changes in the degree of undercooling a_3 for a system size of 256 × 256 during cooling. (a)(i) Microstructure for the austenite (square) phase. The red color indicates zero order parameter e_3.(ii) Initial random plastic strain e_1^p. Colors ranging from −500(black) to 500(yellow). (b)(i) Order parameter strain e_3 for degree of undercooling $a_3 = 1.8$. Colors ranging from −1.5(black) to 1.5(yellow). Rhombic phase starts to appear at sporadic spatial locations indicated by dark yellow and dark blue colors. (ii) Plastic strain e_1^p for the same a_3. Colors range from −4(black) to 2(yellow). The initial random plastic strain is now smoothened to a large extent. (c) e_3 for $a_3 = 1$. Colors ranging from −1.5(black) to 1.5(yellow). Rhombic phases starts to grow and traces of twinned microstructure is now visible. (ii) Plastic strain e_1^p for the same a_3. Colors range from −4(black) to 5(yellow). (d)(i) e_3 for $a_3 = 0.2$. Colors ranging from −1.5(black) to 1.5(yellow). The system has now completely transformed into the twinned martensite. (ii) The plastic strain e_1^p profile has not however undergone much change.

martensite finish temperature. However, on heating the system back to the austenite phase, the system retains it's martensitic microstructure almost completely as far as a temperature of about $a_3 = a_2 = 1$, after which the martensites become unstable and the twinned microstructure slowly dissolves into the austenite phase. The transformation from the martensite to austenite is complete when the degree of undercooling reaches a value of about 2.4. What is clearly evident from Fig.2 is that the martensites appear more rapidly and dissolve slowly.

The gradual evolution of martensitic microstructure with decrease in temperature is shown in Fig.3. We begin with an austenite (square) phase wherein the order parameter strain e_3 is zero everywhere and random plastic strain e_1^p is distributed throughout the system as shown in Fig.3(a)(i) & (ii). The system is then made to evolve according to Eqn.5, Eqn.6 and Eqn.7. Initially, the dynamics smoothens the profile of initial random e_1^p except at some arbitrary points in space which at later times act as the nucleation sites for the martensites [22]. This is evident in Fig.3(b)(ii). At this stage, product phases start to appear in the matrix of the austenite phase (Fig.3(b)(ii)). There are, however, no signs of twinned microstructure yet. The untwinned rhombic phases begin to grow in size on further decrease of temperature and traces of twinned martensite gradually appear as shown in Fig.3(c)(i). However, most of the system is still in the austenite (square) phase. Note that the new domains start forming at finite distances from each other. This is consistent with the autocatalytic nucleation mechanism known to occur in martensitic transitions [5, 22]. On decreasing the degree of undercooling a_3 to as low as 0.2, the austenite (square) phase transforms completely into martensites. The plastic strain profile (Fig.3(d)(ii)) however doesn't undergo much noticeable change.

However, on gradual heating of the martensitic phase, the system retains the twinned microstructure for a large temperature regime before slowly transforming into the austenite (square) phase. The system first loses the twin structure as can be seen from Fig.4(a)(i). Plastic

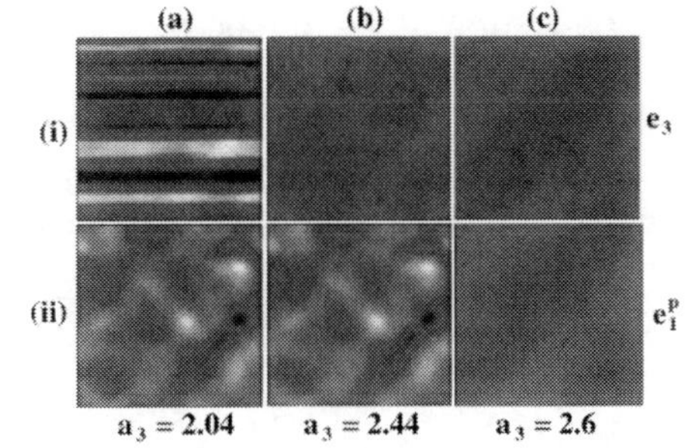

Figure 4. Plot of e_3 profile and e_1^p profile with changes in the degree of undercooling a_3 for a system size of 256×256 during heating. (a)(i) Order parameter strain e_3. Colors ranging from -1.5(black) to 1.5(yellow). The twins starts to dissolve. (ii) Plastic strain e_1^p. Colors range from -2(black) to 1(yellow). The profile is almost same as that of e_1^p for $a_3 = 1$ or $a_3 = 0.2$. (b)(i) Order parameter strain e_3 for degree of undercooling $a_3 = 2.44$. Colors ranging from -1.5(black) to 1.5(yellow). Rhombic phases has almost transformed into the square austenite phase. (ii) Plastic strain e_1^p for the same a_3. Colors range from -2(black) to 1(yellow). The plastic strain continues to diffuse. (c) e_3 for $a_3 = 2.6$. Colors range from -1.5(black) to 1.5(yellow). The system is now in the austenite (square) phase. (ii) e_1^p for $a_3 = 2.6$. Almost the whole of plastic strain has diffused.

strains are not produced anew anywhere because no new twins are formed that can account for large stresses at the twin boundaries required for plastic strain production. The plastic strain e_1^p gradually diffuses with time (see Fig.4(a)(ii)&(b)(ii)). Further increase of a_3 to 2.44 leads to the transformation of the rhombic phase to the square phase almost completely. This is in accordance with the result shown in Fig.2 where the area-fraction is almost zero for $a_3 = 2.44$. On heating the system more, there are no microstructural changes observed as can be seen from Fig.4(c)(i). The plastic strain has now diffused away completely (Fig.4(c)(ii)). Note that if the system is cooled again, it is now more difficult to nucleate martensites because most of the plastic strain, which facilitates nucleation, has diffused away (annealed). However, in real systems, martensites reappear if the system is cooled again. Our Elasto-plastic Theory fails to reproduce this feature because of the absence of the effect of defect reorganization and defect pinning in the plastic strain dynamics. In reality, defects would organize into grain boundaries which provide sites for martensite nucleation.

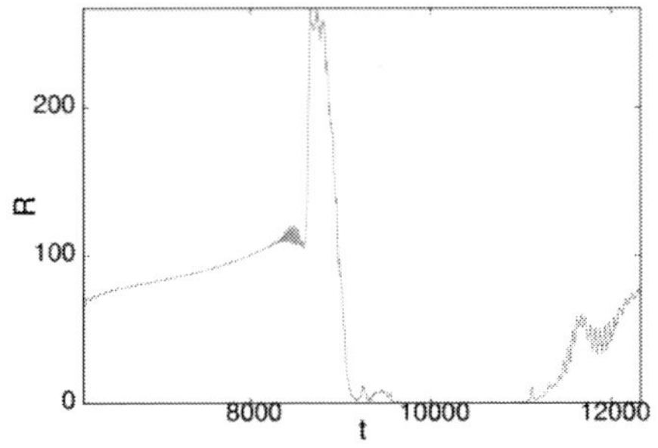

Figure 5. Typical plot of dissipation over a complete thermal cycling. The transformation from the high temperature austenite to the low temperature martensite phase proceeds through bursts of energy release.

3.2. Avalanches and Power Law Statistics

The transformation from the high temperature austenite phase to the low temperature martensite phase proceeds through dissipation of energy. The physical origin of this dissipation can be attributed to phonon viscosity, dislocation drag, electronic damping, frictional sliding of the twins, friction encountered by the growing product phase against the parent phase backdrop, etc. The signature of energy dissipation is captured by acoustic emission measurements in experiments. Measurement of amplitude and time-duration of avalanches in acoustic emission signals during martensitic transformations of single crystal Cu-Zn-Al under thermal cycling yields a power law [21]. A typical plot of dissipation obtained during thermal cycling from our Elasto-plastic Theory is presented in Fig.5. It can be seen in Fig.5 that the transformation proceeds through avalanches. The energy is released through bursts and thus is consistent with acoustic emission studies [21]. Initially, the transition advances through small amplitude avalanches. After some time, however, once the twins start forming and growing, the dissipation rises sharply. The largest peak corresponds to the time when the twins hit the system boundary after which the dissipation decreases sharply to almost zero. The most prominent contribution to the dissipation can hence be attributed to the growth and frictional sliding of the twins. Dissipation thus ceases when the twin reaches the boundary thereby terminating further growth of the twins. It nevertheless rises again when the martensite starts transforming into the austenite on heating.

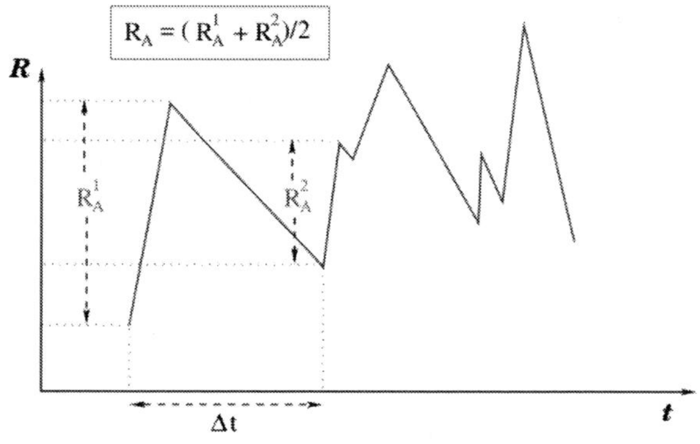

Figure 6. The blue line represents a schematic diagram of the avalanches in dissipation. The time-duration Δt of an avalanche is evaluated by locating two consecutive minima in $\mathcal{R}$ and finding out the time-difference between these two minima as illustrated in the figure. The amplitude of the avalanche R_A corresponding to Δt is evaluated by averaging the difference in the maximum of $\mathcal{R}$ and the two minima of $\mathcal{R}$ (located at the two ends of Δt).

The occurrence of these avalanches en-route to transition needs to be characterized. We evaluated the amplitude and time duration of each avalanche during both cooling and heating. The time duration of avalanches is computed by numerically evaluating the time in-between two adjacent minima of $\mathcal{R}$ in the dissipation plot of Fig.5. To calculate the amplitude of the avalanche R_A corresponding to a time duration, say Δt, we find the average difference in the maximum of $\mathcal{R}$ and the two minima of $\mathcal{R}$ (located at the two ends of Δt) in Δt. The procedure for evaluating R_A and Δt is illustrated in a schematic diagram in Fig.6.

Fig.7(a) & (b) plots the logarithmic plot of the probability distribution of the amplitude and time duration of avalanches respectively during cooling. We have obtained data of these avalanches for 12 realizations of the initial state of the system. The distribution of amplitude R_A of avalanches follows a power-law statistics - $P(R_A) \sim R_A^{-\alpha_R^c}$ where $\alpha_R^c = 1.3$ for almost three

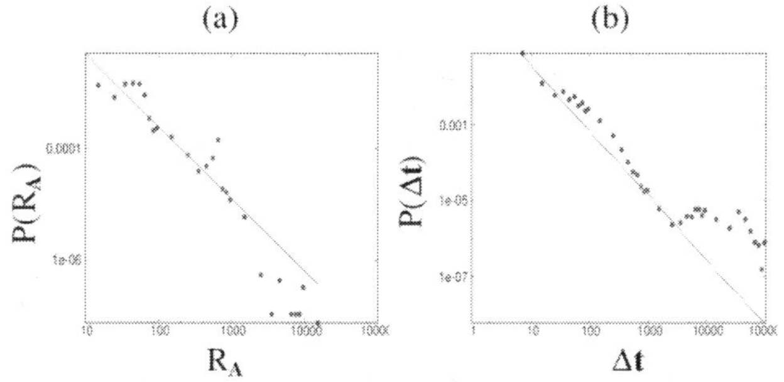

Figure 7. (a) Logarithmic plot of the probability distribution of the amplitude of avalanches during cooling. It has a power-law behavior for almost three decades with an exponent of 1.3. (b) Logarithmic plot of the probability distribution of the time duration of avalanches during cooling. It has a power-law behavior for more than two decades with an exponent of 1.7. The straight lines are a fit to the data.

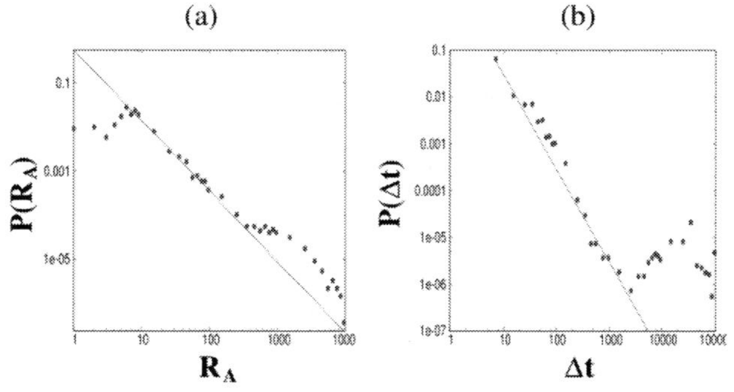

Figure 8. (a) Logarithmic plot of the probability distribution of the amplitude of avalanches during heating. It has a power-law behavior for more than two decades with an exponent of 1.6. (b) Logarithmic plot of the probability distribution of the time duration of avalanches during heating. It has a power-law behavior for more than two decades with an exponent of 2. The straight lines are a fit to the data.

decades. Also the distribution of time duration Δt of avalanches follows a power-law statistics - $P(\Delta t) \sim \Delta t^{-\tau_R^c}$ where $\tau_R^c = 1.7$ for more than two decades. The exponents obtained during heating and cooling are however different. This is also experimentally observed [21, 24] in martensitic transformation, the reason generally attributed to different dissipative mechanism in the growth and shrinkage of domains. Fig.8(a) & (b) show the probability distributions of the amplitude and time duration of avalanches respectively during heating. Both the probabilities showcase a power law statistics for at least two decades with coefficients given by $\alpha_R^h = 1.6$ and $\tau_R^h = 2$ respectively. Experimentally observed [21, 24] values of the exponents during cooling and heating is respectively $\alpha_R^c = 3.2 \pm 0.2, \tau_R^c = 1.6 \pm 0.3$ and $\alpha_R^h = 3.0 \pm 0.2, \tau_R^h = 1.6 \pm 0.3$. The large discrepancy in α_R's obtained from experiments and our Elasto-plastic Theory may be because of the difference in spatial dimension of the two observations. The experiments were done in three dimensions whereas our model system is in two dimensions. Another reason for the discrepancy may be because of the way we characterize the amplitude and time-duration of avalanches. There is however reasonable agreement in the τ_R's.

4. Conclusion

We have explored martensitic transformation under a single transformation protocol - temperature induced martensitic transformation. The martensite phase nucleates at defect cites determined by the presence of large non-affine strain. The transformation proceeds through avalanches in energy dissipation. The amplitude and the time-duration of these avalanches follow a power law statistics with different exponents during heating and cooling in temperature induced transformation. We have also looked into the possibility of reversibility of the microscopic patterns of the OP strain produced during the transformation. Unfortunately, such a reversibility at the very smallest length scales do not exist in our model. The reason for this may be attributed to the very simple phenomenological dynamics (Eqn.7) that we have used. In reality e_1^p, which contains contribution arising from defects, would tend to reorganize and form patterns of its own, over a somewhat larger timescale. These patterns will then be correlated with that of the OP strain since defects provide sites for nucleation of new martensite plates. A more detailed study involving such phenomena is planned for the future.

Acknowledgments

We are grateful to Rajiv Ahluwalia for interesting discussions. Support from the Dept of Science and Technology, Govt. of India is gratefully acknowledged.

References

[1] A. Yethiraj and A. van Blaaderen, Nature **421**, 513(2003); A. Yethiraj *et al.*, Phys. Rev. Lett. **92**, 058301 (2004).
[2] A. N. Vasi'lev *et al*, Phys. Rev. B **59**, 1113 (1999); K. Oikawa *et al*, Appl. Phys. Lett. **81**, 5201 (2002).
[3] E. Collet *et al*, Science **25**, 612 (2003).
[4] K. Otsuka *et al*, Acta Metallurgica **24**, 207 (1976).
[5] A. G. Khachaturyan, *Theory of Structural Transformations in Solids*, (Wiley, NewYork, 1983).
[6] W. I. F. David *et al*, Euro. Phys. Lett. **18**, 219, (1992).
[7] C. Godrèche(ed.) *Solids far from equilibrium*, (Cambridge University Press, New York, 1992).
[8] R. W. Cahn and J. Haasen, *Physical Metallurgy*, 4th Edition, (Elsevier, Amsterdam, 1996).
[9] R. Phillips, *Crystals, Defects and Microstructures: Modeling Across Scales*, (Cambridge University Press, Cambridge, 2001)
[10] J. W. Cahn and J. E. Hilliard, J. Chem. Phys **28**, 258 (1957)
[11] K. Bhattacharyya, *Microstructure of martensite* (Oxford University Press, Oxford, 2003).
[12] J. Bhattacharya, A. Paul, S. Sengupta and M. Rao, J. Phys.: Condens. Matter **20**, 365210 (2008).
[13] A. Paul, J. Bhattacharya, S. Sengupta and M. Rao, J. Phys.: Condens. Matter **20** (2008) 365211.
[14] G. R. Barsch, B. Horovitz and J. A. Krumhansl, Phys. Rev. Lett. **59**, 1251 (1987); B. Horovitz, G. R. Barsch and J. A. Krumhansl, Phys. Rev. B **43**, 1021 (1991); A. E. Jacobs, Phys. Rev. B **31**, 5184 (1985).
[15] K. Ø. Rasmussen *et al.*, Phys. Rev. Lett. **87**, 055704 (2001); T. Lookman *et al.*, Phys. Rev. B **67**, 024114 (2003), and references therein.
[16] L. D. Landau and E. M. Lifshitz, *Theory of Elasticity* (Pergamon, Oxford, 1986).
[17] V. A. Lubarda, *Elastoplastic theory* (CRC Press, Boca Raton, 2002); J. Lubliner, *Plasticity Theory* (Macmillan Publishing Co., New York, 1990)
[18] M. Baus and R. Lovett, Phys. Rev. Lett. **65**, 1781 (1990); **67**, 406 (1991); Phys. Rev. A **44**, 1211 (1991)
[19] D. M. Hatch, T. Lookman, A. Saxena and H. T. Stokes, Phys. Rev. B **64**, 060104(R) (2001).
[20] V. A. Chernenko *et al*, Phys. Rev. B **57**, 2659 (1998).
[21] F-J Perez-Reché, B. Tadić, L. Mañosa, A. Planes and E. Vives, Phys. Rev. Lett. **93**, 195701 (2004); E. Vives, L. Mañosa, R. Ishmael, R. Pérez-Magrané and A. Planes, Phys. Rev. Lett. **72**, 1694, (1994).
[22] P. L. Ferraglio and K. Mukherjee, Acta Metall. **22**, 835 (1974).
[23] S. Limkumnerd and J. P. Sethna, Phys. Rev. Lett. **96**, 095503 (2006).
[24] J. Baram and M. Rosen, Acta Metall. **30**, 655 (1992)

Phase field crystal model of drying induced ordering of colloidal droplet

Nirmalendu Ganai[1], **Arnab Saha**[2], **Surajit Sengupta**[2,3]

[1] Department of Physics, Nabadwip Vidyasagar College, Nabadwip, Nadia 741302, India
[2] S. N. Bose National Centre for Basic Sciences, Salt Lake, Kolkata 700098, India
[3]Centre for Advanced Materials, Indian Association for the Cultivation of Science, Jadavpur, Kolkata 700032 , India

E-mail: camss@iacs.res.in

Abstract. Controlled evaporation of solvent from a suspension of colloidal particles can induce interesting patterns of the residue on a substrate. Under certain circumstances, the colloidal particles can self-assemble in large crystalline arrays. The conditions needed for obtaining such a homogeneous pattern are poorly understood because of the complex interactions between many forces. Here we report on a simple phase-field crystal modeling of the phenomenon which shows that the ability to crystallize depends both on the initial density of the suspension as well as the drying rate.

1. Introduction

Self-assembly of colloidal particles into ordered arrays is interesting for various technological applications [1, 2] and also because it can involve ordering in systems driven far from equilibrium[3]. Many qualitatively different transitory structures can be formed by the colloidal particles, while evaporating the colloidal suspension. Ordered arrays of colloidal particles are often produced by evaporating a drop of colloidal suspension on a substrate[4, 5, 6]. Ordering upon evaporation is not well understood because of the complex interaction of many forces viz. lateral capillary forces, convection and floatation, as well as colloid-colloid interactions, and contact line pinning[7, 8, 9, 10, 11, 12].

Evaporation assisted ordering is usually a two-stage mechanism. In the pinned state, the contact line between the drop and the substrate is pinned. The area of the drop is fixed. Evaporation occurs primarily at the periphery [10] due to the increased local curvature. Therefore, solvent flows towards the periphery to replenish the liquid lost due to evaporation [8, 9] . It advects the particles with it and the build-up of particles at the periphery causes self-assembly due to attractive capillary forces, that has been investigated in the past [11, 9]. However, the contact line is eventually depinned and drying recommences when the drop area decreases [12], pushing the colloidal particles inward. The life time of these two states can be controlled by varying the contact angle between the droplet and the substrate. Here we try to understand the final structure at the end of the evaporation through depinned state via simple phenomenological continuum models. For simplicity we will not consider the effect of the substrate, hydrodynamics and polydispersity of the solute.

2. The Phase Field Crystal Model

Our approach is coarse-grained over the fluctuations at length-scale and time-scale of particle dynamics. We model the drying process as a two dimensional (2d) liquid-gas phase transition. The local density of the solvent $\phi(r, t)$ at position r, satisfies[13],

$$\dot{\phi} = \Gamma \frac{\delta \mathcal{F}_1}{\delta \phi} \tag{1}$$

with

$$\mathcal{F}_1[\{\phi\}] = \int d^2\mathbf{r} \left[-a\phi^2 + b\phi^4 + c(\nabla\phi)^2 + h\phi\right] \tag{2}$$

where h is the chemical potential of the solvent controlling the rate of drying and a, b, c are parameters and $a, b, c > 0$. The free energy density has minima at ϕ_1 and ϕ_2 where $\phi_1 > 0$ and $\phi_2 < 0$. We denote the liquid $\phi = \phi_1$ and vapour $\phi = \phi_2$ phases of the solvent. The resulting partial differential equation is solved by an explicit Euler discretization scheme starting with an initial $\phi(0, \mathbf{r}) = \phi_1$ within a circle of radius R. For $h > 0$, R decreases at subsequent times as the drop evaporates and $\phi \to \phi_2$ everywhere.

We consider Phase Field Crystal(PFC) model for the solute particles. The simplest PFC model defined by the dimensionless free energy functional [14, 15] is

$$\mathcal{F}_2[\{\psi\}] = \int d^2\mathbf{r} \left\{\frac{\psi}{2}[\epsilon + (1+\nabla^2)^2]\psi + \frac{\psi^4}{4}\right\} \tag{3}$$

where ψ corresponds to dimensionless density field [14] of the solute. Note that it contains next order term in gradient expansion of the free energy ($|\nabla^2\psi|^2$) besides the lowest order ($|\nabla\psi|^2$) term of the expansion to ensure the correct crystalline minima, which is triangular in 2d. A more detailed discussion is given in [14]. The dimensionless free energy functional in Eq.3 can be obtained by suitable rescaling of a dimensional free energy functional which is also related to classical density functional theory [14, 15, 16].

In our case the solvent evaporates continually from the surface of the droplet and thus the liquid-gas interface shrinks, pushing the solute particles inward. So, the dynamics of the solute particle will be influenced by the evaporation. More specifically, the evaporating solvent will act as a time-dependent confinement to the solute particles. Therefore, there will be a term in $\mathcal{F}_2[\{\psi\}]$ which will couple with solvent density $\phi(r, t)$. For simplicity, we assume the coupling to be linear and thus the dimensionless free energy functional of the solute becomes,

$$\mathcal{F}_2 = \int d^2\mathbf{r} \left\{\frac{\psi}{2}[\epsilon + (1+\nabla^2)^2]\psi + \frac{\psi^4}{4} + \phi\psi\right\} \tag{4}$$

The dimensionless equation of motion will be then

$$\frac{\partial\psi}{\partial t} = \vec{\nabla}.(\mu\vec{\nabla}\frac{\delta\mathcal{F}_2}{\delta\psi}) = \vec{\nabla}.\left\{\mu\vec{\nabla}([\epsilon + (1+\nabla^2)^2]\psi + \psi^3 + \phi)\right\} \tag{5}$$

where the mobility μ is given by

$$\mu = (\tanh(\gamma\phi) + 1)/2 \tag{6}$$

and μ vanishes in the limit $\phi \to \phi_2$. The choice of the functional form of μ is motivated by the fact that during evaporation if some solute particles go out of the droplet, the motion of those particles cease. The parameter γ will determine how sharply the velocity of the particles vanishes while crossing the perimeter of the drop. Here the control parameters are the average density of the particles in the drop, ψ_{av}, and the drying rate which is controlled by varying h and Γ.

According to the model described here, there are two important time scales, which are involved in 'drying induced ordering' phenomena. One (τ_1) corresponds to the drying rate of the solvent and other (τ_2) corresponds to the relaxation of the solute to its crystalline minima. If initially ψ_{av} is much lower than the critical value (i.e. initially solute is in disordered state) and $\tau_1 < \tau_2$, no ordering of the solute is expected, simply because in this case, before it finds the crystalline minima, the solvent will be completely evaporated. Only if $\tau_1 \geq \tau_2$, then ordering is expected. Such interplay between various time scales is important to understand the aforementioned phenomena.

From the phase diagram of a system obeying Eq.3 [14], we know that for $\epsilon = -0.375$, triangular phase will be obtained when ψ_{av} is in between -0.32143 to -0.1543, as well as in between 0.32143 to 0.1543. Beyond the range, striped phase will be obtained and below the range we obtain liquid like disordered phase. In our case, when the drop evaporates with a drying rate suitable for ordering (i.e. $\tau_1 \geq \tau_2$), the shrinking circular contact line pushes the solute inside the drop. Consequently, ψ_{av} increases and when it is in the aforementioned range (in our case the range is between -0.32143 and -0.1543), ordered phase will be formed. But if the drying is slow enough or the initial density of the solute is high enough so that, final ψ_{av} goes beyond the range, striped phase will be obtained according to the phase diagram given in [14]. In figure 2a and in 2b we show a typical striped and a typical ordered phase. In both the phases density is inhomogeneous over space but in the ordered state density peaks form a crystalline array (triangular) which is clearly absent in the striped phase. If the drying rate is such that the final ψ_{av} remains below the range, we get a typical liquid-like disordered state, which is shown in figure 2c. The density in bulk of this state is homogeneous over space, as it should be and the ring pattern near the boundary (see figure 2c and 3a) appears due to the sharp change of density at the periphery of the drying droplet. When $\phi = \phi_2$ everywhere all the dynamics ceases (because, $\mu(\phi_2) = 0$).

A further simplification may be achieved by solving for the dynamics of the solvent analytically. For a given ψ_{av}, initial condition for solvent-field is

$$\begin{aligned}
\phi(\mathbf{r}, t=0) &= \phi_1 \; for \; |\mathbf{r}| = R \\
&= \phi_1 + \frac{(\phi_1 - \phi_2)}{\xi}(R - \mathbf{r}) \; for \; R \leq |\mathbf{r}| \leq (R + \xi) \\
&= \phi_2 \;\; otherwise
\end{aligned} \tag{7}$$

Here $\xi = \sqrt[2]{\frac{c(\phi_1 - \phi_2)^3}{\mathcal{I}_s}}$ is the interfacial width for $\frac{\xi}{R} \ll 1$, with $\mathcal{I}_s = |\frac{-a(\phi_1^3 - \phi_2^3)}{3} + \frac{b(\phi_1^5 - \phi_2^5)}{5} + \frac{h(\phi_1^2 - \phi_2^2)}{2}|$. The rate of change of $\phi(\mathbf{r}, t)$, $\frac{d\phi}{dt}$, is then obtained as,

$$\frac{d\phi}{dt} = \frac{dR}{dt}\,\frac{d\phi}{d\mathbf{r}}\Big|_{\mathbf{r}=R} \tag{8}$$

which gives, after some straightforward algebra,

$$\frac{dR}{dt} = \dot{R} = \frac{\xi}{(\phi_1 - \phi_2)^2}[-a(\phi_1^2 - \phi_2^2) + b(\phi_1^4 - \phi_2^4) + h(\phi_1 - \phi_2)] + \frac{2c}{R} \tag{9}$$

Where we have assumed that $\frac{\xi}{R} \ll 1$. Integrating Eq.9, we get

$$\mathbf{r} = R_0 - \frac{2c}{D}\ln(\frac{DR_0 + 2c}{D\mathbf{r} + 2c}) + Dt \tag{10}$$

where $D = \frac{\xi}{(\phi_1-\phi_2)^2}[-a(\phi_1^2-\phi_2^2)+b(\phi_1^4-\phi_2^4)+h(\phi_1-\phi_2)]$ and R_0 is the initial radius of the drop.

We see that Eq.10 is a transcendental equation which is solved using bisection at every time step. Typical solutions for $R(t)$ are shown in Fig.1

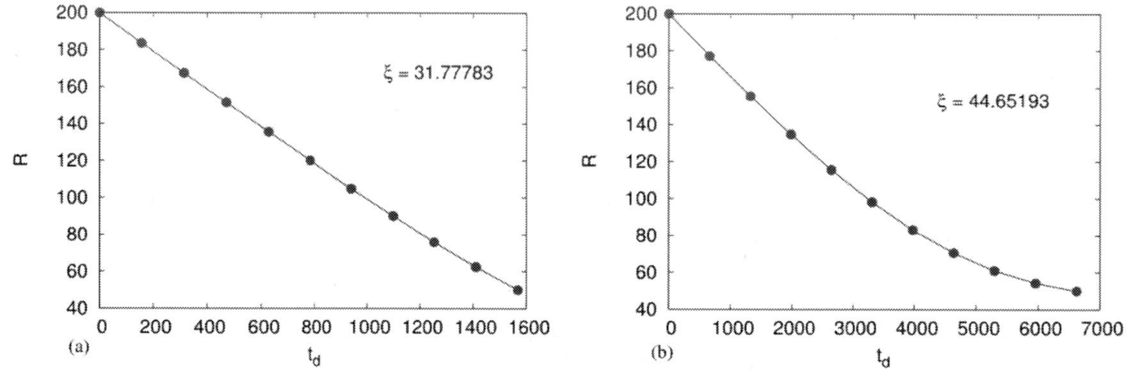

Figure 1. (a) and (b) show how the radius of the drop (R) decreases with time for total drying time $t_d = 1568.15$ (fast drying) for $h = 0.006$ and $t_d = 6620.32$ (slow drying) for $h = 0.002$ respectively. In the both cases the final radius of the drop is $R = 50.0$ which is greater than the values of ξ for the both cases.

3. Results

In our PFC simulations we choose $a = 0.008, b = 0.004, c = 1.0$ and $h = 0.006, 0.002$. Corresponding to these values, the dimensionless free energy $\mathcal{F}_1$ has minima at $\phi_1 = 0.651$, $\phi_2 = -1.15$ (for $h = 0.006$) and $\phi_1 = 0.930$, $\phi_2 = -1.06$ (for $h = 0.002$). The interfacial width, $\xi = 31.77783$, for $h = 0.006$ and $\xi = 44.65193$, for $h = 0.002$. A system of size $512\Delta x \times 512\Delta x$ where $\Delta x = 1.0$ is chosen with periodic boundary conditions. Here we do not explicitly solve the dynamics of the solvent rather we use the solution given in Eq.10, which is valid for $\frac{\xi}{R} \ll 1$. At time $t = 0$ we take the radius of the drop $R_0 = 200$ and $\Gamma = 1.0$..

Initially $\psi(\mathbf{r}, t = 0)$ is chosen randomly (because initially the solute is in disordered state) keeping its average density constant. The quantitiy $[\epsilon+(1+\nabla^2)^2]\psi$ in Eq.5, (with $\epsilon = -0.375$), is calculated using fast fourier transformation (FFT) and Eq.5 is solved using Euler discretization scheme with $\gamma = 10.0$ in Eq.6. For $\gamma = 10.0$, $\mu(\phi_1) = 1.0$ and $\mu(\phi_2) = 0$.

We show a typical configurations for ψ in Fig.2. While Fig.2(a) shows a striped pattern, ψ shows ordered and liquid-like final configurations in Fig.2(b) and (c). Here the initial average value of ψ is largest for Fig.2(a) but is smaller in the subsequent figures (b) and (c) with $\psi_{(a)} > \psi_{(b)} > \psi_{(c)}$. In all these cases total drying time (t_d) i.e. drying rate , is taken to be same. This implies that for a fixed drying rate, for small as well as for large initial densities of solute, the typical final configurations are not in crystalline phase. For small initial densities, they are in liquid-like disordered phase and for large initial densities they are in stripe phase.

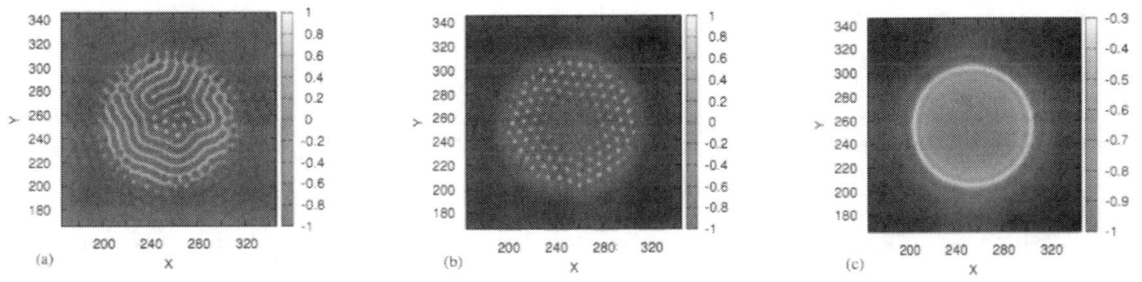

Figure 2. (a),(b) and (c) are the final configuration of $\psi(\mathbf{r},t)$ for three different initial, average solute densities which correspond to $\psi_{(a)} = -0.7$, $\psi_{(b)} = -1.0$ and $\psi_{(c)} = -1.2$ respectively showing the striped phase, crystal and disordered liquid configurations respectively. Total drying time is 1568.15 (For $h = 0.006$) for all three cases.

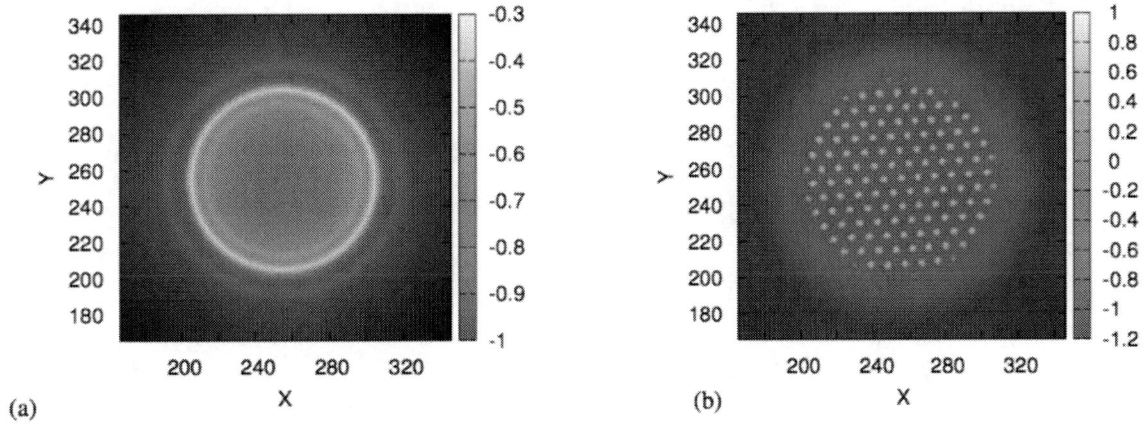

Figure 3. (a) and (b) are the final configuration of $\psi(\mathbf{r},t)$ for $\psi_{av} = -1.2$ when total drying time $t_d = 1568.15$ (fast drying) for $h = 0.006$ and $t_d = 6620.32$ (slow drying) for $h = 0.002$ respectively. In (a) we get a typical liquid-like disordered configuration and in (b) a typical ordered (crystalline) configuration finally.

In intermediate initial densities we get crystalline phase. We also show a typical liquid-like disordered configuration for fast drying (i.e. $\tau_1 < \tau_2$) and a typical ordered (triangular phase) configuration for slow (i.e. $\tau_1 \geq \tau_2$) drying in Fig.3 (a) and (b) respectively. In both the cases the initial solute density is taken to be same and below the critical density. This implies that for slow drying, as was expected, we get crystalline order and for fast drying, we get liquid-like disordered structure. We also show how the radius of the droplet typically shrinks during evaporation in Fig.1 for fast as well as for slow drying.

4. Conclusions

In reality, drying induced pattern formation in colloidal drops is a very complex phenomena due to the interplay of various forces involved. One of the essential feature of the problem is dragging of the solute particles together by the solvent during evaporation in depinned state. This originates due to the surface tension between solute and the solvent. Here we have tried to capture this effect by simple continuum models. According to the model described here, evaporation-assisted ordering depends on the drying rate as well as on the initial concentration of the solute. There is a critical density at which solute particles can be organised in space to form ordered arrays. For slow drying, the solute can reach to the critical density by the end of the evaporation and form the arrays. For fast drying, it may happen that before it reaches to the critical density, solvent is evaporated out completely and we have disordered structures at the end. A full description of this process together with Brownian dynamics simulations will be published elsewhere[6]

Acknowledgements. The authors acknowledge extensive discussions with D.D. Sarma. One of us (SS) acknowledges support from the DST, Govt. of India.

References

[1] Mirkin C A and Rogers J A (Eds.) 2001 *MRS Bull.* **26** 506
[2] Xia Y, Gates B, Yin Y and Lu Y 2000 *Adv.Mater.* **12** 693-713
[3] Rabani E, Reichman D R, Geissler P L and Brus L E 2003 *Nature* **426** 271-274
[4] Brezesinski T, Groenewolt M, Gibaud A, Pinna N, Antonietti M and Smarsly B 2006 *Adv. Mater.* **18** 2260-63
[5] Rakers S, Chi L F and Fuchs H 1997 *Langmuir* **13** 7121-24
[6] Mukherjee S, Saha A, Santra P K, Sengupta S and Sarma D D 2011 *arXiv*: 1106.4716 (preprint).
[7] Bhardwaj R, Fang X and Attinger D 2009 *New J. Phys.* **11** 075020
[8] Deegan R D, Bakajin O, Dupont T F, Huber G, Nagel S R and Witten T A 1997 *Nature* **389** 827-29
[9] Yan Q, Gao L, Sharma V, Chiang Y and Wong C C 2008 *Langmuir* **24** 11518-22
[10] Hu H and Larson R G 2002 *J. Phys. Chem. B* **106** 1334-44
[11] Denkov N D, Velev O D, Kralchevsky P A, Ivanov I B, Yoshimura H and Nagayama K 1992 *Langmuir* **8** 3183-90
[12] Soolaman D M and Yu H 2005 *J. Phys. Chem. B* **109** 17967-73
[13] Chaikin P M and Lubensky T C 1995 *Principles of Condensed Matter Physics* (New York: Cambridge University Press)
[14] Elder K R and Grant M 2004 *Phys. Rev. E* **70** 051605
[15] Elder K R, Katakowski M, Haataja M, and Grant M 2002 *Phys. Rev. Lett.* **88** 245701.
[16] Elder K R, Provatas N, Berry J, Stefanovic P and Grant M 2007 *Phys. Rev. B* **75** 064107.

Continuum Models and Discrete Systems Symposia (CMDS-12) IOP Publishing
Journal of Physics: Conference Series **319** (2011) 012017 doi:10.1088/1742-6596/319/1/012017

Renormalization Group as a Probe for Dynamical Systems

Amartya Sarkar
S. N. Bose National Centre for Basic Sciences, Kolkata-700098, India
E-mail: amarta345@bose.res.in

J. K. Bhattacharjee
S. N. Bose National Center for Basic Sciences, Kolkata-700098, India
E-mail: jkb@bose.res.in

Abstract. The use of renormalization group (RG) in the analysis of nonlinear dynamical problems has been pioneered by Goldenfeld and co-workers [1]. We show that perturbative renormalization group theory of Chen *et al* can be used as an effective tool for asymptotic analysis for various nonlinear dynamical oscillators. Based on our studies [2] done on two-dimensional autonomous systems, as well as forced non-autonomous systems, we propose a unified methodology — that uses renormalization group theory — for finding out existence of periodic solutions in a plethora of nonlinear dynamical systems appearing across disciplines. The technique will be shown to have a non-trivial ability of classifying the solutions into limit cycles and periodic orbits surrounding a center. Moreover, the methodology has a definite advantage over linear stability analysis in analyzing centers.

1. Introduction

In the early 1990s Chen, Goldenfeld and Oono proposed a new method based on renormalization group (RG) [1, 3] to analyze nonlinear dynamical problems. Their method was able to extract the asymptotic behaviour of solutions to differential equations. In quantum field theory (QFT), it's quite well known that RG equations [4] have the peculiar capability of improving the "global" nature of functions obtained in the perturbation theory (which is essentially "local"). The RG flow equations may be interpreted as the following: any physical quantity $\mathcal{Q}(\alpha, \beta, \mu)$ should have no dependence whatsoever, on the renormalization point μ which is arbitrarily chosen and is absent from the original Lagrangian of the problem, i.e.

$$\frac{d\mathcal{Q}}{d\mu} = 0. \tag{1}$$

The concept of such a floating renormalization point was first introduced in a classic paper by Gell-Mann and Low [5]. Chen et al relying heavily on analogies and ideas drawn from RG method for QFT, formulated their own brand of RG which can be used effectively to analyze dynamical systems. They showed how RG can predict the asymptotic behaviour of solutions to various nonlinear differential equations and in the process demonstrated that RG is equivalent to various singular perturbation techniques used in nonlinear dynamics like the multiple scale analysis, boundary layer technique, asymptotic matching, WKB method etc. However the chief advantage of RG over traditional perturbative techniques lies in the fact that it does not require any *ad hoc* assumptions, rather it simply uses a naive perturbative expansion. The mathematical reasons as to why the RG method of Chen et al works has been discussed by Kunihiro [6]. The study of nonlinear differential equations in a two dimensional dynamical system is of considerable interest to researchers across disciplines. And since RG method of Goldenfeld *et al* extracts the asymptotic behavior of solutions to differential equations, makes it particularly handy while analyzing periodic solutions to nonlinear dynamical problems which will be the focus of this study.

2. The Methodology

How does one apply the RG principle to a problem in dynamics? We begin by observing the fact that any periodic solution in 2-dimensions can be expressed as a Fourier series with amplitude A and phase of the lowest harmonic θ determining the amplitude and phase of the higher order ones. The amplitude and phase are the quantities that will feature in the renormalization *flow* equations. The renormalization procedure takes advantage of the fact that for a given phase path (see Fig. 1), the initial condition can be taken at any point, τ, along the path without affecting the final outcome at t i.e. $x(t)$ remains the same and independent of τ. This condition ultimately leads to the RG flow equations. In this respect it is akin to the Bogoliubov-Krylov method [7]. But as mentioned earlier, the advantage is that RG uses a naive perturbation theory;

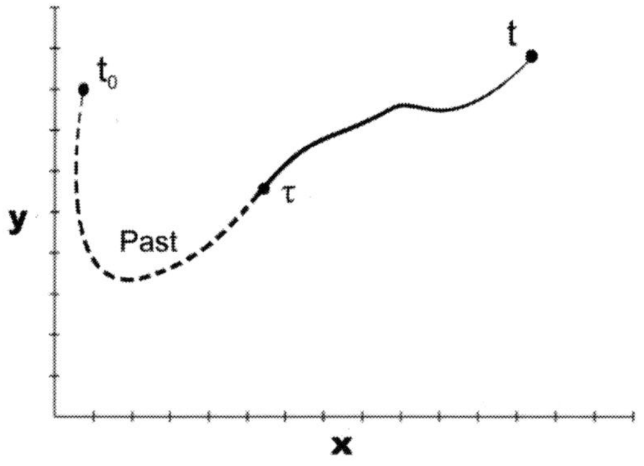

Figure 1. The initial condition can be at any point on the path

and we do not need to anticipate scales (as in multiple scales method) or make an assumption about slowly varying amplitudes and phases (Bogoliubov-Krylov).

Suppose we have an ordinary differential equation of the form: $\ddot{x} + \omega^2 x = \epsilon F(x, \dot{x})$. If one attempts to solve the equation using an expansion, $x(t) = x_0 + \epsilon x_1 + \epsilon^2 x_2 + \cdots$, it results in breakdown of the perturbation theory at times t such that $\epsilon(t - t_0) > \mathcal{O}(1)$ due to the presence of secular terms. Thus a naive expansion of the dynamical variable leads, in the asymptotic limit, to an unphysical answer. If t be the time at which we want to know $x(t)$ and t_0 the initial time, then the perturbative answer $x(t)$ diverges when $t - t_0 \to \infty$. This is completely similar to divergence in field theories where a physical quantity (e.g. two point correlation function) diverges as the renormalization cutoff $\Lambda \to \infty$. If we are dealing with physical variables, we must always have quantities that are finite and while this is achieved in field theory by constructing running coupling constants, it is done for the differential equation by introducing an arbitrary time scale τ and letting the amplitude and phase depend on τ. To regularize the perturbation series, RG technique uses the arbitrary time τ to split $t - t_0$ as $(t - \tau) + (\tau - t_0)$ and absorbing the terms containing $\tau - t_0$ into the respective *renormalized* counterparts A and θ of A_0 and θ_0. At the end of the process one arrives at the RG-flow-equations for A and θ:

$$\frac{dA}{d\tau} = f(A, \theta) \tag{2}$$

$$\frac{d\theta}{d\tau} = g(A, \theta) \tag{3}$$

For autonomous systems, f and g are generally function of A alone. We propose to use flow equations (2) and (3) to differentiate between oscillators which are of the *center* variety and *limit cycles*. The center type oscillation consists of a continuous family of closed orbits in phase space, each orbit being determined by its own initial condition. This implies that the amplitude A gets fixed, once the initial condition is set. This must lead to,

$$\frac{dA}{d\tau} = 0. \tag{4}$$

This statement is exact and is not tied to any perturbation theory argument. On the other hand for the limit cycle ,

$$\frac{dA}{d\tau} = f(A), \tag{5}$$

such that the flow equation (5) has a fixed point. The fixed point corresponding to where $f(A^*) = 0$ $(A^* \neq 0)$, has to be stable for the limit cycle to be stable and A^* gives the radius of the limit cycle. And if $A^* = 0$ is a fixed point of equation (5), then we have either a focus or a node.

This simple prescription, though not proved rigorously, appeals to one's intuition when one notes that (i) $A = 0$ means the *assumed* periodic solution has zero amplitude and hence hints at an attractor (a node or a focus), (ii) $f^{(A)} = 0 \, \forall \, A \geq 0$ hints at a family of non-isolated periodic orbits surrounding the fixed point and so a center solution is implied, and (iii) vanishing of $dA/d\tau$ at $A = A^* \neq 0$ logically indicates that an isolated periodic orbit of amplitude A^* happens to be surrounding the fixed point.

The calculation of $f(A)$ requires the use of perturbation theory. Application of perturbation theory is possible only if one can locate a linear center (the basic periodic state) about which to perturb. Locating a center can sometimes be straightforward, for example: $\dot{x}_1 = x_2$, $\dot{x}_2 = -\partial V/\partial x_1$, where V is a general anharmonic potential, $V = x_1^2/2 + \lambda_1 x_1^3/3 + \lambda_1 x_1^4/4$, where $(x_1, x_2) = (0, 0)$ is a linear center around which perturbation theory can be done; similarly for the Van der Pol oscillator $\ddot{x} + \epsilon\dot{x}(x^2 - 1) + x = 0$, the origin($x = 0$, $\dot{x} = 0$) is a center for $\epsilon = 0$. For the Lotka-Volterra population dynamics model — $\dot{x}_1 = x_1 - x_1 x_2$, $\dot{x}_2 = -x_2 + x_1 x_2$ the origin is a saddle but the other fixed point, $(1, 1)$, is a center. In such cases shifting the origin to the center is the first step in the process of determining $f(A)$.

A more complicated situation can arise in few cases such as the Belushov-Zhabotinsky reaction [8, 9] system (discussed later). In that case, a transfer of origin to the fixed point has to be followed by a proper setting of parameters to make the origin a center which is the starting point of our perturbative method. This raises the problem that the given dynamical system may not have a relevant parameter at all, e.g. the well known paradigm for a limit cycle,

$$\dot{z} = (1 + i)z - \beta |z|^2 z, \tag{6}$$

where $z = x + iy$ is the complex variable and $\beta > 0$. The only fixed point is the origin and it is an unstable focus for all β. However, we can overcome this difficulty by considering the more general system

$$\dot{z} = (\alpha_1 + i\alpha_2)z - \beta|z|^2 z. \tag{7}$$

The origin is now a stable focus for $\alpha_1 < 0$, unstable focus for $\alpha_1 > 0$ and a center for $\alpha_1 = 0$. It is this center about which one can set up a perturbation theory. The perturbative determination of $f(A)$ and $g(A)$ thus involves the following steps:

(i) Find all the fixed points of the system and identify the linear centers.

(ii) If there are no linear centers, look to extend the parameter space and see if a linear center can be located as the parameters are changed.

(iii) For every linear center, thus located, we need to check the existence of a limit cycle by perturbatively constructing $f(A)$ and $g(A)$.

(iv) If linear center is absent even after extension of the parameter space this methodology can't handle the problem.

3. Examples

As a test case let us consider a general 2D-dynamical equation represented by,

$$\ddot{x} + x = -k\dot{x} - \epsilon f(x, \dot{x}) + F\cos\Omega t, \tag{8}$$

where $f(x, \dot{x})$ is a nonlinear function of x and $\dot{x}$. For simplicity we will first consider only the autonomous cases, i.e. where $F = 0$. We will deal with non-autonomous case later on.

3.1. Duffing Oscillator

In order to elucidate the RG methodology we take a fairly simple and well studied nonlinear oscillator, the Duffing oscillator ($f(x, \dot{x}) = x^3$)), as our starting point and do the RG calculation in detail. The equation of motion of this damped nonlinear oscillator is given by,

$$\ddot{x} + k\dot{x} + \omega^2 x + \lambda x^3 = 0. \tag{9}$$

We immediately see that a linear center exists for $k = \lambda = 0$ and around this limit we can do our perturbative calculation. We naively expand x as $x = x_0 + kx_1' + \lambda x_1 + k^2 x_2' + \lambda^2 x_2 + k\lambda x_2'' + \cdots$. Then substituting this into Eq.(9), we obtain the following equations at different orders of λ and k:

$$\ddot{x}_0 + \omega^2 x_0 = 0, \tag{10}$$
$$\ddot{x}_1 + \omega^2 x_1 = -x_0^3, \tag{11}$$
$$\ddot{x}'_1 + \omega^2 x_1' = -\dot{x}_0. \tag{12}$$

We set the initial condition as $x(t = 0) = A_0$ and $\dot{x}(t = 0) = 0$. We then write the solution of Eq.(10) as $x_0 = A_0 \cos\omega t$, so that x_0 picks up the initial condition, $x_0(t = 0) = A_0$. Hence for the subsequent orders, the initial condition becomes $x_i(t = 0) = \dot{x}_i(t = 0) = 0$ for all $i \geq 1$. Using the zeroth order solution we rewrite Eq.(11) and Eq.(12) as:

$$\ddot{x}_1 + \omega^2 x_1 = -\frac{A_0^3}{4}\left(\cos 3\omega t + 3\cos\omega t\right), \tag{13}$$
$$\ddot{x}'_1 + \omega^2 x_1' = \omega A_0 \sin\omega t. \tag{14}$$

Keeping in mind the initial conditions, we can immediately write the solutions to the above equations,

$$x_1 = -\frac{3A_0^3}{8\omega} t \sin \omega t + \frac{A_0^3}{32\omega^2} (\cos 3\omega t - \cos \omega t), \tag{15}$$

$$x_1' = -\frac{A_0}{2} t \cos \omega t + \frac{A_0}{2\omega} \sin \omega t. \tag{16}$$

Upto this order, the displacement of the oscillator can be written as,

$$\begin{aligned} x(t) &= A_0 \cos \omega t - \frac{3\lambda A_0^3}{8\omega}(t - \tau + \tau) \sin \omega t + \frac{\lambda A_0^3}{32\omega^2}(\cos 3\omega t - \cos \omega t) \\ & -\frac{kA_0}{2}(t - \tau + \tau) \cos \omega t + \frac{kA_0}{2\omega} \sin \omega t. \end{aligned} \tag{17}$$

Here we have split the interval 0 to t as 0 to τ and τ to t in accordance with our prescription for renormalization. In order to remove the divergences, we introduce two renormalization constants, $\mathcal{Z}_1(0, \tau)$ and $\mathcal{Z}_2(0, \tau)$, in the following manner:

$$A_0 = A_0(t_0 = 0) = A(\tau)\mathcal{Z}_1(0, \tau), \tag{18}$$

$$0 = \theta_0(t_0 = 0) = \theta(\tau) + \mathcal{Z}_2(0, \tau). \tag{19}$$

The renormalization constants have the expansions,

$$\mathcal{Z}_1(0, \tau) = 1 + a_1\lambda + a_1' k + \cdots, \tag{20}$$

$$\mathcal{Z}_2(0, \tau) = b_1\lambda + b_1' k + \cdots, \tag{21}$$

so that the constants a_i and b_i can be chosen order by order to remove divergences arising at each order. One needs to note here that the presence of nonlinear perturbations means that A_0 and θ_0 do no longer remain constants of motion. We now absorb the $\tau - t_0(= 0)$ containing terms into the renormalized counterparts A and θ of A_0 and θ_0 so that both are now functions of τ. We can now rewrite Eq.(17) as

$$\begin{aligned} x(t) &= A\cos(\omega t + \theta) + (a_1\lambda + a_1' k)A\cos(\omega t + \theta) - (b_1\lambda + b_1' k)A\sin(\omega t + \theta) \\ &- \frac{3\lambda A^3}{8\omega}(t - \tau + \tau)\sin(\omega t + \theta) + \frac{\lambda A^3}{32\omega^2}\Big(\cos 3(\omega t + \theta) - \cos(\omega t + \theta)\Big) \\ &- \frac{kA}{2}(t - \tau + \tau)\cos(\omega t + \theta) + \frac{kA}{2\omega}\sin(\omega t + \theta). \end{aligned} \tag{22}$$

Then choosing $a_1' = kA\tau/2$, $a_1 = 0$, $b_1' = 0$ and $b_1 = -3\lambda\tau/8\omega$, one gets rid of the $\tau - 0$ containing terms and we write Eq.(22) as

$$\begin{aligned} x(t, \tau) &= A\cos(\omega t + \theta) - \frac{3\lambda A^3}{8\omega}(t - \tau)\sin(\omega t + \theta) + \frac{\lambda A^3}{32\omega^2}\Big(\cos 3(\omega t + \theta) - \cos(\omega t + \theta)\Big) \\ & -\frac{kA}{2}(t - \tau)\cos(\omega t + \theta) + \frac{kA}{2\omega}\sin(\omega t + \theta). \end{aligned} \tag{23}$$

Since τ does not appear in the original problem it should not be in the final solution. Therefore we impose the condition that the solution has to be independent of τ i.e. $(\partial x/\partial \tau)_t = 0$, for any t and this yields (to the lowest order) the flow equations:

$$\frac{dA}{d\tau} = -\frac{kA}{2}, \tag{24}$$

$$\frac{d\theta}{d\tau} = \frac{3\lambda A^2}{8\omega}, \tag{25}$$

which integrate to $A(t) = A_0 e^{-k\tau/2}$ and $\theta = \theta_0 + \frac{3\lambda A^2}{8\omega}\tau$. Finally we set $\tau = t$ to get rid of remaining τ–dependence, to get

$$x(t) = A\cos[\Omega t + \theta_0] + \frac{\lambda A^3}{32\Omega^2}\Big(\cos 3(\Omega t + \theta_0) - \cos(\Omega t + \theta_0)\Big) + \frac{kA}{2\Omega}\sin(\Omega t + \theta_0), \tag{26}$$

where $\Omega = \omega + 3\lambda A^2/8\omega$.
When there is no damping term, i.e. $k = 0$, we have the conservative anharmonic oscillator,

$$\ddot{x} + \omega^2 x + \lambda x^3 = 0, \tag{27}$$

for which the fixed point $(0, 0)$ in the $x - \dot{x}$ plane (*i.e.* $x - y$ plane) is a center. And as expected the flow equation in this case reduces to,

$$\frac{dA}{d\tau} = 0, \tag{28}$$

meaning we have a center. The periodic solution correct to first order in λ is given by

$$x(t) = A\cos\Omega t + \frac{\lambda A^3}{32\Omega^2}[\cos 3\Omega t - \cos\Omega t] + \mathcal{O}(\lambda^2) \quad , \tag{29}$$

where Ω is the corrected frequency given by the expression:

$$\Omega = \omega + \frac{3\lambda A^2}{8\omega} + \mathcal{O}(\lambda^2). \tag{30}$$

The standard results [10] for the oscillator have, thus, been correctly captured and we find that the emergence of $x = \dot{x} = 0$ as a center is confirmed by the fact that $dA/d\tau = 0$.

3.2. Van der Pol Oscillator

As our second example we turn to the well known paradigm for limit cycle — the Van der Pol oscillator i.e. $f(x, \dot{x}) = \epsilon\dot{x}\left(x^2 - 1\right)$ and $F = 0$ in Eq. (8). Thus the equation of motion is,

$$\ddot{x} + \epsilon\dot{x}\left(x^2 - 1\right) + \omega^2 x = 0. \tag{31}$$

We look at it as a second order dynamical system: $\dot{x} = y$, $\dot{y} = -\epsilon y(x^2 - 1) - \omega^2 x$, in order to identify the fixed points. Clearly the only fixed point of the system is at the origin which is a stable focus for $\epsilon < 0$ and unstable focus for $\epsilon > 0$. For $\epsilon = 0$ Eq. (31) is simply the well known equation for a simple harmonic oscillator and thus we have located a center about which we can base the perturbation expansion around. Expanding x as $x(t) = x_0(t) + \epsilon x_1(t) + \epsilon^2 x_2(t) + \cdots$, we obtain at different orders of ϵ,

$$\ddot{x}_0 + \omega^2 x_0 = 0, \tag{32}$$
$$\ddot{x}_1 + \omega^2 x_1 = -\dot{x}_0(x_0^2 - 1). \tag{33}$$

We work with initial condition $x = A_0$ at $t = 0$ and $\dot{x} = 0$ at $t = 0$. Proceeding in an identical manner as we did for the Duffing oscillator we can arrive at the following flow equations:

$$\frac{dA}{d\tau} = \frac{\epsilon A}{2}\left(1 - \frac{A^2}{4}\right) + \mathcal{O}(\epsilon^2), \tag{34}$$
$$\frac{d\theta}{d\tau} = 0 + \mathcal{O}(\epsilon^2). \tag{35}$$

Journal of Physics: Conference Series **319** (2011) 012017 doi:10.1088/1742-6596/319/1/012017

The flow equations obtained are in accordance with what our prescription suggests has to be the structure of the RG flow in case of a limit cycle solution. One can immediately see that the flow has a stable fixed point at $A^2 = 4$ which in fact is the usual answer for Van der Pol limit cycle for small ϵ. The solution upto this order of perturbation is given by

$$x(t) = A\cos(\omega t + \theta) - \frac{\epsilon A^3}{96\omega^2}\Big(\sin 3(\omega t + \theta) - 3\sin(\omega t + \theta)\Big) + \mathcal{O}(\epsilon^2). \tag{36}$$

Its well known that in order to obtain the correct results for the Van der Pol oscillator one must anticipate the hidden multiple scale in the problem and include it in the perturbation expansion. But RG starts with a simple expansion and the hidden multiple scales are automatically identified. To illustrate this point we need to calculate the flow equations upto the next higher order. The resultant flow equation turns out to be,

$$\frac{dA}{d\tau} = \frac{\epsilon A}{2}\left(1 - \frac{A^2}{4}\right) + \mathcal{O}(\epsilon^3), \tag{37}$$

$$\frac{d\theta}{d\tau} = -\frac{\epsilon^2}{8}\left(1 - \frac{A^4}{32}\right) + \mathcal{O}(\epsilon^3). \tag{38}$$

One can immediately see that the usual multiple scales $T_1 = \epsilon t$, $T_2 = \epsilon^2 t$ and so on, used in the analysis of Van der Pol oscillator appear naturally in the flow equations.

3.3. Belushov-Zhabotinsky reaction

An similar approach is effective for the so called Belushov-Zhabotinsky reaction system (chlorine dioxide-iodine-malonic acid reaction). A relatively recent two variable model [8, 9] based on the fact that the concentrations of the intermediaries $I^-(x)$ and $ClO_2^-(y)$ vary on a much faster time scale than ClO_2, I_2 and $Malonic\ acid$ is given by,

$$\dot{x} = a - x - \frac{4xy}{1+x^2}, \tag{39}$$

$$\dot{y} = bx\left(1 - \frac{y}{1+x^2}\right). \tag{40}$$

The constants 'a' and 'b' are parameters which depend on the rate constants and the approximately constant concentrations of the other reactants. First we note that there is only one fixed point $x = a/5$ and $y = 1 + x^2 = 1 + a^2/25$. And thus our first step is to shift the origin to $(a/5, 1 + a^2/25)$ i.e. use the shifted variables $X(= x + a/5)$, $Y(= y + 1 + a^2/25)$. A linear stability analysis of the resulting system about the fixed point $X = Y = 0$ reveals that it's a center for some $b^* = 3a/5 - 25/a$. Further that the origin is an unstable focus for $b < b^*$ and stable for $b > b^*$. As discussed we need to fix the parameters to proceed. So we pick a value of 'a' and choose $b = b^* - \delta$, where $\delta \ll b^*$. One carries out a perturbation analysis for the variables X and Y by assuming that the amplitude is small for small δ. The amplitude flow works out to be

$$\frac{dA}{d\tau} = -\frac{a}{5}\delta\Omega A + \frac{\Omega A^3}{\left(1+\frac{a^2}{25}\right)^2}\left[\frac{3a^4}{125} - 3a^2 - 315 + \frac{1875}{a^2}\right], \tag{41}$$

where $\Omega^2 = a\left(1 + \frac{a^2}{25}\right)\left(\frac{3a}{5} - \frac{25}{a}\right)$. From the flow equation we can conclude that a limit cycle exists for positive values of δ. It is apparent that as we measure the value of 'a' for which limit cycles can exist, there is a cyclic-fold bifurcation at $a = a_c \simeq \sqrt{191.43}$ — obtained by setting the expression inside square bracket to zero.

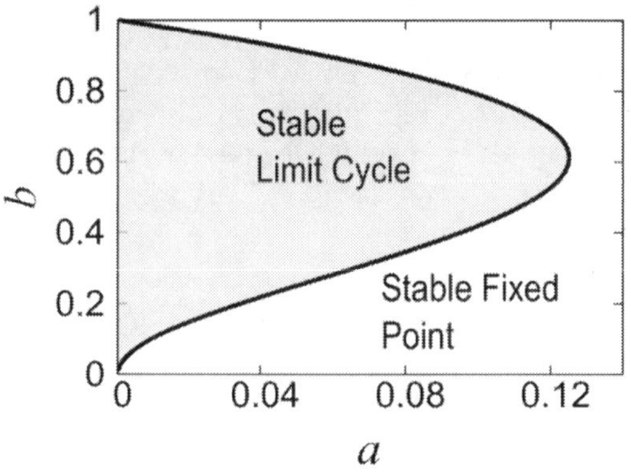

Figure 2. The figure shows the phase diagram of the glycolytic oscillator given by Eqs. (42) and (43). The curve (solid line) separates the figure into two regions. For parameters in the shaded region, one gets limit cycle while unshaded region corresponds to parameters giving rise to a stable focus. Linear stability analysis predicts center for parameters on the curve.

3.4. Glycolytic oscillator

We now turn to a rather interesting example which clearly illustrates the usefulness of shifting of origin. Further it requires a determination of the locus of Hopf bifurcation points, to set up the perturbation theory, in order to locate the limit cycle. This example is drawn from biology – the glycolytic oscillator. Selkov [11] gave a simple mathematical model describing this oscillator as a 2-dimensional dynamical system. The variable x is the concentration of ADP (adenosine diphosphate) and y that of F6P (fructose-6-phosphate). The dynamics is given by

$$\dot{x} = -x + (a + x^2)y, \tag{42}$$
$$\dot{y} = b - (a + x^2)y, \tag{43}$$

where 'b' is the rate of fructose production and 'a' the rate at which fructose decomposes (converts to ADP). It should be noted here that the presence of ADP catalyzes this conversion and hence 'a' is augmented to '$a + x^2$'. The only fixed point of the system is at $x = b$, $y = b/(a + b^2)$. It turns out to be a stable focus for a certain parameter range and an unstable focus for certain others. The crossover from stable to unstable focus occurs on a curve which is a locus of points in the a-b plane where a 'Hopf bifurcation' occurs i.e. the fixed point for those values of (a, b) is a center. The curve is given by $2a = \sqrt{1 + 8b^2} - (1 + 2b^2)$ and is shown in Fig. 2. We shift the origin to the fixed point and use the new coordinates X, Y given by $X = b - x$ and $Y = \frac{b}{a+b^2} - y$. To use perturbation theory, we chose (a, b) close to the boundary. Setting $b = \sqrt{3/8}$ (the turning point of the curve), we take $a = 1/8 - \delta$ to consider a point inside the boundary but close to it. Clearly, δ is small and positive. To $\mathcal{O}(\delta)$, the equation of motion reads

$$\dot{X} = \frac{1}{2}(X + Y) + \mathcal{N}(X, Y), \tag{44}$$
$$\dot{Y} = -\frac{3}{2}X - \frac{Y}{2} - \mathcal{N}(X, Y), \tag{45}$$

where

$$\mathcal{N}(X, Y) = \delta(3X - Y) + \sqrt{\frac{3}{8}}X(X + Y) + X^2Y. \tag{46}$$

$\mathcal{N}$ has to be expanded in amplitude and the parameter δ. On solving the above equation we

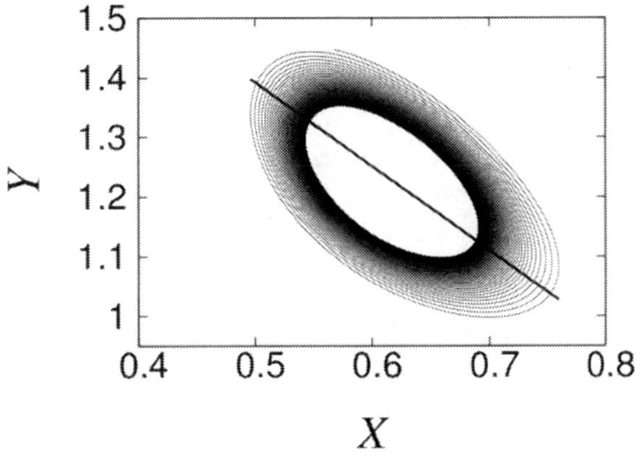

Figure 3. Limit cycle in glycolytic oscillator for $a = 0.124$, $b = \sqrt{0.375}$ and $\delta = 0.001$.

find the amplitude flow,

$$\frac{dA}{d\tau} = 2\delta A - \frac{3A^3}{8} + \mathcal{O}(\delta^2). \tag{47}$$

Thus the amplitude emerges to go as $\delta^{1/2}$ for small δ. The frequency changes from the zeroth order value of $\frac{1}{\sqrt{2}}$ according to the flow

$$\frac{d\theta}{d\tau} = -\frac{\delta}{\sqrt{2}} + \frac{A^2}{4\sqrt{2}} + \mathcal{O}(\delta^2). \tag{48}$$

The stable fixed point $A^2 = 16\delta/3$ gives us the size of the limit cycle for $\delta \ll 1$. A typical small-δ orbit obtained numerically, is shown in Fig. 3 which bears out the correctness of the above flow.

3.5. Advantage over linear stability analysis

Lets reconsider the glycolytic oscillator given by eqs. (42) and (43). As discussed earlier according to linear stability analysis we have a loci of Hopf bifurcation points on the solid curve in Fig. 2. So for parameter values corresponding to the ones on the curve (e.g. $a = 1/8$, $b = \sqrt{3/8}$), we are supposed to have a center (periodic solution). But a simple numerical check establishes otherwise; it turns out for parameter values on the curve the solution is actually a *focus*. It is well known that linear stability analysis can predict wrongly the existence of a center [12] when the fixed point in question is actually a focus. However our RG method can clearly distinguish between an attractor and a center. If one looks at the amplitude flow equation (47), we notice for $\delta = 0$ it reduces to,

$$\frac{dA}{d\tau} = -\frac{3A^3}{8}. \tag{49}$$

Which according to our prescription suggests that the fixed point in question is an stable attractor and not an center. So we see that the linearized version of a nonlinear dynamical system may not reproduce qualitatively correct picture of the phase portrait near a fixed point because due to the linearization, the fixed point is shielded from full bombardment of the non-linear terms. On the other hand RG manages to capture the true nature of the fixed point.

3.6. Non-autonomous system

The RG technique can also be used to deal with non-autonomous systems. A very simple example ($f(x,\dot{x}) = 0$ and $F \neq 0$ in Eq.(8)) will suffice to show how. We consider a damped driven oscillator given by,

$$\ddot{x} + \omega^2 x + k\dot{x} = F \cos \Omega t, \tag{50}$$

which we rewrite as,

$$\ddot{x} + \Omega^2 x = -k\dot{x} + F \cos \Omega t + (\Omega^2 - \omega^2)x. \tag{51}$$

We treat k, F and $\Omega^2 - \omega^2$ as small to perturb about a center ($k = F = \Omega^2 - \omega^2 = 0$). Proceeding as explained earlier, to the first order in all these small parameters, we obtain:

$$\frac{dA}{d\tau} = -\frac{kA}{2} - \frac{F \sin \Theta}{2\Omega} \tag{52}$$

$$\frac{d\Theta}{d\tau} = -\frac{F \cos \Theta}{2\Omega A} + \Delta\omega, \tag{53}$$

where $\Delta\omega \equiv \omega - \Omega$. Since we are dealing with a forced oscillator, i.e. Ω is maintained externally, it cannot change, implying $d\Theta/d\tau = 0$. Also, existence of a periodic solution requires that $dA/d\tau = 0$. Therefore, setting both Eqs. (52) and (53) to zero and solving we find that the fixed point corresponds to the amplitude $A = f/[k^2+4(\Delta\omega)^2]^{1/2}$ (where $f = F/\Omega$) and the phase $\theta = \tan^{-1}[-k/2(\Delta\omega)]$. This is exactly in accordance with the literature of forced oscillators. The stable non-zero fixed point in the evolution of A corresponds to a limit cycle in accordance with what we have claimed has to happen. So in case of non-autonomous systems one needs to set both the flow equations to zero and solve for the fixed points in order to extract information about the periodic solution. This process can also be repeated for cases when $f(x,\dot{x}) \neq 0$ and obtain the correct results.

3.7. Oscillator without any linear terms

As we have pointed out earlier in this article, for the RG method to work one needs to find a linear center (a basic periodic state) about which to perturb. We have also showed that in a few cases if initially there isn't a linear center in the problem one can suitably expand the parameter space or introduce new parameters to create one. But now we ask the question what if the dynamical system does not have a linear part nor is it amenable for parameter expansion via which one can locate a center. For example consider the oscillator: $\ddot{x} + \lambda x^3 = 0$ or $\ddot{x} + \beta\lambda x\dot{x} + \lambda^2 x^3 = 0$. Such cases can't be handled by the prescription we have laid down here nevertheless a little tweaking of the method can handle a class of such problems. To illustrate our point lets consider the so called Riccati equation given by,

$$\ddot{x} + \beta\lambda x\dot{x} + \lambda^2 x^3 = 0. \tag{54}$$

The trick we use here is motivated from the so called "equivalent linearization" method of nonlinear dynamics. We rewrite the above equation as,

$$\ddot{x} + \beta\lambda x\dot{x} + \lambda^2\alpha\langle x^2\rangle x + \lambda^2\left(x^3 - \alpha\langle x^2\rangle x\right) = 0$$

$$\text{or} \quad \ddot{x} + \Omega^2 x = -\beta\lambda x\dot{x} - \lambda^2\left(x^3 - \alpha\langle x^2\rangle x\right) \tag{55}$$

where $\langle x^2\rangle$ is the average of $x(t)^2$ over a cycle. In the above equation α is a constant number which we need to find out. If we ignore the term in parenthesis, we have an equivalent linear oscillator with frequency

$$\Omega^2 = \alpha\lambda\langle x^2\rangle. \tag{56}$$

So now we have a linear system about which to perturb and we can carry out the normal RG calculation. A detailed account of the problem and calculations can be found in the reference [13]. The RG calculation yields the flow equations,

$$\frac{dA}{d\tau} = 0, \tag{57}$$

$$\frac{d\theta}{d\tau} = \lambda\left(\frac{3}{4} - \frac{\alpha}{2} - \frac{\beta^2}{12}\right)\frac{A^2}{2\Omega}. \tag{58}$$

And the above flow equations correctly predict the salient features of this oscillator [13].

4. Conclusion

In concluding we must revisit the method described here and discuss some of its merits and demerits. We have shown here that the RG method can handle a wide variety of nonlinear dynamical problems in 2-dimensions. It can further differentiate between different kinds of periodic solutions, namely centers and limit cycles. Further where as linear stability analysis fails to distinguish between an attractor and a center in a few cases our RG method successfully distinguishes the two. The major advantage of RG however is the fact that it does not require any *apriori ad hoc* assumptions like the other singular perturbation techniques. It starts with a simple perturbative expansion assuming only the existence of a small parameter in the problem and is able to capture various features of the dynamics. For example the multiple scales involved in a dynamics need not be pre-identified as is done in multiple scale analysis rather the scales appear naturally in the RG calculation.

However being perturbative this method has its limitations. As we have already discussed, for a given 2-D system, application of RG depends on our ability to identify a suitable linear center (periodic state) about which to perturb. While in many cases identifying a linear center is straightforward, there are situations where one needs to employ certain tricks to do so. We have discussed here a few ways in which one can identify a linear center in the problem. But if in a problem one is not able to identify such a basic periodic state about which to perturb we can't proceed with the RG prescription. Nevertheless a wide range of dynamical problems can be handled with our prescription. And it has been our intention here to illustrate our method in a manner so that people from various disciplines may find it useful while analyzing periodic solution in two dimensions.

References

[1] Chen L Y, Goldenfeld N and Oono Y 1994, *Phys. Rev Lett.* **73**, 1311
[2] Sarkar A, Bhattacharjee J K, Chakraborty S and Banerjee D 2010 *arXiv:1005.2858v1*
[3] Chen L Y, Goldenfeld N and Oono Y 1996, *Phys. Rev. E* **54**, 376.
[4] Wilson K G 1971, *Phys. Rev. B* **4**, 3174; **4**, 3184
[5] Gell-Mann M and Low F E 1954, *Phys. Rev.* **95**, 1300
[6] Kunihiro T 1997, *Prog. of Theor. Phys.* **97**, 179
[7] Jordan D W and Smith P A 1999, *Nonlinear Ordinary Differential Equations: An Introduction to Dynamical Systems* (Oxford University Press, New York).
[8] Lengyel I and Epstein I R 1992, *Proc. Nat. Acad. Sci., USA* **89**, 3977.
[9] Lengyel I, Rabai G and Epstein I R 1990, J. Am. Chem. Soc. **112**, 9104.
[10] Bhattacharjee J K, Malik A K and Chakraborty S 2007, *Indian Jour. Phys.*, **81**, 1115
[11] Sel'kov E E 1968, *Eur. J. Biochem.* **4**, 79.
[12] Strogatz S H 1994, *Nonlinear Dynamics and Chaos: With Applications to Physics, Biology, Chemistry and Engineering*, (Westview Press, USA).
[13] Sarkar A and Bhattacharjee J K 2010, *Europhys. Lett.*, **91** 60004.

Journal of Physics: Conference Series **319** (2011) 012018 doi:10.1088/1742-6596/319/1/012018

Linear elastic trusses leading to continua with exotic mechanical interactions

Pierre Seppecher, Jean-Jacques Alibert and Francesco Dell Isola
IMATH, University of Toulon, France
E-mail: seppecher@imath.fr

Abstract. We study the statics of some trusses, i.e. networks of nodes linked by linear springs. The trusses are designed in such a way that a few number of floppy modes are present and remain when considering the homogenized limit of the truss. We then obtain linear elastic materials with exotic mechanical interactions which cannot be described in the classical framework of Cauchy stress theory. For aim of simplicity, the structures described here are two-dimensional. The extension to the 3D case does not present any difficulty.

1. Introduction

The theory of Cauchy stress is so efficient in describing mechanical interactions in usual materials that its foundations have seldom been reconsidered (see [3, 20, 19, 26, 28, 27, 37, 38, 16, 17, 21, 22, 23, 24, 25, 29, 30, 31, 32, 33, 40, 41]. Rapid progress in nanotechnologies makes now possible the design of new materials having exotic mechanical properties. One basic assumption of Cauchy stress theory is that the mechanical contact interaction between two parts of the material is only due to a surface density of contact forces concentrated on the "cut", i.e. on the dividing surface [44, 34, 35]. In this paper we describe some microscopic structures leading at the macroscopic level to materials for which this basic assumption is no longer valid.

Microscopic (atomic) structures leading to high order gradient materials have already been described [1, 6, 42, 43]. The advantage of the structures presented here is that the non classic properties are preponderant and are not limited to producing thin boundary layers.

The microscopic structures we consider are simple is the sense that they consist only of linear elastic springs joining nodes (what is usually called "trusses"). The idea is to design structures with a few number of floppy modes, i.e. global deformations which do not change the length of any spring of the structure and so do not need any energy.

In the different sections we consider structures with an increasing degree of complexity. We compute the deformation energy in terms of the displacement of a network of nodes. This energy is generally self-explanatory : writing the homogenized energy which corresponds to a very dense network of nodes is straightforward even if its rigorous mathematical derivation is not so easy [5, 7, 9, 10, 11, 12, 13] (for a rigorous derivation of the homogenization of the structure presented in section 3.1 see [2]). It is not the aim of this paper to give such a rigorous mathematical analysis, we prefer to emphasize the mechanical features of the proposed structures.

2. The flexion beam and the flexion truss

2.1. The flexion beam

Let us begin by considering the very simple reticulated structure described in figure 1. When

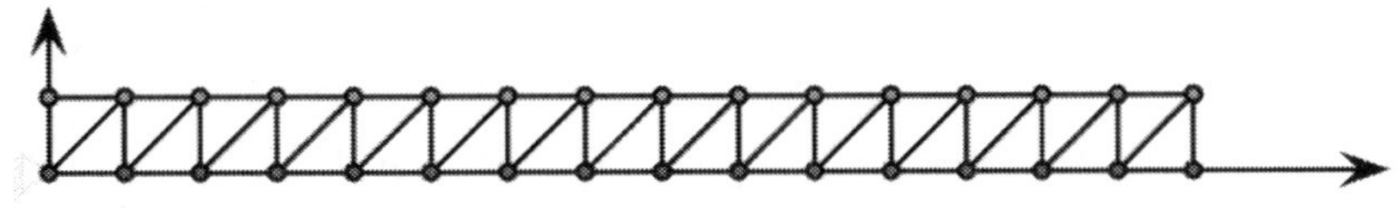

Figure 1. A beam truss.

the stiffness of elastic bars is suitably scaled, the homogenized behavior of the structure is one-dimensional and corresponds to a non-extensional flexion beam. This is not very difficult to check but the computation is much more easy when considering a slightly different structure (figure 2) in which we assume that the horizontal bars have infinite stiffness while all other bars have the same finite stiffness k.

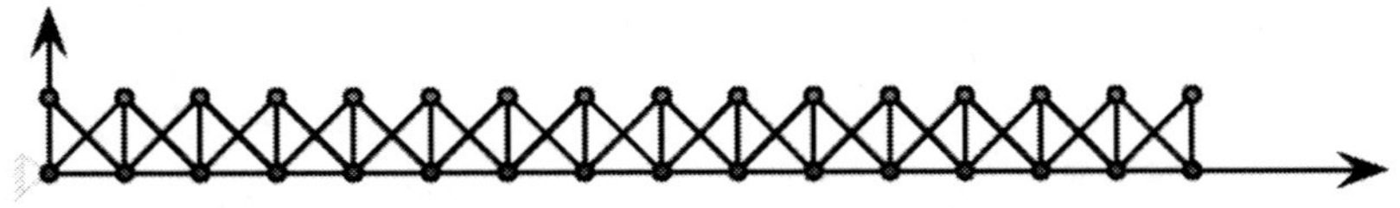

Figure 2. A alternative beam truss. The horizontal bars have infinite stiffness. All other bars have the same finite stiffness. The diagonal bars do not interact : this can be achieved either by using a 3D structure or by adding an extra node at the junction.

Indeed this alternative structure is made of a periodic superposition of basic elements described in figure 3. Thus its elastic energy E is easy to write in terms of the displacements

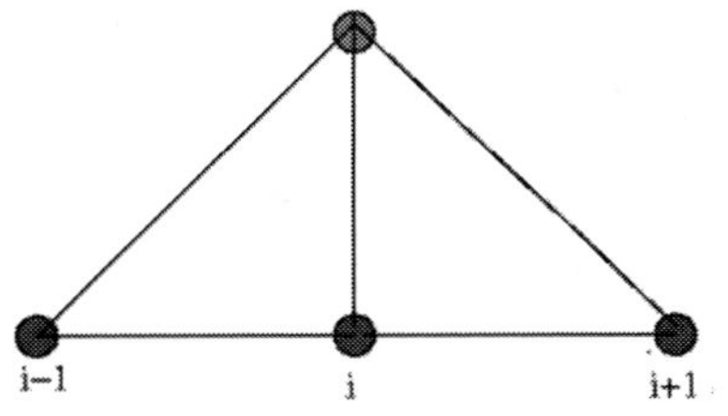

Figure 3. An element of the beam of fig.2.

of the nodes. Assuming that forces are applied on the lowest (blue) nodes only, we write the energy in terms of the displacements $u[i] = (u_1[i], u_2[i])$ $(i \in \{1, \ldots N\})$ of the blue nodes. (The N nodes are numbered from the left to the right). The longitudinal displacement of the blue nodes has to vanish and the elastic energy contained in each element depends only on the transverse displacement u_2. It reduces to $(k/16)(u_2[i-1] - 2u_2[i] + u_2[i+1])^2$. The elastic energy of the beam then reads

$$E(u) = \sum_{i=2}^{N-1} (k/16)(u_2[i-1] - 2u_2[i] + u_2[i+1])^2$$

Let us assume that the total length of the structure is ℓ and let us denote $\epsilon = \ell/(N-1)$ the distance between two consecutive blue nodes. If the stiffness of the bars scales as ϵ^{-3} : $k = 16K\epsilon^{-3}$, then

$$E(u) = \epsilon \sum_{i=2}^{N-1} K \left(\frac{u_2[i-1] - 2u_2[i] + u_2[i+1]}{\epsilon^2} \right)^2 .$$

When the beam is made of a large number of small cells, i.e. when ϵ tends to zero, then one can recognize in the expression of E the finite difference approximation for the continuous energy

$$\tilde{E}(u) = \int_0^\ell K \left(\frac{\partial^2 u_2}{\partial x_1^2} \right)^2 .$$

This is the energy of the well known linear flexion beam. Nevertheless let us make some remarks before using this structure to build a new 2D material : First, we have assumed that all bars are linear elastic bars. They have spring-like behavior or correspond to long range linear interactions : the possibility of buckling is not considered here. Secondly, external forces are exerted only on the lowest (blue) nodes, the upper (red) nodes are considered as "internal". Finally note that, even when the first node is fixed (which corresponds to a Dirichlet condition $u(0) = 0$ in the continuous model) the truss is not isostatic. A floppy mode (rotation around node 1) remains. For instance, in order to counterbalance a vertical force applied in the middle of the beam, it is necessary to apply a suitable force on the second node (which corresponds in the continuous model to a torque applied at $x = 0$).

2.2. The flexion 2D material

Now let us consider many (M) parallel beam structures similar to the one described in the previous section (each one with first node fixed) and let us link them by adding vertical bars which link each blue node with the blue node just above; the added vertical bars have stiffness k'; see figure 4. Note that, for sake of clearnesss, figure 4 shows a slightly different structure than the one we just described.

This structure clearly has a floppy mode : the deformation shown in figure 4 needs no energy as no bar is extended. It is the only floppy mode. Indeed it is easy to check that fixing an extra node makes the structure isostatic. At a first insight, one could think that we have here designed a degenerated classical material with a vanishing shear stiffness. It is easy to reject this interpretation by considering the dimension of the space of floppy modes in both cases.

The elastic energy of the whole structure is simply the sum of the elastic energies of the beams we just derived plus the energy concentrated in the additional vertical junctions. It writes

$$E(u) = \sum_{j=1}^{M} \sum_{i=2}^{N-1} (k/16)(u_2[i-1,j] - 2u_2[i,j] + u_2[i+1,j])^2 + \sum_{j=1}^{M-1} \sum_{i=1}^{N} k'(u_2[i,j] - u_2[i,j-1])^2$$

where $u[i,j]$ denotes the displacement of the i-th (blue) node of the j-th beam. With the previous scaling for k and a suitable scaling for k' : $k' = K'\epsilon$ we obtain by passing to the limit $\epsilon \to 0$ the following continuous energy :

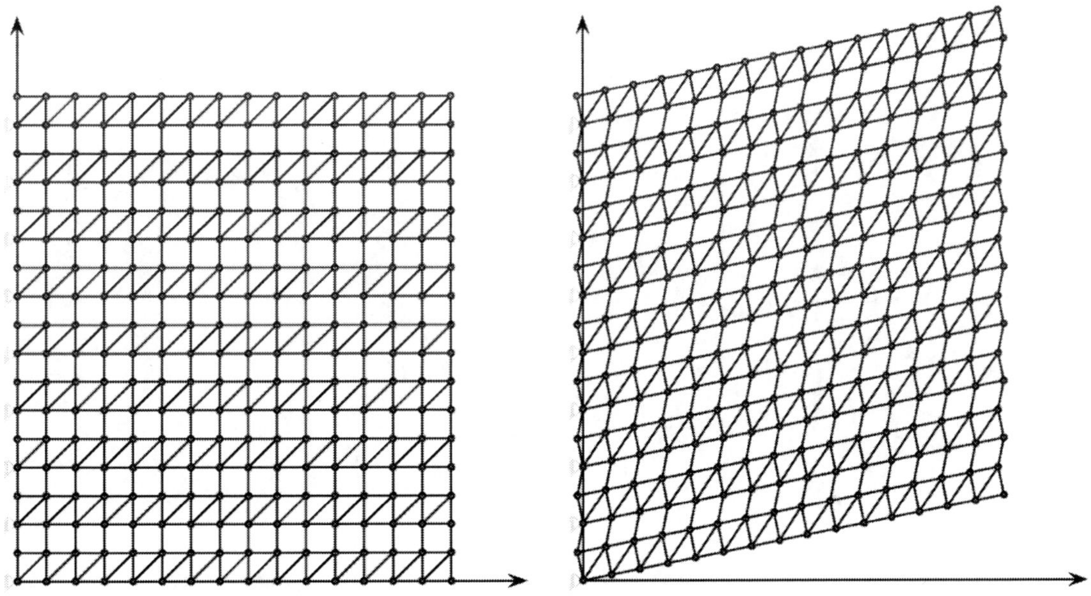

Figure 4. A 2D flexion structure. For the clearness of the figure, any red node has been joined with the blue one just above, but the computation is straightforward only when assuming that the blue nodes actually join the blue nodes just above. For the same reason we draw the first version of the flexion beam while the alternative version is used in the computation. The truss admits a floppy mode.

$$\tilde{E}(u) = \int_\Omega K \left(\frac{\partial^2 u_2}{\partial x_1^2}\right)^2 + K' \left(\frac{\partial u_2}{\partial x_2}\right)^2$$

subjected to the constraint $u_1 = 0$ in the whole domain $\Omega := (0, \ell) \times (0, \ell')$ and to the Dirichlet boundary condition $u = 0$ on the side $x_1 = 0$.

In order to understand the behavior of the continuum described by this elastic energy, let us consider the response of the system to some test load. We assume that one exerts a vertical force f concentrated on a vertical surface S in the middle of the domain (that is an academic example since applying such a concentrated force inside the material is not really easy). An equilibrium solution minimizes the total energy $\tilde{E} - \int_S f u_2 dx_2$. In a classical Cauchy material, an equilibrium state exists with (approximately) a constant shear deformation in the left hand part of the domain and no deformation in the right hand part (cf. figure 5). The situation is completely different for our new material. First, due to the aforementioned floppy mode, the Dirichlet condition is not sufficient for ensuring the existence of equilibrium. If the second node of all beams is also fixed (or alternatively if a suitable vertical force is applied on them) then a solution exists. This corresponds in the continuous model to adding an extra clamping constraint $\frac{\partial u_2}{\partial x_1} = 0$ on the side $S_l := \{x_1 = 0\}$ (or alternatively to applying a suitable surface density of torques on it, that is adding in the energy a term of the form $\int_{S_l} g \frac{\partial u_2}{\partial x_1} dx_2$). The equilibrium state is then very different from the classical case : the shear deformation varies linearly in the left hand part of the domain and remains constant but not vanishing in the right hand part (cf. figure 5).

Up to now the reader may be deceived. Indeed the interactions in the described material are

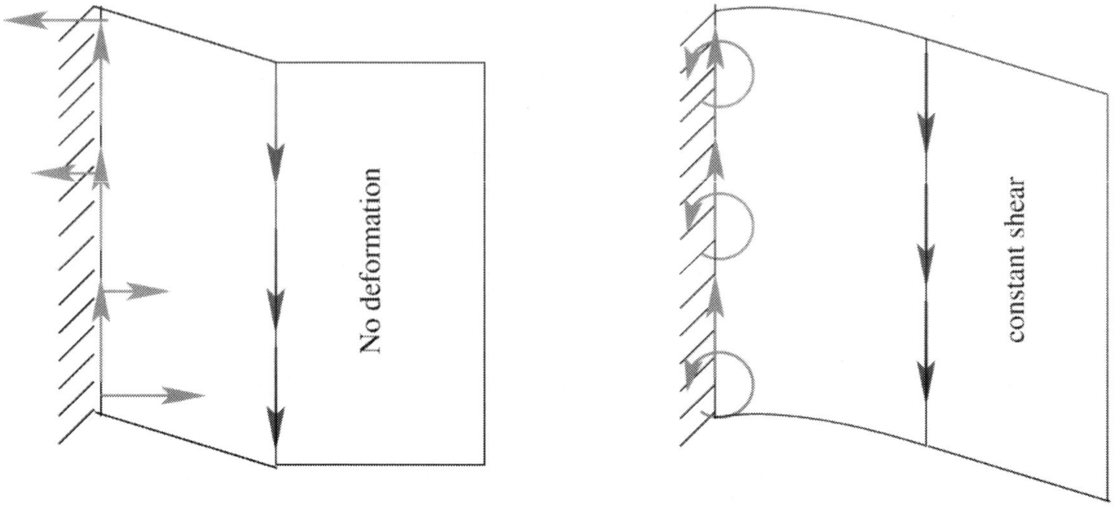

Figure 5. Comparison of the equilibrium states of classical (left hand drawing) and flexion (right hand drawing) materials subjected to a Dirichlet condition on the left hand side S_l and a vertical force on a vertical middle surface S. In the classical case, the balance of forces is ensured by a suitable surface density of reaction forces on S_l while, in the flexion case, a suitable density of torque has to be applied on S_l.

limited to surface densities of forces and torques which are not very exotic actions. The point is that at the microscopic level the displacement of the considered truss is a "local rotation", the truss behaves like a Timochenko continuum [39] or a Cosserat continuum [14, 15] in the sense cleared up by Bleustein [8] (see also [18]). In the next section we build a structure endowed with local deformation instead of local rotation. Interactions will then be of a completely different nature.

3. The pantograph beam and the pantograph truss

3.1. The pantograph beam

Now we build a structure based on the basic element described in figure 6 which possesses the same floppy mode as scissors. Indeed, for the clearness of the forthcoming figures we will simply draw a pair of scissors for representing this structure but remember that this is only a symbolic notation for a structure simply made of nodes joined by elastic bars.

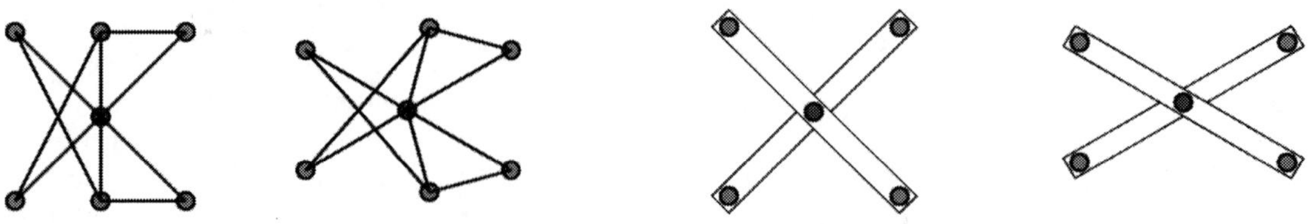

Figure 6. Basic element of the pantographic structures. We assume that the bars linking blue and red nodes have very high (infinite) stiffness corresponding to pure flexion (no extension) for the scissors. The structure is symbolized by two crossing rods. Its floppy mode plays an essential role.

By linking many such cells we get the pantographic structure described in figure 7. The

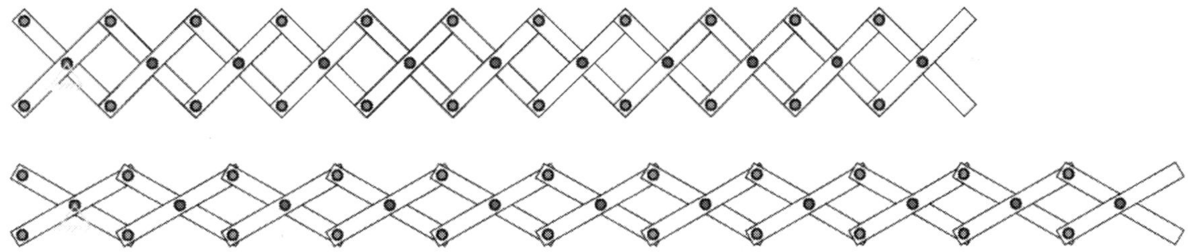

Figure 7. The pantographic structure and its extensional floppy mode.

remarkable feature of this structure is its extensional floppy mode (see fig. 7). Note that fixing the deformation of the first cell tends to fix the deformation of all cells. In particular, fixing the two first (blue) nodes makes the structure isostatic.

Computing the elastic energy in terms of the displacement of the blue nodes is straightforward. We get

$$E(u) = \sum_{i=2}^{N-1} k(u_2[i-1] - 2u_2[i] + u_2[i+1])^2 + k'(u_1[i-1] - 2u_1[i] + u_1[i+1])^2$$

We now recognize in $\epsilon^{-2}(u_1[i-1] - 2u_1[i] + u_1[i+1])$ the finite difference approximation of the second derivative of u_1 with respect to x_1. With a suitable scaling for k and k', the continuous limit ($N \to \infty$) model reads

$$\tilde{E}(u) = \int_0^\ell K\left(\frac{\partial^2 u_2}{\partial x_1^2}\right)^2 + K'\left(\frac{\partial^2 u_1}{\partial x_1^2}\right)^2$$

The transverse displacement u_2 and the axial one u_1 decouple. With respect to the transverse displacement, the behavior of the beam is exactly the behavior of the flexion beam described in the previous section. Owing to the symmetry of the terms involving u_1 and u_2 in the expression of the energy, every result valid for the transverse displacement can be transposed to the axial displacement, only the mechanical interpretation differs.

For instance, when the two first nodes are fixed, the displacement at the equilibrium under the action of a single axial force F at a middle point point $x_1 = \ell/2$ minimizes $\tilde{E}(u) - Fu_1(\ell/2)$ under the constraints $u_1(0) = \frac{\partial u_1}{\partial x_1}(0) = 0$: the deformation increases linearly from $x_1 = 0$ to $x_1 = \ell/2$ and remains constant but non zero for $x_1 > \ell/2$. Fixing the two first nodes corresponds in the continuous limit to imposing the constraint $\frac{\partial u_1}{\partial x_1}(0) = 0$. This can be replaced by the dual condition : two opposite axial forces applied at the two first nodes (see figure 8) which corresponds in the continuous limit to applying a new type of action. Two opposite axial forces are often said to have no action on a body but it is clearly not the case here as two such forces tend to modify the deformation of the first cell and as this deformation tends to propagate in the whole beam. Such "double force" is a new type of interaction [21, 22, 36] which needs a new symbol (⟛ see figure 8).

3.2. The pantographic truss

Consider now an array of many (M) horizontal pantographic beams (see figure 9), and link them using vertical pantographic beams. We fix the first node of each horizontal beam. Then we get a structure with four floppy modes (see figure 10).

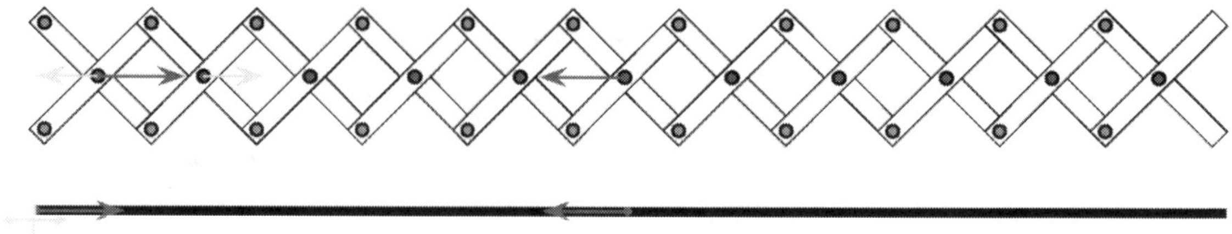

Figure 8. An axial force is applied at the middle point of a pantographic beam. On the left hand side of the pantograph, in addition to the reaction force, two opposite forces on the right hand side of the pantograph must be applied in order to enable the equilibrium of the structure. In the continuous limit (below) such additional action is a point-wise "double force".

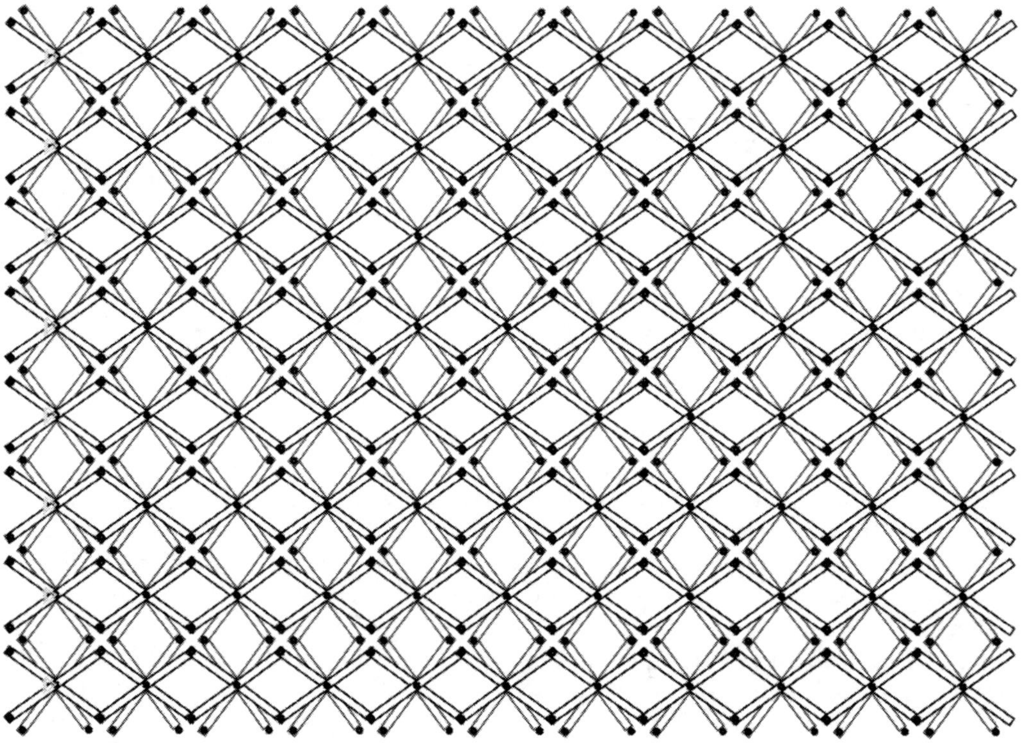

Figure 9. The pantographic 2D structure.

The elastic energy of the structure is the sum of the energies of all pantotographic beams:

$$E(u) = \sum_{j=1}^{M}\sum_{i=2}^{N-1} k(u_1[i-1,j] - 2u_1[i,j] + u_1[i+1,j])^2 + k'(u_2[i-1,j] - 2u_2[i,j] + u_2[i+1,j])^2$$
$$+ \sum_{j=2}^{M-1}\sum_{i=1}^{N} k(u_1[i,j-1] - 2u_1[i,j] + u_1[i,j+1])^2 + k'(u_2[i,j-1] - 2u_2[i,j] + u_2[i,j+1])^2$$

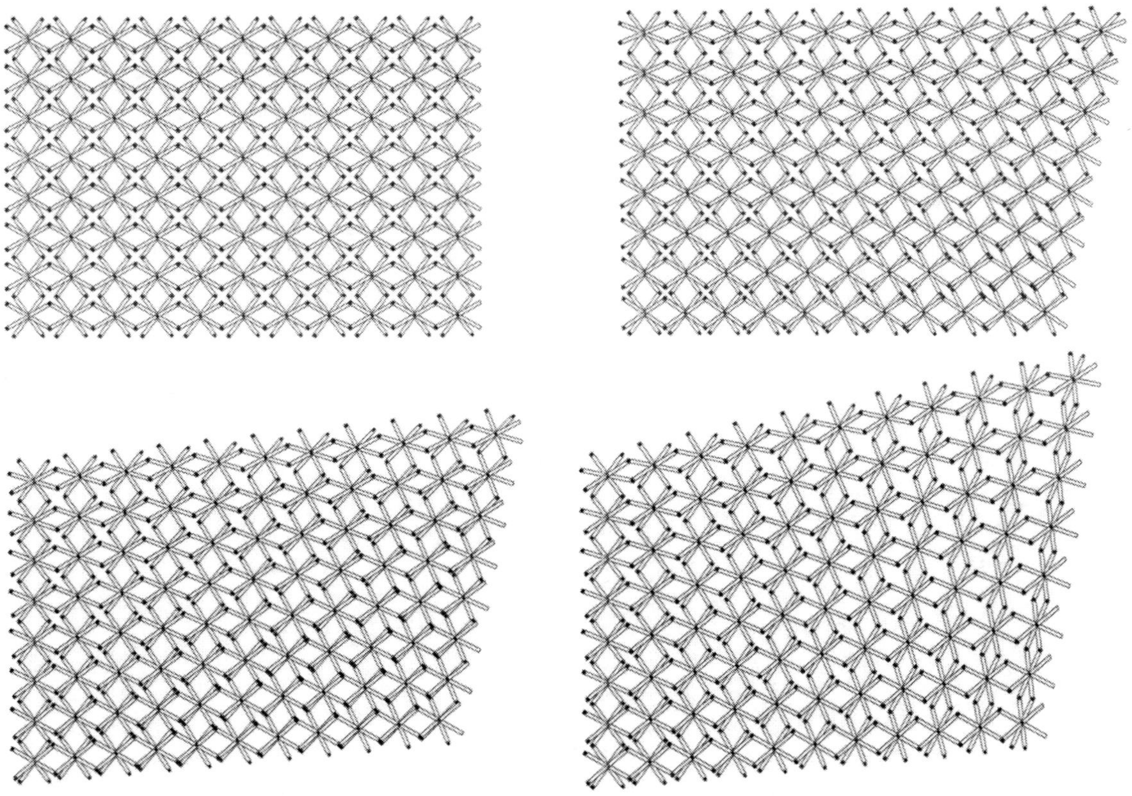

Figure 10. When imposing a Dirichlet condition on one side, the pantographic 2D structure has four floppy modes.

This clearly leads to the continuum model

$$\tilde{E}(u) = \int_\Omega K\left(\frac{\partial^2 u_2}{\partial x_1^2}\right)^2 + K'\left(\frac{\partial^2 u_2}{\partial x_2^2}\right)^2 + K'\left(\frac{\partial^2 u_1}{\partial x_1^2}\right)^2 + K\left(\frac{\partial^2 u_1}{\partial x_2^2}\right)^2$$

The floppy modes are of the form $u(x_1, x_2) = (ax_1x_2 + bx_1 + cx_2 + d, ex_1x_2 + fx_1 + gx_2 + h)$ which, taking into account the Dirichlet condition $u = 0$ on the boundary $x_1 = 0$, reduces to a four dimensional space of floppy modes:

$$u(x_1, x_2) = ((ax_2 + b)x_1, (ex_2 + f)x_1),$$

(see figure 10). Note that one can still act on the material at the fixed surface $x_1 = 0$ by applying a density $\mathcal{G}$ of torques (or fixing the "rotation $\frac{\partial u_2}{\partial x_1}$) or by applying a density $\mathcal{F}$ of "double-forces" (or fixing the "dilatation $\frac{\partial u_1}{\partial x_1}$). For instance, let us consider the extension to the 2D case of the situation described in figure 8. We apply a surface density of horizontal forces f on the middle surface $S = \{x_1 = \ell/2\}$ adding in the energy the term $\int_S -fu_1dx_2$. The existence of an equilibrium is ensured only if we apply on the surface $x_1 = 0$ a surface density of horizontal forces $-f$ plus a surface density of double forces adding in the energy the terms $-\int_{S_l} fu_1dx_2$ and $-\int_{S_l} \mathcal{F}\frac{\partial u_1}{\partial x_1}dx_2$ with $\mathcal{F} = f\ell/2$).

4. Third gradient structure

Up to now the continuous materials we have designed are second gradient material (see [21])that is materials with an elastic energy depending only on the first and second gradient of the displacement. It is possible, to the expense of some extra complexity, to design trusses leading to higher gradient material and consequently supporting more exotic mechanical interactions.

4.1. Third gradient flexion beam

We first consider the basic element described in figure 11. Note that we build on the structure we already introduced (fig. 6) and remember that such a structure is only made of nodes and linear springs. The lowest horizontal bars are assumed to have infinite stiffness while all other ones have the same finite stiffness. One can easily check that the elastic energy

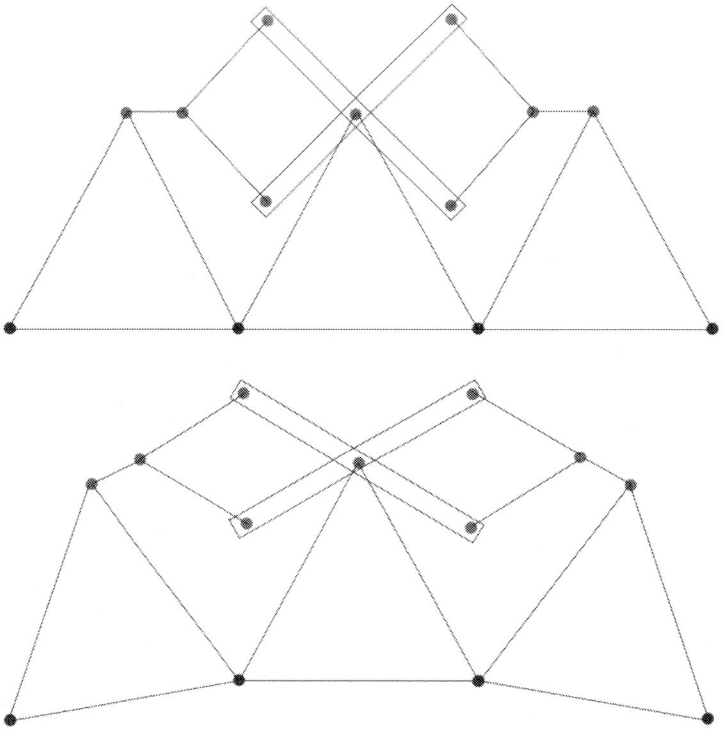

Figure 11. The basic element of the 3rd gradient beam and its major floppy mode.

of this element in terms of the displacements of the 4 lowest (red) nodes is proportional to $(u_2[i] - 3u_2[i+1] + 3u_2[i+2] - u_2[i+3])^2$.

In order to simplify the following drawings, we use the symbol  for representing this basic element. Then we construct a beam by link any four successive nodes of the beam by such basic elements (figure 12). It is non extensible ($u_1 = 0$) and its elastic energy reads

$$E(u) = \sum_{i=1}^{N-3} k(u_2[i] - 3u_2[i+1] + 3u_2[i+2] - u_2[i+3])^2.$$

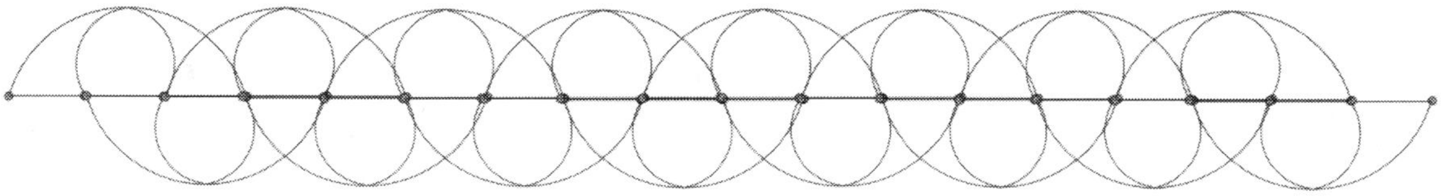

Figure 12. A 3rd gradient beam made by superposition of basic elements of figure 11.

When the two first nodes are fixed, a floppy mode remains which corresponds to deformations with constant curvature along the beam. This curvature is fixed for instance by fixing the displacement of the third node. One can also act on the beam by applying three vertical forces $(f, -2f, f)$ on three successive nodes. In the continuous limit we recognize the finite difference approximation of the third derivative of the transverse displacement and the energy reads

$$\tilde{E}(u) = \int_0^\ell K \left(\frac{\partial^3 u_2}{\partial x_1^3} \right)^3 .$$

4.2. Third gradient 2D structure

Consider now an array of many (M) horizontal 3rd gradient beams (see figure 13), and link them using vertical elastic springs. We fix the two first nodes of each beam. Then we get a structure with one floppy mode. The elastic energy of the whole structure the sum of the elastic energies

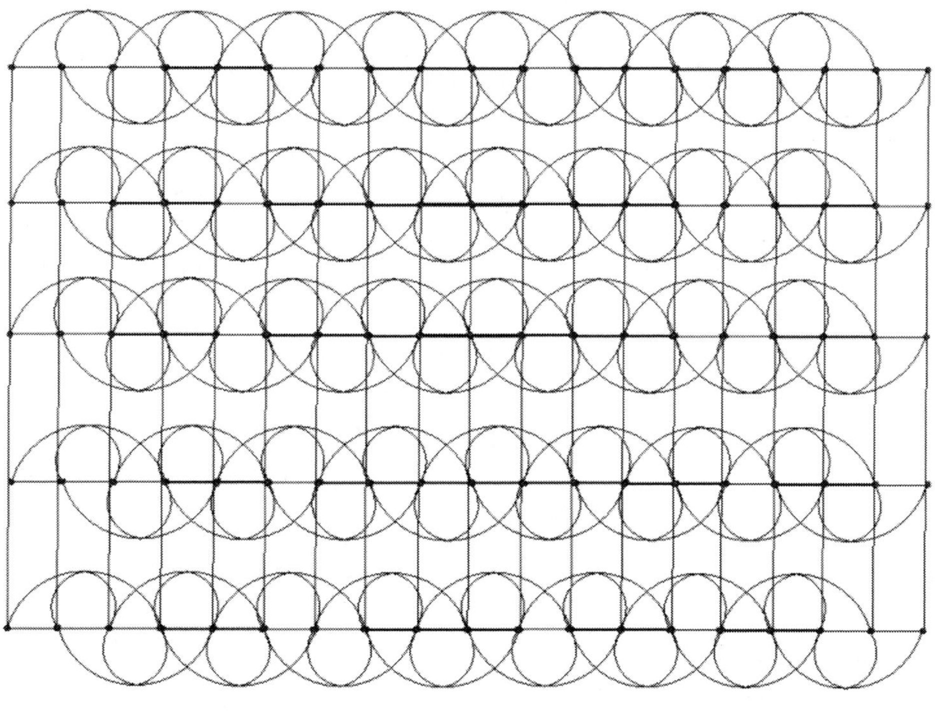

Figure 13. A 3rd gradient 2D structure.

of the beams we just derived plus the energy concentrated in the additional vertical junctions.

It takes the form

$$E(u) = \sum_{j=1}^{M} \sum_{i=1}^{N-3} k(u_2[i,j] - 3u_2[i+1,j] + 3u_2[i+2,j] - u_2[i+3,j])^2 + \sum_{j=1}^{M-1} \sum_{i=1}^{N} k'(u_2[i,j] - u_2[i,j-1])^2$$

and with a suitable scaling for k and k' we obtain the following continuous energy :

$$\tilde{E}(u) = \int_{\Omega} K \left(\frac{\partial^3 u_2}{\partial x_1^3}\right)^2 + K' \left(\frac{\partial u_2}{\partial x_2}\right)^2$$

with the constraints $u_1 = 0$ on the domain Ω and $u_2 = 0$, $\frac{\partial u_2}{\partial x_1} = 0$ on the boundary $x_1 = 0$.

Let us again consider the problem defined in section 2 (figure 5). The equilibrium solution is still different. First such a solution exists only if the third node of all beams is also fixed (or alternatively if a suitable vertical force is applied on them). This corresponds in the continuous model to adding an extra clamping constraint $\frac{\partial^2 u_2}{\partial x_1^2} = 0$ on the side $S_l := \{x_1 = 0\}$ (or alternatively to applying a suitable surface density of "triple-forces" on it, that is adding in the energy a term of the form $\int_{S_l} h \frac{\partial^2 u_2}{\partial x_1^2} dx_2$). This type of triple-forces can also be interpreted as "double-torques" and we use the symbol  for them. In the equilibrium state the curvature increases in the left hand part of the domain and remains constant but not vanishing in the right hand part (cf. figure 14).

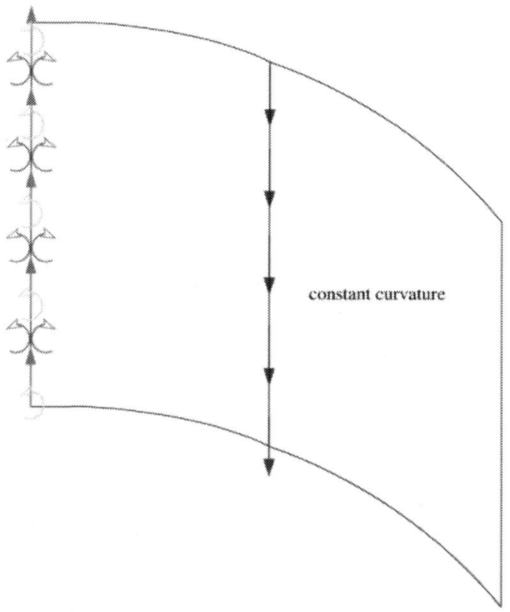

Figure 14. The equilibrium state of a flexion third gradient material clamped on the left hand side S_l and submitted to a vertical force on the vertical middle surface.

5. Conclusion

We could carry on this enumeration of more and more intricate structures. For instance, building on the structural element introduced and symbolized in the previous section, we could consider

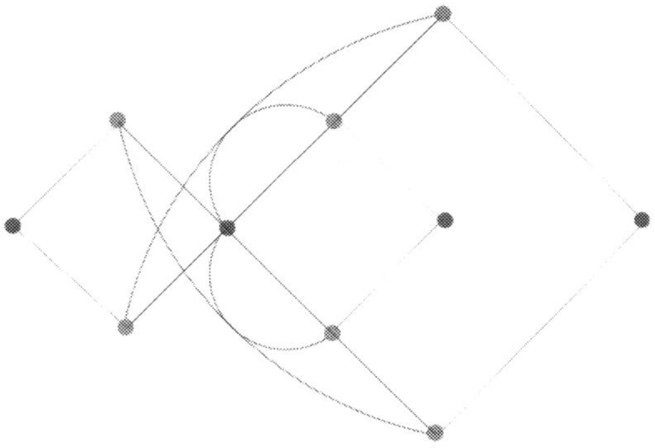

Figure 15. A structural element for a material sensitive to the second derivative of the dilatation.

the element described in figure 15. We let the reader check that it has a floppy mode which mimics the third derivative of u_1 with respect of x_1 (that is the second derivative of the dilatation) and that, introducing a new symbol to denote it, the element described in figure 16 has a floppy mode which mimics the fourth derivative of u_2 with respect of x_1. This enumeration

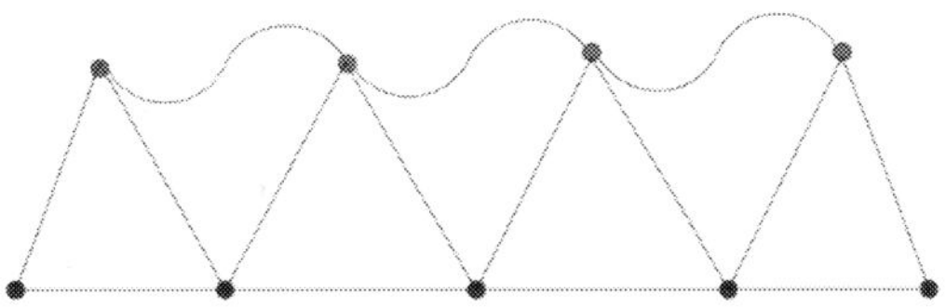

Figure 16. A structural element for a material sensitive to the fourth derivative of the transverse displacement.

would not be of great interest : the reader should yet be convinced by the examples we have presented that homogenization of trusses can lead to any high gradient materials.

In these materials, contact interactions between parts of the domain are far belong those described by the Cauchy stress theory. In our examples we have illustrated the possibility of surfaces densities of torques, "double-forces" or "triple-forces". Our trusses give a clear microscopic interpretation of all these different actions : they are different distributional limits of highly oscillating forces. The denominations that P. Germain [21] gave to them are illustrated.

Let us recall that it has been rigorously proved in [4] that all types of high gradient materials can be achieved by homogenizing trusses.

The highest gradient terms are clearly active when the lower gradient terms do not play any role, that is when the structures possess floppy modes. The reader who prefer to work with non degenerated materials and thus to consider only trusses with no floppy mode (apart those corresponding to rigid motions) can add linear elastic links between each node and its neighbors. This will simply add a classical (first gradient) elastic term in the elastic energy. These supplementary links have to be much weaker than the links we have used for building our structures (the scaling with respect to ϵ is different). That explains why, for natural materials, these new types of interactions scarcely matter : to obtain a material in which a new type of action is preponderant, one must carefully build with very stiff links a structure having a floppy mode and then, possibly add some weaker links.

References

[1] Abeyaratne R, Triantafyllidis N, 1984, *Trans. ASME J. Appl. Mech.* **51**, no. 3, 481–486.
[2] Alibert J-J, dell'Isola F and Seppecher P, 2003, *Math. Mech. Solids* **8** , no. 1, 51–73.
[3] Banfi C, Marzocchi A and Musesti A, 2006, *Ricerche di Matematica* **55** 299–310.
[4] Camar-Eddine M and Seppecher P, 2003, *Arch. Ration. Mech. Anal.*, **170** 3, 211 - 245.
[5] Cioranescu D and Rebnord D A, 1995, *Asymptotic Anal.*, **11**, 2, 107–130.
[6] Bardenhagen S and Triantafyllidis N, 1994, *J. Mech. Phys. Solids* **42** , no. 1, 111–139.
[7] Bensoussan A, Lions J L and Papanicolaou G, 1978, *Asymptotics Analysis for periodic Structures*, North-Holland Amsterdam.
[8] Bleustein J L, 1967, *International Journal of Solids and Structures*, **3** (6), pp. 1053-1057.
[9] Braides A and Gelli M S, 1999, *Preprint SISSA*, 41/99/M.
[10] Cioranescu D, Saint Jean Paulin J, 1989, *C.R.Acad.Sci. Paris I*, **308** no.2, 41-46.
[11] Cioranescu D, Saint Jean Paulin J, 1990, *Lecture notes in Control and Inform.Sci.*, **147**, Springer Verlag Berlin.
[12] Cioranescu D, Saint Jean Paulin J, 1992, *Lecture notes in Control and Inform.Sci.*, **178** Springer Verlag Berlin.
[13] Cioranescu D, Saint Jean Paulin J, 1999, *Applied Mathematical Sciences.*, **136** Springer Verlag New York.
[14] Cosserat E and Cosserat F, 1908, *Note sur la thorie de l'action euclidienne* Paris, Gauthier-Villars.
[15] Cosserat E and Cosserat F, 1909, *Théorie des Corps Déformables*, Paris, Hermann.
[16] Dell'Isola F and Seppecher P, 1995, *Comptes Rendus de l'Academie de Sciences - Serie IIb*, **321**, , pp. 303-308.
[17] Dell'Isola F and Seppecher P, 1997, *Meccanica*, **32**, pp. 33-52.
[18] Di Carlo A Rizzi N and Tatone A, 1990, *Meccanica*, **25**, 168-174.
[19] Degiovanni M, Marzocchi A and Musesti A, 2006, *Annali di Matematica***185**, 81–103.
[20] Degiovanni M, Marzocchi A and Musesti A, 1999, *Arch. Ration. Mech. Anal.* **147** 197–223.
[21] Germain P, 1973, *J. Mécanique*, **12**, 235–274.
[22] Germain P, 1973, *SIAM J. Appl. Math.*, **25** 3.
[23] Green A E and Rivlin R S, 1964, *Arch. Rational Mech. Anal.*, **16**, 325–353.
[24] Green A E and Rivlin R S, 1964, *Arch. Rational Mech. Anal.*, **17**, 113–147.
[25] Green A E and Rivlin R S, 1964, *Z. Angew. Math. Phys.*, **15**, 290–292.
[26] Lucchesi M, Silhav M and Zani N, 2008, *J Elasticity* **90**, 209–223.
[27] Marzocchi A and Musesti A, 2001, *Cont. Mech. Thermodyn.* **13**, 149–169.
[28] Marzocchi A and Musesti A, 2003, *Meccanica* **38**, 369–389.
[29] Maugin G and Metrikine A V, 2010, *Mechanics of Generalized Continua, One Hundred Years After the Cosserats*, Springer.
[30] Mindlin R D, 1964, *Arch. Rat. Mech. Analysis*, **16**, 51-78.
[31] Mindlin R D, 1965, *Int. J. Solids and Struct.* **1**, 4, 417-438.
[32] Mindlin R D and Eshel N N, 1968,*International Journal of Solids and Structures*, **4**, 1, 109-124.
[33] Mindlin R D and Tiersten H F, 1962, *Arch.Rat.Mech.Anal.*, **11**, 415-448.
[34] Noll W, 1959, *Proceeding of the Berkeley Symposium on the Axiomatic Method*, Amsterdam, 226–281.
[35] Noll W, 1973 *Arch. Rational Mech. Anal.*,**52** 62–92.
[36] Seppecher P, 1989, *C. R. Acad. Sci. Paris*, **309**, Srie II 497–502.
[37] Silhavy M, 1985, *Arch.Ration. Mech. Anal.*, **90** 195–211.
[38] Silhavy M, 1991, *Arch. Ration. Mech. Anal.*, **116** 223–255.
[39] Timoshenko S, 1976, *Strenght of Materials*, Krieger Publishing.
[40] Toupin R A, 1962 *Arch. Rat. Mech. Anal.*, **11**, 385-414.
[41] Toupin R A, 1964 *Arch. Rational Mech. Anal.*, **17** 85–112.
[42] Triantafyllidis N and Bardenhagen S, 1993, *J. Elasticity*, **33**.
[43] Triantafyllidis N and Bardenhagen S, 1996, *J. Mech. Phys. Solids*, **44** , no. 11, 1891–1928.
[44] Truesdell C A, 1977, *A First Course in Rational Continuum Mechanics, Vol. I General Concepts*, Academic Press.

Stochastic population oscillations in spatial predator-prey models

Uwe C. Täuber

Department of Physics, Virginia Tech, Blacksburg, VA 24061-0435, USA

E-mail: tauber@vt.edu

Abstract. It is well-established that including spatial structure and stochastic noise in models for predator-prey interactions invalidates the classical deterministic Lotka–Volterra picture of neutral population cycles. In contrast, stochastic models yield long-lived, but ultimately decaying erratic population oscillations, which can be understood through a resonant amplification mechanism for density fluctuations. In Monte Carlo simulations of spatial stochastic predator-prey systems, one observes striking complex spatio-temporal structures. These spreading activity fronts induce persistent correlations between predators and prey. In the presence of local particle density restrictions (finite prey carrying capacity), there exists an extinction threshold for the predator population. The accompanying continuous non-equilibrium phase transition is governed by the directed-percolation universality class. We employ field-theoretic methods based on the Doi–Peliti representation of the master equation for stochastic particle interaction models to (i) map the ensuing action in the vicinity of the absorbing state phase transition to Reggeon field theory, and (ii) to quantitatively address fluctuation-induced renormalizations of the population oscillation frequency, damping, and diffusion coefficients in the species coexistence phase.

1. Introduction

Over the past decade, mathematical and computational tools from statistical physics have been increasingly and quite successfully applied to ecological problems, including attempts at a quantitative understanding of biodiversity [1]–[4]. In this context, physicists typically consider simplified idealized models that hopefully capture the essential features of interacting biosystems; leaving aside some of the biological complexity allows the consistent incorporation of stochastic fluctuations and spatio-temporal correlations, whose crucial importance has long been recognized in the field [5], but is still often neglected.

Predator-prey models defined via reaction-diffusion systems on a regular lattice, whose rate equations in the well-mixed mean-field limit reduce to the classic coupled Lotka–Volterra ordinary differential equations, constitute paradigmatic examples of the dynamics of two competing populations [6]–[8]. Monte Carlo simulations of these models, specifically in two dimensions, display a remarkable wealth of intriguing features (for a fairly recent overview, see, e.g., Ref. [9]): In contrast to the regular non-linear oscillations of the deterministic Lotka–Volterra model for which the population densities invariably return to their initial values (c.f. figure 1 below), computer simulations display persistent, but eventually decaying stochastic population oscillations (figure 2) [10]–[17]. In the absence of spatial degrees of freedom, these erratic population oscillations may be understood through a resonant stochastic amplification mechanism [18] that drastically extends the transient time interval before any finite system ultimately reaches its absorbing stationary state, where the predator population becomes extinct [19]. In spatially extended systems, it is well-known that the mean-field Lotka–Volterra reaction-diffusion equations allow for traveling wave solutions [20]–[22]. In the corresponding stochastic spatial realizations, spreading activity fronts (figure 4, [23]) induce short-ranged but significant positive correlations of either species, and anti-correlations between the predator and prey populations, which have the effect of further enhancing the amplitude and life time of local population oscillations [9, 24]. We have investigated various different variants of stochastic spatial Lotka–Volterra models for competing predator-prey populations, and found these intriguing spatio-temporal structures to be remarkably robust against rather drastic changes of the detailed microscopic interaction rules [24, 25], and even the introduction of quenched spatial disorder in the reaction rates [26].

In this brief communication, I will provide an overview of our Monte Carlo simulation results, specifically contrasting model variants with and without restrictions on the number of particles per lattice site. The former describe ecological systems with finite local carrying capacity, and display a continuous non-equilibrium phase transition from an active species coexistence state to an absorbing phase wherein the predators become extinct. Numerical evidence supports the general expectation [27]–[32] that this extinction transition should be governed by the directed-percolation universality class [7, 8], [11]–[14], [16, 17]. I will then demonstrate how field-theoretic tools based on the Doi–Peliti representation of the master equation for stochastic interacting particle systems [33]–[35] (for recent reviews, see Refs. [36, 37]), augmented with a means to incorporate restricted site occupation numbers [38], can be employed to gain a comprehensive understanding of fluctuation and correlation effects in Lotka–Volterra predator models. Specifically, the effective action near the extinction transition in model variants with restricted site occupations will be explicitly mapped onto Reggeon field theory which describes the universal scaling of directed-percolation clusters [27, 32, 39, 40]. Moreover, expanding on the treatment in Ref. [41], I shall report a computation of the fluctuation-induced renormalizations of the population oscillation frequency, damping, and diffusion coefficients in the species coexistence phase to lowest order in a perturbation expansion with respect to the predation rate [42].

2. Stochastic lattice Lotka–Volterra models

2.1. Model variants and mean-field description

We consider a two-species system of diffusing particles (with diffusion constant D) that undergo the following stochastic reactions:

$$\begin{aligned} A &\to \emptyset && \text{with rate } \mu, \\ A + B &\to A + A && \text{with rate } \lambda, \\ B &\to B + B && \text{with rate } \sigma. \end{aligned} \tag{1}$$

The 'predators' A decay or die spontaneously at rate $\mu > 0$, whereas the 'prey' B produce offspring with rate $\sigma > 0$. In the absence of the binary 'predation' interaction with rate λ, the uncoupled first-order processes would naturally lead to predator extinction $a(t) = a(0)\, e^{-\mu t}$, and Malthusian prey population explosion $b(t) = b(0)\, e^{\sigma t}$; here $a(t)$ and $b(t)$ respectively indicate the A / B concentrations or population densities. The binary predation reaction induces species coexistence through the non-linear interaction of both particle species.

In the simplest spatial realization of this stochastic reaction-diffusion model, both particle species are represented by unbiased random walkers on a d-dimensional hypercubic lattice, and we allow an arbitrary number of particles per lattice site [24]. All reactions (1) can then be implemented strictly on-site: Offspring particles are placed on the same lattice point as their parents, and the predation reaction happens only if an A and a B particle meet on the same lattice site. If we then assume the populations to remain well mixed, and consequently ignore both spatial fluctuations and correlations, we can approximately describe the coupled reactions (1) through the associated mean-field rate equations for spatially homogeneous concentrations $a(t) = \langle a(\vec{x},t)\rangle$, $b(t) = \langle b(\vec{x},t)\rangle$, where $a(\vec{x},t)$ and $b(\vec{x},t)$ respectively denote the local predator and prey densities. One then arrives at the classic Lotka–Volterra equations [4], a coupled set of two ordinary non-linear differential equations:

$$\dot{a}(t) = \lambda\, a(t)\, b(t) - \mu\, a(t)\ , \quad \dot{b}(t) = \sigma\, b(t) - \lambda\, a(t)\, b(t)\ . \qquad (2)$$

The rate equations (2) display three stationary states (a_s, b_s), namely the empty absorbing state with total population extinction $(0,0)$, which is obviously linearly unstable if $\sigma > 0$; a predator extinction absorbing state wherein the prey population diverges $(0,\infty)$, which for $\lambda > 0$ is also linearly unstable; and finally a species coexistence state $(a_u = \sigma/\lambda, b_u = \mu/\lambda)$, which however represents only a marginally stable fixed point with purely imaginary eigenvalues $\pm i\sqrt{\mu\,\sigma}$ of the associated Jacobian stability matrix: Linearizing eqs. (2) near (a_u, b_u) results in the coupled differential equations $\delta\dot{a}(t) = \sigma\,\delta b(t)$, $\delta\dot{b}(t) = -\mu\,\delta a(t)$, which are readily solved by $\delta a(t) = \delta a(0)\, \cos\left(\sqrt{\mu\,\sigma}\,t\right) + \delta b(0)\,\sqrt{\sigma/\mu}\, \sin\left(\sqrt{\mu\,\sigma}\,t\right)$ and $\delta b(t) = -\delta a(0)\,\sqrt{\mu/\sigma}\,\sin\left(\sqrt{\mu\,\sigma}\,t\right) + \delta b(0)\,\cos\left(\sqrt{\mu\,\sigma}\,t\right)$, describing harmonic oscillations about the center fixed point (a_u, b_u) with frequency $\omega = 2\pi\, f = \sqrt{\mu\,\sigma}$. Indeed, the phase space trajectories for the full non-linear coupled differential equations (2) are determined by $da/db = a\,(\lambda\, b - \mu)/b\,(\sigma - \lambda\, a)$, with a conserved first integral

$$K(t) = \lambda[a(t) + b(t)] - \sigma \ln a(t) - \mu \ln b(t) = K(0)\ . \qquad (3)$$

Consequently, as depicted in figure 1, the solutions of the deterministic mean-field Lotka–Volterra model are closed orbits in phase space, i.e., regular periodic non-linear population oscillations whose amplitudes are fixed by the initial configuration. Naturally, the precise periodic return to the initial concentration values does not appear to be a very realistic feature. In addition, the neutral cycles of the coupled mean-field rate equation system (2) indicates that this deterministic mathematical model is fundamentally unstable with respect to slight modifications [4].

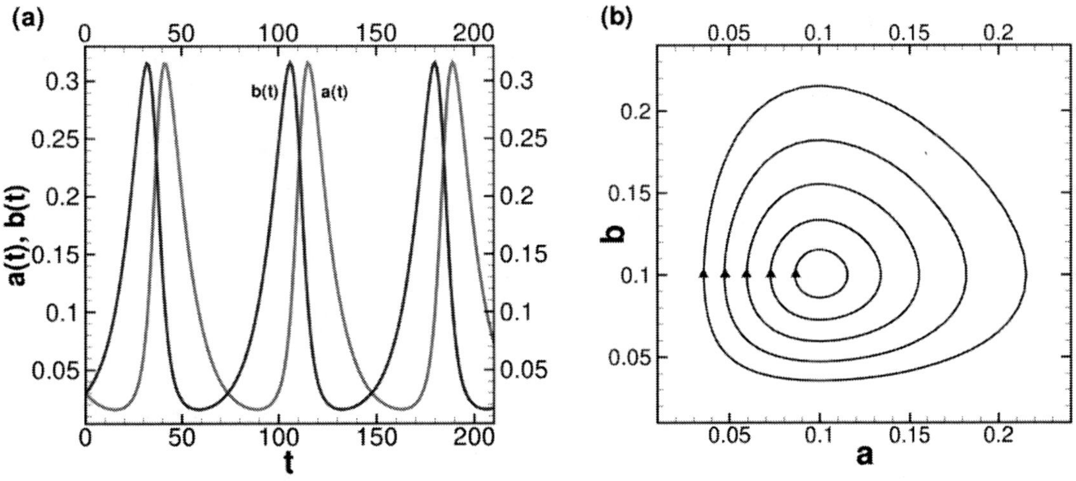

Figure 1. Solutions of the coupled Lotka–Volterra mean-field rate equations (2): (a) Non-linear predator $a(t)$ (red) and prey $b(t)$ (blue) population density oscillations; (b) periodic orbits in the a-b phase plane. For small amplitudes the oscillations become harmonic (circular orbits) with frequency $\omega = \sqrt{\mu\sigma}$. (Reproduced with permission from Ref. [24], p. 4.)

One such modification that aims at rendering the Lotka–Volterra system more relevant biologically is to introduce a finite carrying capacity (total particle density) $\rho > 0$ that limits the prey population growth, modeling, e.g., the effect of limited food resources [4]. Within the mean-field rate equation approximation, the second differential equation in (2) is then replaced with

$$\dot{b}(t) = \sigma\, b(t)\, [1 - b(t)/\rho] - \lambda\, a(t)\, b(t) \ . \tag{4}$$

The non-trivial stationary states in this restricted Lotka–Volterra model are predator extinction and prey saturation $(0, \rho)$, linearly stable for $\lambda < \lambda_c = \mu/\rho$; and species coexistence (a_r, b_r) with $b_r = \mu/\lambda$ and $a_r = (1-\mu/\rho\lambda)\sigma/\lambda$, which both exists and is linearly stable provided the predation rate is sufficiently large, $\lambda > \lambda_c$. The eigenvalues of the Jacobian now acquire negative real parts, $\epsilon_\pm = -\mu\,\sigma \left[1 \pm \sqrt{1 - 4\rho\lambda\,(\rho\lambda/\mu - 1)\,/\sigma}\right] /2\rho\lambda$, which implies an exponential approach to the stable fixed point (a_r, b_r), replacing the neutral cycles of the unrestricted model (2). Moreover, for $\sigma > \sigma_s = 4\lambda\,\rho\,(\rho\lambda/\mu - 1) > 0$, or $\mu/\rho < \lambda < \lambda_s = \left(1 + \sqrt{1 + \sigma/\mu}\right) \mu/2\rho$, the eigenvalues are real, indicating a nodal stable fixed point, whereas for $\sigma < \sigma_s$ or $\lambda > \lambda_s$, i.e., deep in the species coexistence phase, the eigenvalues $\epsilon_\pm$ turn into a complex conjugate pair, and (a_r, b_r) becomes a stable spiral singularity which is approached in a damped oscillatory manner. Adding spatial degrees of freedom, finite local carrying capacities can be implemented in a lattice model through limiting the maximum occupation number per site for each species. Most drastically, one can permit at most a single particle per lattice site [9]; the binary predation reaction then has to occur between predators and prey on adjacent nearest-neighbor sites, and new offspring needs to be placed on neighboring positions. In that case, one can in fact entirely dispense with hopping processes, since all particle production reactions entail population spreading as well.

In summary, already within the mean-field rate approximation, a finite prey carrying capacity ρ, which can be viewed as the average result of local restrictions on the prey density originating from limited resources, crucially changes the phase diagram: There emerges an extinction threshold (at λ_c for fixed μ) for the predator population, which in a spatially extended system becomes a genuine continuous active-to-absorbing non-equilibrium phase transition in the thermodynamic limit of infinite system size and time.

2.2. Monte Carlo simulation results

Various authors have studied stochastic lattice predator-prey models that in the well-mixed mean-field limit reduce to the classical Lotka–Volterra system [6]–[8], [10]–[17]. In this section, I shall briefly discuss the pertinent results from our own individual-based Monte Carlo simulation studies, performed mostly on two-dimensional square lattices with periodic boundary conditions. Technical details and more precise descriptions of the algorithms we have employed can be found in Refs. [9] and [24].

Figure 2 shows typical simulation data for the temporal evolution of the total predator and prey particle densities in a two-dimensional stochastic lattice model with (almost) arbitrarily large site occupation numbers and on-site reactions [24]. One observes long-lived but clearly damped population oscillations that are actually quite independent of the initial state; neither are they caused by the constancy of the first integral K, eq. (3), that follows from the deterministic rate equations: As is apparent from the numerical data, in the stochastic spatial model $K(t)$ is manifestly time-dependent, and in fact traces the overall population oscillations. We note that as the system size increases, the relative oscillation amplitudes become smaller; in the thermodynamic limit, the quasi-periodic population fluctuations eventually die out entirely. From the marked peaks in the Fourier-transformed concentration signals, $a(f) = \int a(t)\, e^{2\pi i f t}\, dt$ for the predators, and similarly for the prey density, we may infer a characteristic oscillation frequency f. As illustrated in figure 3, the typical population oscillation frequencies thus obtained roughly follow the square-root dependence on the rates μ and σ as predicted by

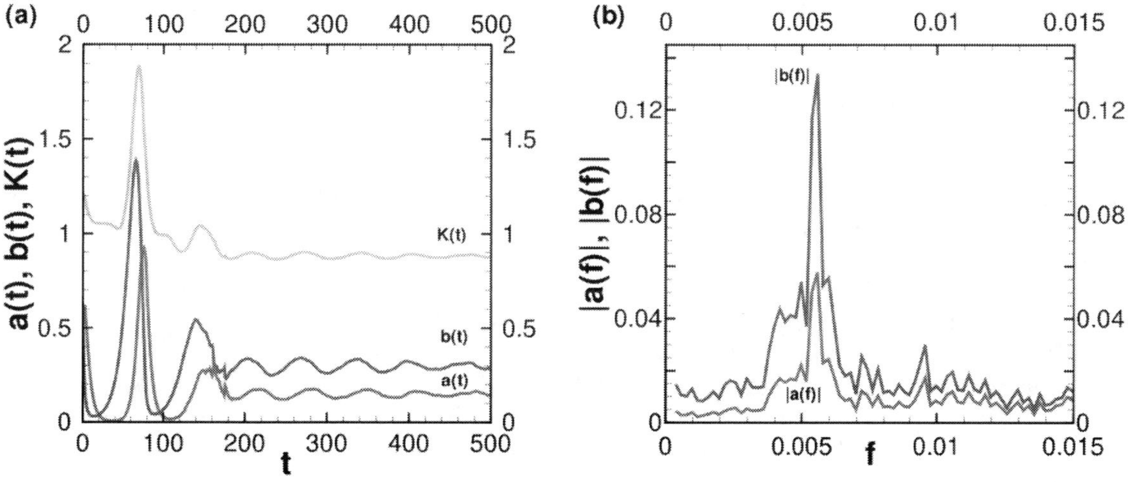

Figure 2. Monte Carlo simulation data for a stochastic spatial Lotka–Volterra system on a 1024×1024 square lattice with periodic boundary conditions, in the absence of site occupation number restrictions: (a) Temporal evolution for the predator $a(t)$ (red) and prey $b(t)$ (blue) population densities, and the quantity $K(t)$ (green) for $\sigma = 0.1$, $\mu = 0.2$,, and $\lambda = 1.0$; (b) Fourier-transformed population density signals $|a(f)|$ and $|b(f)|$ for $\sigma = 0.03$, $\mu = 0.1$,, and $\lambda = 1.0$. (Reproduced with permission from Ref. [24], pp. 9 and 10.)

the linearized mean-field approximation, but with measurable deviations both for low and high rates. Yet the numerical frequency values are reduced by about a factor of four in the stochastic spatially extended system, an apparent considerable downward renormalization caused by fluctuations and reaction-induced spatio-temporal correlations [24]. Note also that figure 3(a) shows a remarkably similar functional dependence of f on the rates μ and σ.

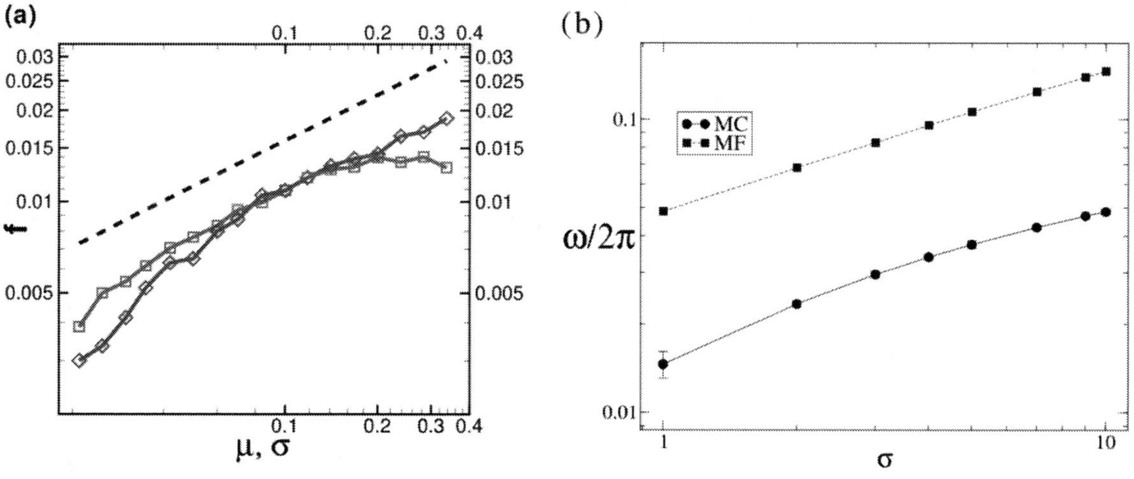

Figure 3. Characteristic peak frequencies obtained form $|a(f)|$ or $|b(f)|$, (a) as functions of the rates σ (red squares) and μ (blue diamonds), with otherwise $\sigma = 0.1 = \mu$ and $\lambda = 1.0$ held fixed on a 1024 lattice without site occupation restrictions; (b) as function of σ with $\mu = 0.1$, $\lambda = 1.6$, $D = 0$ on a 128×128 square lattice with at most a single particle per lattice site (black line). The dashed black lines represent the oscillation frequency from linearized mean-field theory, $f = \sqrt{\mu\sigma}$. (Reproduced with permission from Ref. [24], p. 10 and Ref. [9], p. 469.)

Very similar features are found in stochastic spatial Lotka–Volterra models that incorporate stringent site occupation number restrictions (allowing only at most one particle on each site), deep in the species coexistence phase, i.e., for large predation rates, corresponding to a stable focal mean-field fixed point (stability matrix eigenvalues with negative real and non-vanishing imaginary parts). Both in the absence and presence of local density limitations, the coexistence phase is governed by remarkably strong spatio-temporal fluctuations: Striking spreading activity waves of prey closely followed by predators periodically sweep the system; any small surviving clusters of prey subsequently serve as sources for resurgent expanding prey-predator fronts [9]. An average over these weakly coupled local oscillations then yields the total population time traces depicted in figure 2. These spreading activity fronts appear especially sharp for the site-restricted model variants, as displayed in figure 4, whereas in realizations with arbitrarily many particles per site, the fronts look more diffuse [23]. In either situation, one can employ stationary-state correlation functions to measure the spatial width $\sim 10\ldots20$ lattice sites of the spreading activity regions. At roughly the same length scale, the cross-correlations of the A and B particles peak at a positive value before slowly decaying to zero; at shorter distances, the prey are naturally anti-correlated with the predators [9, 24].

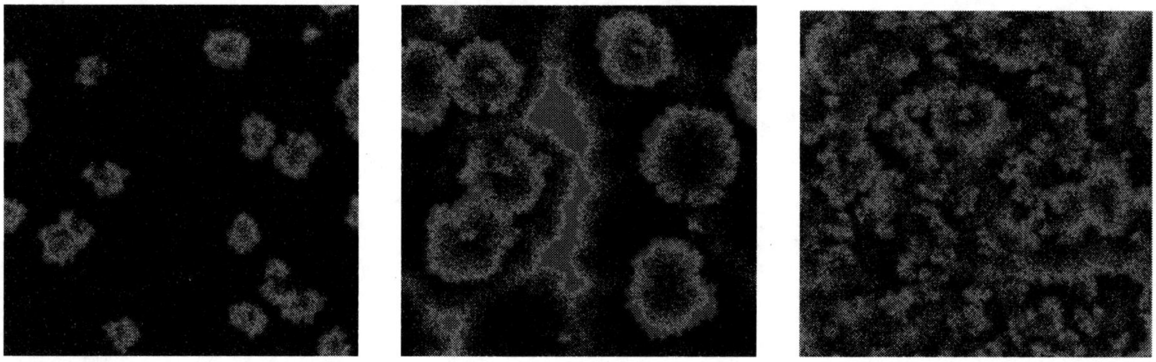

Figure 4. Snapshots illustrating the evolution in time (from left to right) of a two-dimensional stochastic lattice Lotka–Volterra model (with 512×512 sites), incorporating local occupation number restrictions in the species coexistence phase, with rates $\sigma = 4.0$, $\mu = 0.1$, $\lambda = 2.2$, and $D = 0$, when the fixed point is a focus; the initial densities are $a(0) = 1/3 = b(0)$. The red, blue, and black dots respectively represent the prey, predators, and empty lattice sites. (Reproduced with permission from Ref. [9], p. 464.)

In the absence of spatial degrees of freedom, the observed persistent population oscillations can be mathematically understood by performing a systematic van-Kampen expansion about the absorbing steady state [18]. The fluctuation corrections may then essentially be described through a damped harmonic oscillator driven by white noise that will on occasion resonantly incite large-amplitude excursions away from the stable fixed point in the phase plane. In our spatial systems, we may also interpret the persistent population oscillations in the species coexistence regime through a similar mechanism, as suggested by the bottom spiraling trajectory (for large predation rate $\lambda = 1.0$) in the phase portrait depicted in figure 5, which was obtained in simulation runs for stochastic lattice Lotka–Volterra models with restricted site occupancy [9]. As the rate λ is reduced (with all other parameters held constant) and the predators become less efficient, the stochastic lattice systems with site occupation restrictions qualitatively display the same scenarios as revealed by the mean-field analysis for eqs. (4) with finite prey carrying capacity: First, the focal stationary points in the phase plane are replaced by stable nodes (real stability matrix eigenvalues); the population oscillations then cease, and no interesting spatial structures aside from localized activity clusters with meek fluctuations are seen (c.f. the

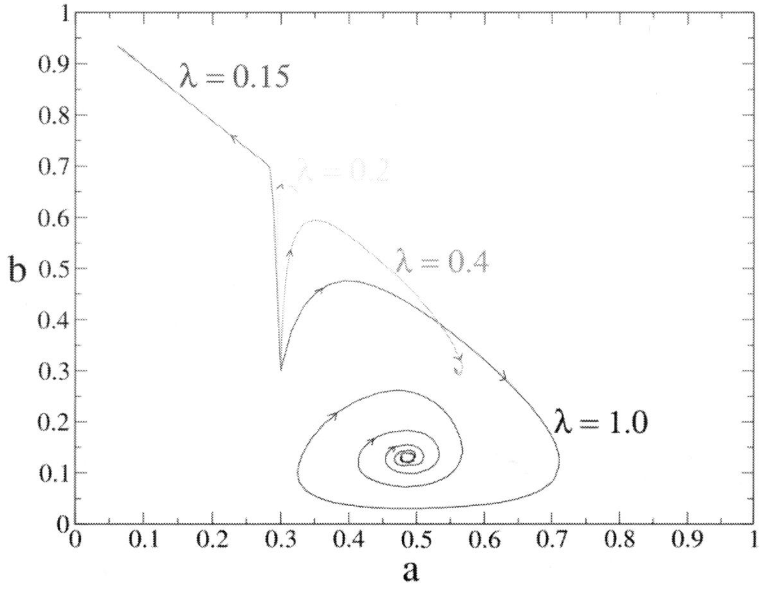

Figure 5. Typical trajectories in the predator-prey phase space for a 512 × 512 stochastic lattice Lotka–Volterra system all initialized with $a(0) = 1/3 = b(0)$ and fixed rates $\sigma = 4.0$, $\mu = 0.1$, and $D = 0$, but different predation rates $\lambda = 0.15, 0.20, 0.40, 1.0$. For small values of λ (typically $\lambda < 0.4$) the fixed point is a stable node, whereas for higher values of λ one observe the characteristic spirals that indicate a focus in phase space. (Reproduced with permission from Ref. [9], p. 463.)

trajectories for $\lambda = 0.4$ and 0.2 in figure 5). At a sufficiently small critical value λ_c (≈ 0.1688 here, see figure 6), a predator extinction threshold is encountered, and for $\lambda < \lambda_c$ ultimately the prey population fills the entire lattice.

For the predator population, the extinction threshold in stochastic spatial Lotka–Volterra models with local particle density restrictions represents a genuine continuous non-equilibrium phase transition in the thermodynamic and infinite-time limit. Since no conserved quantities or disorder are present, one expects this active to absorbing state phase transition to be described by the scaling exponents of critical directed percolation [27]–[32]. Heuristically, one may reason as follows: The prey density is essentially uniform and constant $b \approx \rho$ near the critical point. The Lotka–Volterra reactions (1) then basically reduce to $A \to \emptyset$ and $A \to A + A$; but since the A population cannot multiply to arbitrarily large density (due to prey depletion), we need to add a growth-limiting reaction such as $A + A \to A$, whereupon we arrive at the simplest microscopic reaction-diffusion model realization for the directed-percolation universality class [29, 31, 32, 37]. This assertion is indeed supported by careful analysis of Monte Carlo simulation data [7]–[9], [11]–[14], [16, 17]. We performed dynamical Monte Carlo simulations starting from a single active site with a predator particle in a lattice otherwise filled with prey, choosing reaction rates in the vicinity of the extinction threshold. The survival probability of predators at criticality is expected to decay algebraically as $P(t) \sim t^{-\delta'}$, while the number of active sites with predators should grow according to the power law $N(t) \sim t^{\theta}$ [29, 31, 32], with $\delta' \approx 0.451$ and $\theta \approx 0.230$ for directed percolation in two dimensions [29, 31]. Figure 6 shows the corresponding effective exponents as functions of inverse time as measured in our Monte Carlo simulations for various values of λ with $\sigma = 4.0$, $\mu = 0.1$, and $D = 0$ held fixed [9]. From these data we infer $\lambda_c \approx 0.1688$ as best estimate for the critical predation rate (compare with the mean-field prediction $\lambda_c = \mu = 0.1$), and the extrapolation to $t \to \infty$ yields very good agreement of the asymptotic critical exponents with the accepted directed-percolation values.

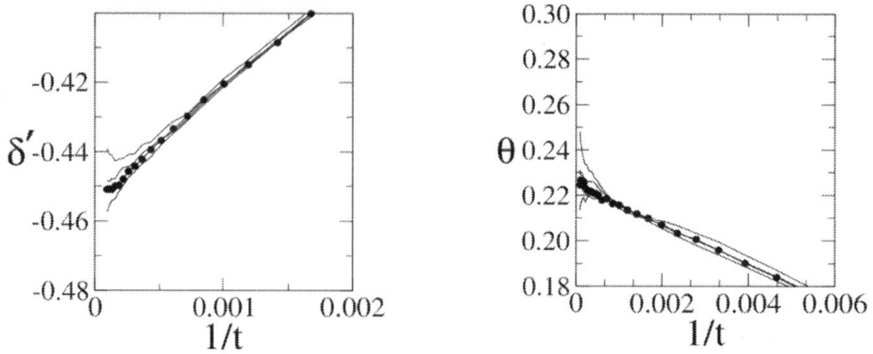

Figure 6. Dynamical Monte Carlo simulation data to estimate the critical point and scaling exponents for the predator extinction threshold of the two-dimensional stochastic lattice Lotka–Volterra model with site occupation restrictions (on a 512×512 lattice): The effective scaling exponents $-\delta'(t)$ vs. $1/t$ (left) and $\theta(t)$ vs. $1/t$ (right) are depicted for four values of λ (from top to bottom): 0.1690, 0.1689, 0.1688, and 0.1687, at fixed $\sigma = 4.0$, $\mu = 0.1$, and $D = 0$. (Reproduced with permission from Ref. [9], p. 470.)

Simulations in one spatial dimension (on a circular domain) yield remarkable differences between model variants with and without site occupation number restrictions: In the former situation, the A and B particles quickly segregate into distinct domains, with the predation reactions occurring only at the boundary. The subsequent time evolution is governed by very slow coarsening induced by merging predator domains [9]. Without site occupation restrictions, in contrast we always observe the active coexistence state [24]. All the above statements of course pertain to sufficiently large lattices. In principle, any finite system with an absorbing steady state will eventually terminate in it; however, the associated survival times are expected to grow with system size according to a power law [19], and for our lattices are much longer than the duration of the simulations.

3. Field-theoretic analysis

3.1. Field theory representation

In the remainder of this paper, I shall describe how stochastic fluctuations, internal reaction noise, and emerging correlations in spatial predator-prey models can be systematically captured by means of a field-theoretic representation of the associated classical master equation, which is then amenable to analytic approximations. Since the two-species Lotka–Volterra model (1) is defined via a diffusion-limited stochastic reaction system, we may employ the by now standard Doi–Peliti framework to map the associated master equation onto a field theory action [33]–[37]. This approach is based on the fact that at any time the configurations in such systems are completely enumerated through specifying the occupation numbers of each species per lattice site, and that all occurring stochastic processes merely modify these local integer occupation numbers. It is therefore natural to use bosonic creation and annihilation operators to formally represent the system's temporal evolution which is given in terms of a stochastic master equation. Subsequently the continuum limit can be taken, which in the time domain is most conveniently accomplished through a coherent-state path integral representation for the evolution operator.

For the diffusion-limited reactions (1) in d spatial dimensions one thus arrives at the following action [9] (see also Ref. [41])

$$S[\hat{a}, a; \hat{b}, b] = \int d^d x \int dt \left[\hat{a} \left(\partial_t - D_A \nabla^2 \right) a + \hat{b} \left(\partial_t - D_B \nabla^2 \right) b \right.$$
$$\left. + \mu \, (\hat{a} - 1) \, a + \sigma \left(1 - \hat{b} \right) \hat{b} \, b + \lambda \left(\hat{b} - \hat{a} \right) \hat{a} \, a \, b \right], \quad (5)$$

with e^{-S} providing the statistical weight for any observables that must be functions of the local 'density' fields $a(\vec{x},t)$ and $b(\vec{x},t)$. The top line here obviously accounts for nearest-neighbor hopping processes in the continuum limit (through the inverse diffusion propagators with diffusivities D_A and D_B for the predators and prey, respectively). The bottom line in (5) contains the stochastic reactions: spontaneous predator death with rate μ, prey birth with rate σ, and predation with rate λ. Note that each reaction process is represented by two contributions, originating from the gain and loss terms in the master equation. One can easily reconstruct these contributions in the action by noting that the second one directly reflects the reaction process itself through the annihilation operators a, b and creation operators $\hat{a}$, $\hat{b}$, whereas the first one encodes the 'order' of the corresponding reaction (i.e., which powers of the concentrations $\hat{a}\,a$ and $\hat{b}\,b$ enter the rate equations). It is important to realize that the Doi–Peliti action faithfully contains all stochastic fluctuations associated with the underlying microscopic processes, namely discrete finite-number fluctuations and internal reaction noise [37]. Following van Wijland's analysis [38], restricted site occupation numbers or finite local carrying capacities ρ for the prey species can be incorporated in this bosonic formalism through the replacement $\sigma \to \sigma\, e^{-\hat{b}b/\rho}$ in the B particle reproduction term. (Alternatively, a growth-limiting reaction such as $B + B \to B$ could have been added.)

The associated classical field equations follow from the stationarity conditions $\delta S/\delta a = 0 = \delta S/\delta b$, always solved by $\hat{a} = 1 = \hat{b}$ (actually just reflecting probability conservation [37]), and $\delta S/\delta\hat{a}(\vec{x},t) = 0 = \delta S/\delta\hat{b}(\vec{x},t)$, which yields precisely the mean-field rate equations augmented by diffusion terms. Indeed, for $\rho = \infty$ one arrives at eqs. (2), while expanding to first order in ρ^{-1} recovers eq. (4). It is then convenient to perform a field shift according to $\hat{a} = 1 + \tilde{a}$, $\hat{b} = 1 + \tilde{b}$, whereupon the action becomes, again to lowest order in the inverse carrying capacity,

$$S[\tilde{a},a;\tilde{b},b] = \int d^dx \int dt \Big[\tilde{a}\left(\partial_t - D_A\nabla^2 + \mu\right) a + \tilde{b}\left(\partial_t - D_B\nabla^2 - \sigma\right) b \\ -\sigma\,\tilde{b}^2\, b + \sigma\,\rho^{-1}\,(1+\tilde{b})^2\,\tilde{b}\,b^2 - \lambda\,(1+\tilde{a})\,(\tilde{a}-\tilde{b})\,a\,b \Big] . \qquad (6)$$

In the following, the 'microscopic' field theory action (6) will serve as the starting point (i) for further manipulations to identify the universality class of the continuous active to absorbing state phase transition at the predator extinction threshold, and (ii) to compute the fluctuation-induced renormalization to lowest order in the predation rate for the population oscillation frequency and damping, as well as the diffusion coefficient in the two-species coexistence phase.

3.2. Extinction transition and directed percolation

Our goal is to construct an effective field theory [9] that describes the universal scaling properties near the non-equilibrium phase transition at $\lambda_c \approx \mu/\rho$ where the predators go extinct, and the prey fill the entire lattice: $a_s = 0$, $b_s \approx \rho$. Consequently we transform the action (6) to new fluctuating fields $c = b_s - b$ with $\langle c \rangle = 0$, and $\tilde{c} = -\tilde{b}$:

$$S[\tilde{a},a;\tilde{c},c] = \int d^dx \int dt \Big[\tilde{a}\left(\partial_t - D_A\nabla^2 + \mu - \lambda\, b_s\right) a + \tilde{c}\left(\partial_t - D_B\nabla^2 + (2b_s/\rho - 1)\,\sigma\right) c \\ +\sigma\, b_s(2b_s/\rho - 1)\,\tilde{c}^2 - \sigma\,\rho^{-1}\, b_s^2\,\tilde{c}^3 - \sigma\,(4b_s/\rho - 1)\,\tilde{c}^2\, c \\ -\sigma\,\rho^{-1}\,(1+\tilde{c}^2)\,\tilde{c}\,c^2 + 2\sigma\,\rho^{-1}\,\tilde{c}^2\,(c + b_s\,\tilde{c})\,c \\ -\lambda\, b_s\left(\tilde{a}^2 + (1+\tilde{a})\,\tilde{c}\right) a + \lambda\,(1+\tilde{a})\,(\tilde{a}+\tilde{c})\,a\,c \Big] . \qquad (7)$$

Next we note that the birth rate is a relevant parameter in the renormalization group sense, which scales to infinity under scale transformations; this observation simply expresses the fact that fluctuations of the nearly uniform prey population become strongly suppressed through the 'mass' term $\propto \sigma$ for the c fields. It is therefore appropriate to introduce rescaled fields $\phi = \sqrt{\sigma}\, c$ and $\tilde{\phi} = \sqrt{\sigma}\, \tilde{c}$, and subsequently take the limit $\sigma \to \infty$, which yields the drastically reduced effective action

$$S_\infty[\tilde{a}, a; \tilde{\phi}, \phi] = \int d^d x \int dt \left[\tilde{a} \left(\partial_t - D_A \nabla^2 + \mu - \lambda\, b_s \right) a - \lambda\, b_s\, \tilde{a}^2\, a + \tilde{\phi}\, \phi + b_s\, \tilde{\phi}^2 \right] . \quad (8)$$

As a final step, one needs to add a growth-limiting process for the predator population, for example through the binary coagulation reaction $A + A \to A$ with rate τ. Since the fields ϕ and $\tilde{\phi}$ only appear as a bilinear form in the action (8), they can readily be integrated out, leaving

$$S_\infty[\tilde{\psi}, \psi] = \int d^d x \int dt \left[\tilde{\psi} \Big(\partial_t + D_A\, (r_A - \nabla^2) \Big) \psi - u\, \tilde{\psi} \left(\tilde{\psi} - \psi \right) \psi + \tau\, \tilde{\psi}^2\, \psi^2 \right] , \quad (9)$$

where $\psi = a\, \sqrt{\lambda\, b_s / \tau}$, $\tilde{\psi} = \tilde{a}\, \sqrt{\tau / \lambda\, b_s}$, $r_A = (\mu - \lambda\, b_s)/D_A$, and $u = \sqrt{\tau\, \lambda\, b_s}$. This new effective non-linear coupling u becomes dimensionless at $d_c = 4$, signifying the upper critical dimension for this field theory. Near four dimensions, the quartic term $\propto \tau$ constitutes an irrelevant contribution in the renormalization group sense and may be omitted for the analysis of universal asymptotic power laws at the phase transition. The action (9) then becomes identical to Reggeon field theory, which is known to describe the critical scaling exponents for directed percolation [27, 32, 39, 40]. This mapping to Reggeon field theory [9] firmly corroborates the expectation that the predator extinction threshold is governed by the directed-percolation universality class [7, 8], [11]–[14], [16, 17], which features quite prominently in phase transitions to absorbing states [27, 28], even in multi-species systems [30]. The universal scaling properties of critical directed percolation are well-understood and quantitatively characterized to remarkable accuracy, both numerically through extensive Monte Carlo simulations and analytically by means of renormalization group calculations (for overviews, see Refs. [29, 31, 32]).

3.3. Fluctuation corrections in the coexistence phase

In order to address fluctuation corrections in the predator-prey coexistence phase [42], we start again from the Doi–Peliti field theory action (6), and introduce the proper fluctuating fields $c = a - \langle a \rangle$ and $d = b - \langle b \rangle$:

$$a = \frac{\sigma}{\lambda} \left(1 - \frac{\mu}{\rho\, \lambda} + A_c \right) + c \, , \quad b = \frac{\mu}{\lambda} \left(1 + B_c \right) + d \, . \quad (10)$$

Here, the mean-field values for the stationary densities have been taken into account already, such that the counter-terms A_c and B_c, which are naturally determined by the conditions $\langle c \rangle = 0 = \langle d \rangle$, contain only fluctuation contributions. The bilinear terms in the ensuing action may then readily be diagonalized by introducing new fields $\varphi_\pm$ and $\tilde{\varphi}_\pm$,

$$\begin{aligned} c &= \frac{1}{\sqrt{2\mu}} \left[\varphi_+ + \varphi_- - \frac{\gamma_0}{i\omega_0} (\varphi_+ - \varphi_-) \right] , \quad d = \sqrt{\frac{\mu}{2}}\, \frac{\varphi_+ - \varphi_-}{i\omega_0} \\ \tilde{a} &= \sqrt{\frac{\mu}{2}}\, \frac{\tilde{\varphi}_+ - \tilde{\varphi}_-}{i\omega_0} , \quad \tilde{b} = \frac{1}{\sqrt{2\mu}} \left[\tilde{\varphi}_+ + \tilde{\varphi}_- + \frac{\gamma_0}{i\omega_0} (\tilde{\varphi}_+ - \tilde{\varphi}_-) \right] , \end{aligned} \quad (11)$$

with the mean-field (or 'bare') oscillation frequency and damping constant (see also Ref. [41])

$$\omega_0^2 = \mu\, \sigma \left(1 - \frac{\mu}{\rho\, \lambda} \right) - \gamma_0^2 \, , \quad \gamma_0 = \frac{\sigma\, \mu}{2\, \rho\, \lambda} \, . \quad (12)$$

Note that $\omega_0^2 = \mu\,\sigma$ and $\gamma_0 \to 0$ as $\rho \to \infty$: There is no damping of the mean-field oscillations in the absence of local carrying capacity restrictions.

In the following, we shall consider equal diffusivities $D_A = D_0 = D_B$; the harmonic propagators in the diagonalized theory then read in Fourier space

$$\langle \widetilde{\phi}_\pm(\vec{q},\omega)\, \phi_\pm(\vec{q}\,',\omega')\rangle_0 = \frac{\pm i\omega_0}{-i\omega + D_0\, q^2 \pm i\omega_0 + \gamma_0}\, (2\pi)^{d+1}\, \delta(\vec{q}+\vec{q}\,')\, \delta(\omega+\omega')\ . \tag{13}$$

Along with two two-point noise sources and several non-linear vertices, these propagators form the building blocks for the Feynman diagrams that graphically represent the different contributions in a perturbation expansion in terms of the non-linear coupling λ [42]. To lowest non-trivial ('one-loop') order, only the noise and three-point vertices are needed to determine the counter-terms A_c and B_c, as well as to compute the fluctuation corrections to the bare propagators (13). From the ensuing one-loop expressions, one may infer renormalized versions of the diffusivity D_R, oscillation frequency ω_R, and damping γ_R. In addition, one finds that in the absence of site occupation restrictions (i.e., for infinite local prey carrying capacity ρ), the stochastic spatial fluctuations generate a damping term, just as seen in the lattice simulations. These perturbational calculations are fairly straightforward, but lengthy and somewhat tedious; details will be reported elsewhere [42]. Here I merely provide the explicit results for the renormalized parameters in several space dimensions.

For $d = 1$ and $d = 2$, the expressions for the renormalized oscillation frequency become singular in the limit $\gamma_0 \to 0$; in the list below, only the leading terms in γ_0 are retained:

$$\begin{aligned}
d=1: \quad & D_R = D_0 + \frac{3\,\lambda}{64\,\sqrt{2}} \sqrt{\frac{D_0}{\omega_0}} \left[1 + \frac{1}{12}\left(\sqrt{\frac{\sigma}{\mu}} - \sqrt{\frac{\mu}{\sigma}}\right) + \frac{3}{4}\left(\frac{\sigma}{\mu} + \frac{\mu}{\sigma}\right)\right] + \mathcal{O}(\lambda^2)\ , \\
& \gamma_R = \frac{\lambda}{8\,\sqrt{2}} \sqrt{\frac{\omega_0}{D_0}} \left[1 + \frac{3}{4}\left(\sqrt{\frac{\sigma}{\mu}} - \sqrt{\frac{\mu}{\sigma}}\right) - \frac{3}{4}\left(\frac{\sigma}{\mu} + \frac{\mu}{\sigma}\right)\right] + \mathcal{O}(\lambda^2)\ , \\
& \omega_R = \omega_0 - \frac{\lambda}{16}\, \frac{\mu\,\sigma}{\sqrt{D_0\,\gamma_0}} \left[1 + \frac{1}{2}\left(\frac{\sigma}{\mu} + \frac{\mu}{\sigma}\right)\right] \\
& \qquad + \frac{\lambda}{8\,\sqrt{2}} \sqrt{\frac{\omega_0}{D_0}} \left[1 - \frac{57}{32}\sqrt{\frac{\sigma}{\mu}} + \frac{25}{32}\sqrt{\frac{\mu}{\sigma}} + \frac{1}{32}\left(\frac{\sigma}{\mu} + \frac{\mu}{\sigma}\right)\right] + \mathcal{O}(\lambda^2)\ .
\end{aligned} \tag{14}$$

$$\begin{aligned}
d=2: \quad & D_R = D_0 + \frac{\lambda}{96\,\pi}\left[1 + 2\left(\frac{\sigma}{\mu} + \frac{\mu}{\sigma}\right)\right] + \mathcal{O}(\lambda^2)\ , \\
& \gamma_R = \frac{\lambda}{64}\,\frac{\omega_0}{D_0}\left[\frac{6}{\pi}\left(\sqrt{\frac{\sigma}{\mu}} - \sqrt{\frac{\mu}{\sigma}}\right) - \left(\frac{\sigma}{\mu} + \frac{\mu}{\sigma}\right)\right] + \mathcal{O}(\lambda^2)\ , \\
& \omega_R = \omega_0 - \frac{\lambda}{32\,\pi}\,\frac{\omega_0}{D_0}\,\ln\frac{\omega_0}{\gamma_0}\left[1 + \frac{1}{2}\left(\frac{\sigma}{\mu} + \frac{\mu}{\sigma}\right)\right] \\
& \qquad + \frac{3\,\lambda}{32\,\pi}\,\frac{\omega_0}{D_0}\left[1 - \frac{\pi}{3}\sqrt{\frac{\sigma}{\mu}} - \frac{1}{4}\left(\frac{\sigma}{\mu} + \frac{\mu}{\sigma}\right)\right] + \mathcal{O}(\lambda^2)\ .
\end{aligned} \tag{15}$$

Notice that the infrared singularities encountered in the limit $\gamma_0 \to 0$ cancel for the renormalized diffusivity D_R and the fluctuation-generated damping γ_R. In dimensions $d < 2$, the leading fluctuation correction to the oscillation frequency diverges as $(\omega_0/\gamma_0)^{1-d/2}$, acquiring a logarithmic dependence in two dimensions; it is negative, and symmetric under formal rate exchange $\mu \leftrightarrow \sigma$ (c.f. the top lines in the above one-loop results for ω_R). If we interpret γ_0 in the above equations as a small, self-consistently determined damping, these features are in remarkable agreement with our earlier Monte Carlo observations displayed in figure 3: Fluctuations and correlations induced by the stochastic reaction processes induce a strong downward numerical renormalization of the oscillation frequency, with very similar functional dependence on the rates μ and σ.

In three dimensions, we may set the bare damping constant to zero (or $\rho \to \infty$) to obtain

$$
\begin{aligned}
d = 3: \quad D_R &= D_0 - \frac{\lambda}{384\sqrt{2}\,\pi}\sqrt{\frac{\omega_0}{D_0}}\left[1 + \frac{9}{4}\left(\sqrt{\frac{\sigma}{\mu}} - \sqrt{\frac{\mu}{\sigma}}\right) - \frac{13}{4}\left(\frac{\sigma}{\mu} + \frac{\mu}{\sigma}\right)\right] + \mathcal{O}(\lambda^2)\,, \\
\gamma_R &= \frac{\lambda}{16\sqrt{2}\,\pi}\left(\frac{\omega_0}{D_0}\right)^{3/2}\left[-1 + \frac{3}{4}\left(\sqrt{\frac{\sigma}{\mu}} - \sqrt{\frac{\mu}{\sigma}}\right) - \frac{1}{4}\left(\frac{\sigma}{\mu} + \frac{\mu}{\sigma}\right)\right] + \mathcal{O}(\lambda^2)\,, \qquad (16) \\
\omega_R &= \omega_0 + \frac{\lambda}{128\sqrt{2}\,\pi}\left(\frac{\omega_0}{D_0}\right)^{3/2}\left[1 - \frac{13}{4}\sqrt{\frac{\sigma}{\mu}} - \frac{19}{4}\sqrt{\frac{\mu}{\sigma}} - \frac{13}{4}\left(\frac{\sigma}{\mu} + \frac{\mu}{\sigma}\right)\right] + \mathcal{O}(\lambda^2)\,.
\end{aligned}
$$

In higher dimensions $d \geq 4$, the fluctuation corrections become formally ultraviolet-divergent, and thus a finite cut-off Λ in momentum space must be implemented; e.g., in four dimensions one finds

$$
\begin{aligned}
d = 4: \quad D_R &= D_0 - \frac{\lambda}{512\,\pi}\frac{\omega_0}{D_0}\left[1 + \frac{1}{\pi}\left(\sqrt{\frac{\sigma}{\mu}} - \sqrt{\frac{\mu}{\sigma}}\right)\ln\left(1 + \frac{\Lambda^4}{\omega_0^2/D_0^2}\right) - \left(\frac{\sigma}{\mu} + \frac{\mu}{\sigma}\right)\right] + \mathcal{O}(\lambda^2)\,, \\
\gamma_R &= \frac{\lambda}{32\,\pi^2}\left(\frac{\omega_0}{D_0}\right)^2\left[1 - \frac{1}{2}\ln\left(1 + \frac{\Lambda^4}{\omega_0^2/D_0^2}\right) + \frac{3\,\pi}{8}\left(\sqrt{\frac{\sigma}{\mu}} - \sqrt{\frac{\mu}{\sigma}}\right) - \frac{1}{4}\left(\frac{\sigma}{\mu} + \frac{\mu}{\sigma}\right)\right] \\
&\quad + \mathcal{O}(\lambda^2)\,, \\
\omega_R &= \omega_0 + \frac{\lambda}{256\,\pi}\left(\frac{\omega_0}{D_0}\right)^2\left[1 - \frac{2}{\pi}\sqrt{\frac{\mu}{\sigma}}\ln\left(1 + \frac{\Lambda^4}{\omega_0^2/D_0^2}\right) - \frac{5}{2\,\pi^2}\left(\sqrt{\frac{\sigma}{\mu}} - \sqrt{\frac{\mu}{\sigma}}\right)\right. \\
&\qquad \left. - \left(\frac{\sigma}{\mu} + \frac{\mu}{\sigma}\right)\right] + \mathcal{O}(\lambda^2)\,. \qquad (17)
\end{aligned}
$$

We finally remark that the effective expansion parameter in this fluctuation perturbation series in d dimensions is $(\lambda/\omega_0)\,(\omega_0/D_0)^{d/2}$.

4. Concluding remarks

In conclusion, in this contribution I have reviewed the most striking features of stochastic predator-prey models on regular lattices that in the well-mixed mean-field limit reduce to the celebrated Lotka–Volterra model. It turns out that the spatially extended stochastic systems display both richer behavior than the associated deterministic rate equations, and are actually also more robust with respect to modifications of model and algorithmic details: Spatial predator-prey systems in the species coexistence phase are generically characterized by the emergence of persistent spatio-temporal structures, namely continually expanding and merging activity fronts, leading to transient oscillations for the total (or mean) particle densities. Fluctuations in the two-species coexistence phase are remarkably and unusually strong; they markedly alter the oscillation frequency as compared to the (linearized) mean-field prediction, and in addition generate damping. Restricting the (local) prey population through a growth-limiting finite carrying capacity induces a genuine continuous non-equilibrium extinction phase transition for the predators. I have also outlined how the Doi–Peliti field theory representation of the associated master equation can be employed to (i) demonstrate that this active to absorbing state transition is governed by the universal scaling exponents of critical directed percolation, and (ii) permits a systematic perturbational approach to compute the fluctuation-induced renormalizations of the population spreading and oscillation parameters in the coexistence phase.

It remains to be elucidated which of the standard mathematical models in ecology, population dynamics, and chemical kinetics, many of which are frequently just discussed on the level of mean-field rate equations, are similarly strongly affected by stochastic fluctuations and intrinsic correlations. Perhaps unexpectedly, stochastic spatial variants of cyclic three-species predator-prey systems that are often referred to as rock-paper-scissors models represent an intriguing counter-example: Lattice simulations of these reaction-diffusion systems hardly show any noticeable fluctuation effects, both for model variants with conserved and non-conserved total particle number, despite the formation of striking spiral structures in the latter, so-called May–Leonard model, see Refs. [43, 44] (and further references therein).

Acknowledgments

The author warmly thanks the organizers of CMDS-12 for their kind invitation to participate at this very stimulating conference. Fruitful collaborations and insightful discussions with Ulrich Dobramysl, Erwin Frey, Ivan Georgiev, Qian He, Swapnil Jawkar, Rahul Kulkarni, Gabriel Martinez, Mauro Mobilia, Tim Newman, Michel Pleimling, Beate Schmittmann, Siddharth Venkat, Mark Washenberger, and Royce Zia are gratefully acknowledged.

References

[1] May R M 1973 *Stability and complexity in model ecosystems*, (Princeton: Princeton University Press)
[2] Maynard Smith J 1974 *Models in ecology* (Cambridge: Cambridge University Press)
[3] Hofbauer J and Sigmund K 1998 *Evolutionary games and population dynamics* (Cambridge: Cambridge University Press)
[4] Murray J D 2002 *Mathematical biology*, Vols. I and II (New York: Springer, 3rd ed.)
[5] Durrett R 1999 *SIAM Review* **41** 677
[6] Matsuda H, Ogita N, Sasaki A and Satō K 1992 *Prog. Theor. Phys.* **88** 1035
[7] Satulovsky J E and Tomé T 1994 *Phys. Rev. E* **49** 5073
[8] Boccara N, Roblin O and Roger M 1994 *Phys. Rev. E* **50** 4531
[9] Mobilia M, Georgiev I T and Täuber UC 2007 *J. Stat. Phys.* **128** 447 [doi: 10.1007/s10955-006-9146-3]
[10] Provata A, Nicolis G and Baras F 1999 *J. Chem. Phys.* **110** 8361
[11] Rozenfeld A F and Albano E V 1999 *Physica A* **266** 322
[12] Lipowski A 1999 *Phys. Rev. E* **60** 5179
[13] Lipowski A and Lipowska D 2000 *Physica A* **276** 456
[14] Monetti R, Rozenfeld A F and Albano E V 2000 *Physica A* **283** 52
[15] Droz M and Pękalski A 2001 *Phys. Rev. E* **63** 051909
[16] Antal T and Droz M 2001 *Phys. Rev. E* **63** 056119
[17] Kowalik M, Lipowski A and Ferreira A L 2002 *Phys. Rev. E* **66** 066107
[18] McKane A J and Newman T J 2005 *Phys. Rev. Lett.* **94** 218102
[19] Parker M and Kamenev A 2009 *Phys. Rev. E* **80** 021129
[20] Dunbar S R 1983 *J. Math. Biol.* **17** 11
[21] Sherratt J, Eagen B T and Lewis M A 1997 *Phil. Trans. R. Soc. Lond. B* **352** 21
[22] de Aguiar M A M, Rauch A M and Bar-Yam Y 2004 *J. Stat. Phys.* **114** 1417
[23] Monte Carlo simulation movies are available at `http://www.phys.vt.edu/ ~tauber/PredatorPrey/movies/`
[24] Washenberger M J, Mobilia M and Täuber UC 2007 *J. Phys. Condens. Matter* **19** 065139 [doi: 10.1088/0953-8984/19/6/065139]
[25] Mobilia M, Georgiev I T and Täuber U C 2006 *Phys. Rev. E* **73**, 040903(R)
[26] Dobramysl U and Täuber U C 2008 *Phys. Rev. Lett.* **101**, 258102
[27] Janssen H K 1981 *Z. Phys. B* **42** 151
[28] Grassberger P 1982 *Z. Phys. B* **47** 365
[29] Hinrichsen H 2000 *Adv. Phys.* **49** 815
[30] Janssen H K 2001 *J. Stat. Phys.* **103** 801
[31] Ódor G 2004 *Rev. Mod. Phys.* **76** 663
[32] Janssen H K and Täuber U C 2005 *Ann. Phys.* **315** 147
[33] Doi M 1976 *J. Phys. A: Math. Gen.* **9** 1465
[34] Grassberger P and Scheunert P 1980 *Fortschr. Phys.* **28** 547
[35] Peliti L 1985 *J. Phys. (France)* **46** 1469; 1479
[36] Mattis D C and Glasser M L 1998 *Rev. Mod. Phys.* **70** 979
[37] Täuber U C, Howard M and Vollmayr-Lee B P 2005 *J. Phys. A: Math. Gen.* **38** R79
[38] van Wijland F 2001 *Phys. Rev. E* **63** 022101
[39] Obukhov S P 1980 *Physica A* **101** 145
[40] Cardy J L and Sugar R L 1980 *J. Phys. A: Math. Gen.* **13** L423

[41] Butler T and Reynolds D 2009 *Phys. Rev. E* **79** 032901
[42] Täuber U C 2011 *manuscript in preparation*
[43] He Q, Mobilia M and Täuber U C 2010 *Phys. Rev. E* **82** 051909
[44] He Q, Mobilia M and Täuber U C 2011 *Eur. Phys. J. B* in press

Journal of Physics: Conference Series **319** (2011) 012020 doi:10.1088/1742-6596/319/1/012020

Higher Order Macro Coefficients in Periodic Homogenization

Carlos Conca

Departamento de Ingeniería Matemática, Facultad de Ciencias Físicas y Matemáticas, Universidad de Chile and Centro de Modelamiento Matemático, UMR 2071 CNRS-UChile, Casilla 170/3 - Correo 3, Santiago, Chile.

Jorge San Martín

Departamento de Ingeniería Matemática, Facultad de Ciencias Físicas y Matemáticas, Universidad de Chile and Centro de Modelamiento Matemático, UMR 2071 CNRS-UChile, Casilla 170/3 - Correo 3, Santiago, Chile.

Loredana Smaranda

Department of Mathematics, Faculty of Mathematics and Computer Science, University of Piteşti, 110040 Piteşti, Str. Târgu din Vale Nr.1, Argeş, Romania.

Muthusamy Vanninathan

TIFR-CAM, Post Bag 6503, GKVK Post, Bangalore - 560065, India.

E-mail: `cconca@dim.uchile.cl`, `jorge@dim.uchile.cl`, `smaranda@dim.uchile.cl`, `vanni@math.tifrbng.res.in`

Abstract. A first set of macro coefficients known as the homogenized coefficients appear in the homogenization of PDE on periodic structures. If energy is increased or scale is decreased, these coefficients do not provide adequate approximation. Using Bloch decomposition, it is first realized that the above coefficients correspond to the lowest energy and the largest scale. This naturally paves the way to introduce other sets of macro coefficients corresponding to higher energies and lower scales which yield better approximation. The next task is to compare their properties with those of the homogenized coefficients. This article reviews these developments along with some new results yet to be published.

1. Introduction

In this paper, we revisit the classical problem of homogenization of fine periodic structures ([1], [2], [3]) and report on some fundamental developments that have taken place recently. In Homogenization Theory, a basic model problem regards such periodic structures as heat conductors with fine heterogeneities in which heat conductivities vary rapidly. The main feature in this problem is that the temperature field exhibits oscillations making realistic computations very difficult. In general, Homogenization Theory [4] is a study of oscillations of physical fields and one seeks tools and methods to describe these oscillations. Approximation to the physical fields and heat conductivities at a macro scale is deduced as a consequence. This process involves averaging of oscillations, **homogenized matrix** and **homogenized equation**. In periodic media, the oscillations of physical fields are somewhat restricted; they are created by periodic distribution of microstructures in the media. Yet, they are rich enough to exhibit a variety of phenomena which require sophisticated theory for their understanding. They provide a fertile ground to study the dependence of physical fields on the coefficients of the equation, and on the microstructure, in particular. Another importance of periodic media is that arbitrary heterogeneous structures can be locally approximated by periodic ones since the homogenization process itself is local in this case.

In the periodic setting, there are many methods of homogenization [5]. Our contribution [6] is a spectral approach based on Bloch Waves. One of the advantages of our method is that it enables us to go beyond the homogenized matrix and introduce other macro coefficients in a systematic manner. We call them **higher order macro coefficients**. While the usual homogenized coefficients correspond to the approximation at the lowest energy and the largest scale, the higher order macro coefficients correspond to higher energy or smaller scales, as we shall see. Naturally, these higher order macro coefficients are useful in the study of finer aspects of oscillations.

Let us now give the plan of this write-up. In Section 2, after formally introducing periodic homogenization problem, we give the conventional definition of the homogenized matrix as well as our principal example of two-phase conducting composite. We realize immediately that it is not clear how to generalize this definition to introduce its higher order version. Our main tool to overcome this difficulty is Bloch waves which are introduced in Section 3. Section 4 establishes the connection between Bloch waves and homogenization and this paves the way to define higher order macro coefficients. Main results of this paper are included in Section 5; they compare the properties of homogenized coefficients and their higher order version. In particular, we will define higher order analogues of two prominent homogenized coefficients in the case of two-phase conductors, namely the Arithmetic Mean and the Harmonic Mean. Section 6 ends the article with conclusions and perspectives.

2. Periodic Structures and Homogenization

Starting with the basic reference cell $Y =]0, 2\pi[^N \subset \mathbb{R}^N$, one can tile the entire space $\mathbb{R}^N$ by integer translations of the cell Y. This defines the periodic structure in unit scale in which the independent variable is denoted by y. We can increase the complexity of the structure by considering the rescaled periodic tiling with period $2\pi\varepsilon$ where $\varepsilon > 0$ is a small parameter. The corresponding independent variable is denoted by x and we note the scaling relation $y = \frac{x}{\varepsilon}$. We would like to view these structures as heat conductors. To this end, we introduce conductivity matrix $a(y) = (a_{kl}(y))$ of order $N \times N$ enjoying the following properties:

$$a(y) \text{ is } Y\text{- periodic } : a(y + 2\pi p) = a(y) \quad \forall p \in \mathbb{Z}^N,$$

$$a(y) \text{ is symmetric},$$

$$\text{For some } \alpha > 0,\ a(y)\eta \cdot \eta \geq \alpha|\eta|^2 \quad \forall \eta \in \mathbb{R}^N \text{ (uniform positive definiteness)},$$

$$\text{For some } M,\ |a_{kl}(y)| \leq M \text{ (uniform boundedness)}.$$

The corresponding ε - scaled conductivity matrix is denoted by $a^\varepsilon(x) = a(\frac{x}{\varepsilon})$. Steady state heat conduction in such media is governed by the following elliptic operators in divergence form:

$$Au \equiv -\frac{\partial}{\partial y_k}\left(a_{kl}(y)\frac{\partial u}{\partial y_l}\right),$$

$$A^\varepsilon u^\varepsilon \equiv -\frac{\partial}{\partial x_k}\left(a_{kl}\left(\frac{x}{\varepsilon}\right)\frac{\partial u^\varepsilon}{\partial x_l}\right).$$

These operators represent periodic structures in the same way as Laplace operator did for homogeneous media.
Example (Two - Phase conductor). Let $\alpha_0 > 0, \alpha_1 > 0$ represent the conductivities and let $0 < \gamma < 1$. We define the conductivity matrix by $a_{kl}(y) = a(y)\delta_{kl}$ with $a(y) = \alpha_1\chi_T(y) + \alpha_0\chi_{Y\setminus T}(y)$. Here $T \subset Y$ is a measurable subset and $\gamma = \frac{|T|}{|Y|}$. In other words, we have two phases (α_0, α_1) in volume proportions $(1-\gamma, \gamma)$ respectively in our heat conductor. We call the arrangement of the subset T inside Y, alternatively, the characteristic function χ_T, as the **microstructure** of the two-phase medium.

One of the fundamental questions of Homogenization Theory can now be phrased as follows: Consider oscillating temperature fields $\{u^\varepsilon\}$ with weak limit(average) u as $\varepsilon \to 0$. Their oscillations are restricted by

$$\int_{\mathbb{R}^N} |u^\varepsilon|^2 + \int_{\mathbb{R}^N} |\nabla u^\varepsilon|^2 \le c,$$

$$A^\varepsilon u^\varepsilon = f \text{ in } \mathbb{R}^N.$$

Here the constant $c > 0$ and the heat source density f are independent of ε. Under the above conditions, one wishes to describe the oscillations in u^ε as $\varepsilon \to 0$; in particular, one is interested in the equation satisfied by u. The real issue involved is to compute the weak limit (average) of the heat fluxes

$$a_{kl}\Big(\frac{x}{\varepsilon}\Big)\frac{\partial u^\varepsilon}{\partial x_l}(x), \quad k = 1, 2, \ldots, N.$$

Taking averages of nonlinear functionals of oscillating quantities being not easy, one can easily appreciate the difficulties involved in this homogenization problem which requires the average of the above quadratic functional. These were first overcome and the equation satisfied by u was found in [7].

The pioneering approach in [4] solves this problem in two steps. In the first, one shows that the limit of Fourier materials is a material of same type. More precisely, one shows that the (weak) limit of the heat flux is of the form $q_{kl}\frac{\partial u}{\partial x_l}$ for some constant matrix $q = (q_{kl})$, called the **homogenized matrix**. This is a general phenomenon going well beyond periodic microstructures. In particular, this step shows the existence of a composite medium which is really a mixture of several individual phases with given volume proportions and with arbitrary micro-geometry of mixing. The underlying mathematical theory is the so-called **compensated compactness**. For a nice account of this, see [4].

In the second step, one computes the matrix q by placing the periodic medium in the temperature field $\eta \cdot y$ and computing the average of the resulting heat flux [4]. This amounts to solving the following cell test problem:

$$Aw = 0 \text{ in } \mathbb{R}^N, \quad w(y) - \eta \cdot y \text{ is } Y\text{-periodic.}$$

Then the homogenized matrix q is given by

$$q\eta = \langle a(y)\nabla w(y)\rangle \quad \forall \eta \in \mathbb{R}^N,$$

where the angle brackets represent average over the periodic cell Y. Alternatively, setting $\chi^{(1)}(y) = w(y) - \eta \cdot y$, we see that the **cell test function** $\chi^{(1)}$ satisfies

$$A\chi^{(1)} = \eta_l \frac{\partial a_{kl}}{\partial y_k}(y), \quad \chi^{(1)} \text{ is } Y\text{- periodic.}$$

Further, the homogenized matrix q can be expressed in terms of $\chi^{(1)}$ as follows:

$$q\eta = \langle a(y)\nabla(\chi^{(1)} + \eta \cdot y)\rangle \quad \forall \eta \in \mathbb{R}^N.$$

Apart from the explicit dependence on the microstrucutre via $a(y)$, the homogenized matrix q depends on the microstructure implicitly through the cell test function $\chi^{(1)}$. In the sequel, it will be convenient to express $\chi^{(1)}$ as a linear superposition of χ_l, $l = 1, 2, \ldots, N$: $\chi^{(1)} = \eta_l \chi_l$, where χ_l is solution of

$$A\chi_l = \frac{\partial a_{kl}}{\partial y_k}(y), \quad \chi_l \text{ is } Y\text{-periodic}.$$

Finally, combining the two previous steps, we conclude that the weak limit u satisfies the **homogenized equation**:

$$Qu \equiv -\frac{\partial}{\partial x_k}\left(q_{kl}\frac{\partial u}{\partial x_l}\right) = f \text{ in } \mathbb{R}^N.$$

3. Bloch Waves

In this section, we briefly sketch the Bloch waves associated with the periodic structure and the operator A on it. Some of the original references in the subject are [8], [9], [10]. For further details, the interested reader is referred to our papers on Bloch waves and their applications cited in the list of References.

Just as plane waves/Fourier waves $\{e^{iy\cdot\eta}; \eta \in \mathbb{R}^N\}$ form an orthonormal basis (in a generalized sense) in homogeneous media, Bloch waves form an orthonormal basis of eigenvectors of A in periodic media. To introduce them as a generalization of plane waves, let us recall the property characterizing the plane waves; namely they are common eigenvectors for all $\mathbb{R}^N-$ translation operators: $e^{i(y+y_0)\cdot\eta} = e^{iy_0\cdot\eta}e^{iy\cdot\eta}$. Replacing $\mathbb{R}^N$ - translations by $\mathbb{Z}^N$ - translations, we are motivated to define **Bloch waves** as eigenvectors of the following eigenvalue problem which seeks simultaneous eigenvectors of A and integer translation operators:

$$\begin{gathered}\psi = \psi(y,\eta),\\ A\psi = \lambda(\eta)\psi \text{ in } \mathbb{R}^N,\\ \psi(\cdot,\eta) \text{ is } (\eta - Y) \text{ periodic},\\ i.e., \psi(y + 2\pi p, \eta) = e^{2\pi i p\cdot\eta}\psi(y,\eta)\ \forall p \in \mathbb{Z}^N.\end{gathered}$$

Since (η, Y) periodicity condition is invariant if we replace η by $\eta + q, q \in \mathbb{Z}^N$, it is enough to restrict η in the **dual cell** $Y' =]-\frac{1}{2}, \frac{1}{2}[^N$. Undoubtedly, the novelty of the above system is the (η, Y) periodicity condition. One can offer one more explanation for it, based on the Theory of Group Representations. In fact, the microstructure being periodic, has invariance group $\mathbb{Z}^N$ which is abelian. According to the general theory [11], all its irreducible unitary representations are one-dimensional and given by its characters. Though this may be a more abstract way of rephrasing (η, Y) periodicity condition, there are advantages: it points to a way of treating general non-periodic microstructures admitting an invariance group.

By means of the change of variables $\psi(y,\eta) = e^{iy\cdot\eta}\phi(y,\eta)$, the above problem is transformed to the following one:

$$\begin{gathered}A(\eta)\phi = \lambda(\eta)\phi \text{ in } \mathbb{R}^N,\\ \phi(\cdot,\eta) \text{ is } Y\text{- periodic},\end{gathered}$$

where we see the appearance of the **translated operator:**

$$A(\eta) \equiv -\left(\frac{\partial}{\partial y_k} + i\eta_k\right)\left\{a_{kl}(y)\left(\frac{\partial}{\partial y_l} + i\eta_l\right)\right\}.$$

Spectral theory of compact self-adjoint operators applies here. It shows that the operator $A(\eta)$ can be diagonalized in an orthonormal basis $\{\phi_m(y,\eta); m \in \mathbb{N}\}$ with the corresponding eigenvalues $\{\lambda_m(\eta); m \in \mathbb{N}\}$.

Finally to get the spectral decomposition of A in $\mathbb{R}^N$, it is enough to put the previous elements together by varying η continuously in the dual cell $Y' =]-\frac{1}{2},\frac{1}{2}[^N$. This process gives rise to what is known as **Bloch decomposition** of functions, analogous to the classical **Fourier decomposition**. Roughly speaking, any square integrable function $f(y)$ defined on $\mathbb{R}^N$ can be expressed as a superposition of basis functions $\{\phi_m(y,\eta)e^{iy\cdot\eta}; \eta \in Y', m \in \mathbb{N}\}$, each of which is an eigenvector for A. The coefficients appearing in this superposition are called **Bloch coefficients** which are analogous to **Fourier coefficients**. Finally, the above decomposition is unitary, i.e., it preserves the respective norms. For details, see for instance [6].

4. Bloch waves, Homogenization and Macro Coefficients

Though not evident, somewhat surprising connections exist between homogenization and Bloch waves and our idea is to exploit them further and introduce macro coefficients at various orders. Roughly speaking, Bloch eigenvalue problem introduced in the last section contains a wealth of information about the periodic structure; in particular, its infinitesimal versions at $\eta = 0$ yield information on homogenization. To make this statement more meaningful, we need to make a few observations.

Consider the parametrized eigenvalue problem for the translated operator $A(\eta)$ written down in the last section. Even though $A(\eta)$ depends quadratically on η, it does not imply any regularity beyond Lipschitz property of its eigenvalues $\{\lambda_m(\eta); m \in \mathbb{N}\}$ with respect to η. This is because eigenvalues may coalesce as η varies and become degenerate. However miracle occurs at $\eta = 0$. Indeed, when $\eta = 0$, the first eigenvalue $\lambda_1(0) = 0$ of $A(0) = A$ is simple with eigenvectors being constants. Under such circumstances, regular perturbation theory of Kato-Rellich ([12], [13]) can be applied to show that for a suitable choice of first eigenvector $\phi_1(y,\eta)$, the map $\eta \mapsto (\lambda_1(\eta), \phi_1(y,\eta))$ is analytic near $\eta = 0$. Further, starting from "initial condition" $(\lambda_1(0), \phi_1(y,0))$, one can compute all Taylor coefficients by differentiating the eigen-equation successively and evaluating at $\eta = 0$. In this way, one shows that (see [6]):

- All odd order derivatives of $\lambda_1(\eta)$ at $\eta = 0$ vanish; thus we have

$$\lambda_1(\eta) = \frac{1}{2!}\lambda_1^{(2)}(0)\eta^2 + \frac{1}{4!}\lambda_1^{(4)}(0)\eta^4 + \frac{1}{6!}\lambda_1^{(6)}(0)\eta^6 + \cdots$$

- Taylor coefficient $\frac{1}{2!}\lambda_1^{(2)}(0)$ coincides with the homogenized matrix q (see [14]).
- First order derivatives of the first eigenvector coincide with the cell test functions introduced in Section 2:

$$\frac{\partial \phi_1}{\partial \eta_l}(y,0) = i\chi_l(y), \quad l = 1,2\ldots,N.$$

Above results lead to the following nice spectral interpretation of homogenization process: since Bloch waves diagonalize A with eigenvalues $\{\lambda_m(\eta); m \in \mathbb{N}, \eta \in Y'\}$, we have a nice decomposition of the operator A on the periodic structure in terms of discrete energy levels $m \in \mathbb{N}$ and continuous (quasi) momentum $\eta \in Y'$. Homogenization approximation corresponds to the lowest energy level $m = 1$ and the lowest momentum $\eta = 0$ or equivalently the largest length scale.

Above interpretation shows two directions to proceed further.

Direction 1.(High Energy Homogenization). One can rise the energy level and fix it at an arbitrary value $m \in \mathbb{N}$ and seek homogenization approximation at that level. This leads entirely to a new set of structural conditions on the periodic structure under which one can perform homogenization [15]. They are as follows:

(a) $\eta = 0$ is the minimizer of $\lambda_m(\eta)$ as η varies over the dual cell and it is unique:

$$\lambda_m(0) = \min\{\lambda_m(\eta); \eta \in Y'\}.$$

(b) $\lambda_m(0)$ is a simple eigenvalue of the operator $A(0) = A$.

(c) The Hessian matrix $\lambda_m^{(2)}(0)$ is positive definite.

When $m = 1$, all the above conditions $(a) - (c)$ are automatically satisfied. When $m > 1$, they need not be satisfied and we need to postulate them. If they are not satisfied, new physical phenomena appear ([16], [17]). For instance, if the condition (c) above is not satisfied, we have loss of convexity of $\lambda_m(\eta)$ at $\eta = 0$. This leads to a leakage of energy from the mth mode to other modes and as a consequence, the homogenized limit will not be a scalar equation, but it is a system coupling all these modes. However, if the above conditions are satisfied, one can prove a homogenization theorem involving only the mth mode and show that the homogenized matrix at the energy level $m \in \mathbb{N}$ is given by $\frac{1}{2!}\lambda_m^{(2)}(0)$. The main difficulty at high energy levels is that the **factorization principle** is not valid; that is, the eigenvector $\phi_m(y, 0)$ need not conserve a constant sign. This principle states roughly that if the eigenvector keeps a constant sign, it can be factorized out, giving place to a simpler problem to homogenize. High energy homogenization problems do not fall in this class. To overcome this difficulty, we exploit Bloch decomposition along with the above structural conditions and the notion of **two-scale convergence** [18] to carry out homogenization process. See [15] for details. It is an open problem to translate the above conditions (a) - (c) concretely in terms of microstructures in the physical space.

Direction 2.(Homogenization with smaller scales) We start with the remark that one approximates $\lambda_1(\eta)$ by $\frac{1}{2}\lambda_1^{(2)}(0)\eta^2$ in the homogenization process. Now, one maintains the energy level at $m = 1$ and decreases the length scale to have a closer look at the effects of microstructure, that is, one increases slightly the value of $|\eta|$ to the extent that $\frac{1}{2}\lambda_1^{(2)}(0)\eta^2$ is no more a good approximation for $\lambda_1(\eta)$ and we need the next term of the Taylor expansion to have a good approximation:

$$\lambda_1(\eta) \approx \frac{1}{2!}\lambda_1^{(2)}(0)\eta^2 + \frac{1}{4!}\lambda_1^{(4)}(0)\eta^4.$$

In such situations, one observes that the homogenized matrix q alone is not good enough to provide a good approximation to the inhomogeneous medium at the energy level $m = 1$ and the chosen length scale. We need the 4th order tensor $d = \frac{1}{4!}\lambda_1^{(4)}(0)$ also. Indeed, let us consider the periodic medium with fine microstrucuture described by A^ε introduced in Section 2. The first Bloch eigenvalue on this structure satisfies the scaling relation: $\lambda_1^\varepsilon(\xi) = \varepsilon^{-2}\lambda_1(\varepsilon\xi)$. Expanding this around $\xi = 0$, we get

$$\lambda_1^\varepsilon(\xi) = \frac{1}{2!}\lambda_1^{(2)}(0)\xi^2 + \frac{1}{4!}\varepsilon^2\lambda_1^{(4)}(0)\xi^4 + \frac{1}{6!}\varepsilon^4\lambda_1^{(6)}(0)\xi^6 + \cdots$$

Consider now the problem of acoustic wave propagation in such a medium with well-prepared initial conditions at energy level $m = 1$:

$$v_{tt}^\varepsilon + A^\varepsilon v^\varepsilon = 0 \text{ in } \mathbb{R}^N \times \mathbb{R},$$

$$v^\varepsilon(x, 0) = \phi_1\left(\frac{x}{\varepsilon}, 0\right) v_0(x),$$

$$v_t^\varepsilon(x,0) = \phi_1\left(\frac{x}{\varepsilon},0\right) v_1(x).$$

Since the size of the microstructure and the wave length of propagating waves are of same order, there is a resonant interaction between them. Thus, homogenization process and Bloch waves play a role in this propagation problem. For waves satisfying $|\xi| = o(\varepsilon^{-\frac{1}{2}})$, we see that $\varepsilon^2|\xi|^4 = o(1)$ which shows that all terms in the expansion except the first one can be neglected and so the usual homogenized medium provides a good description of propagation of such waves. For shorter waves satisfying $|\xi| = o(\varepsilon^{-\frac{2}{3}})$, we have $\varepsilon^4|\xi|^6 = o(1)$ and $\varepsilon^2|\xi|^4 = o(\varepsilon^{-\frac{2}{3}})$ and so we cannot neglect the first term as well as the second one for reasonable approximation. Thus both (q,d) are relevant for the approximate description of propagation of such short waves. To see the consequence, let us focus on the one-dimensional case, for simplicity. According to the above discussion, we have the approximation

$$A^\varepsilon \approx -q\frac{d^2}{dx^2} + \varepsilon^2 d\frac{d^4}{dx^4}.$$

The presence of 4th order derivatives signifies that the short waves under consideration undergo dispersion, a feature not captured by the homogenized operator. This dispersion effect is due to the decrease of scales.

Taking together the two directions described above, we have a general picture of periodic structures: Derivatives of various orders of eigenvalues $\lambda_m(\eta)$ at $\eta = 0$ (if they exist) provide macro coefficients associated with the periodic structure at various orders of scales. Similarly, derivatives of various orders of eigenvectors $\phi_m(y,\eta)$ at $\eta = 0$ (if they exist) yield different cell test functions containing hierarchy of information about the microstructure. Viewing through Bloch decomposition, we see that at each energy level $m \in \mathbb{N}$, there is a sequence of intermediate scales indexed by $l \in \mathbb{N}$ ranging from the macroscopic level (at which the usual homogenization is done) upto the scale corresponding to microstructure. Taking $m = 1$ for instance, for each such scale $l \in \mathbb{N}$, there is a new macro coefficient given by the $(2l)^{\text{th}}$ order derivative of $\lambda_1(\eta)$ at $\eta = 0$ which incorporates the microstructure information through the cell test functions defined by derivatives of $\phi_1(y,\eta)$ at $\eta = 0$ upto order l. The new tensors which emerge out of the above picture, namely $q_m = \frac{1}{2!}\lambda_m^{(2)}(0), d = \frac{1}{4!}\lambda_1^{(4)}(0)$ etc. are called **higher order macro coefficients.**

In the sequel, we pursue the Direction 2 and report on the results obtained. When $m = 1$, because of analyticity, all derivatives of $\lambda_1(\eta)$ and $\phi_1(y,\eta)$ exist at $\eta = 0$. Thus the existence question regarding macro coefficients is settled and we focus on their properties.

Example: Before proceeding to the next section, let us consider the example of homogeneous medium defined by the conductivity matrix $a_{kl}(y) \equiv \delta_{kl}$. In this case, the Bloch waves are naturally indexed by $m \in \mathbb{Z}^N$, they are independent of η and given by $\phi_m(y,\eta) = e^{im\cdot y}$. The associated eigenvalues are quadratic and given by $\lambda_m(\eta) = |m+\eta|^2 \ \forall\ m \in \mathbb{Z}^N$. Further, the cell test functions $\chi^{(1)}, \chi^{(2)}$ are all constants and consequently, the homogenized matrix and their higher order versions are trivially found to be

$$q = I, \quad q_m = I, \quad d = 0.$$

5. Properties of Macro Coefficients

This is the main part of this article, which recalls some properties of the homogenized matrix q and gives parallel developments about the tensor d. For a detailed study of q, see [19], [20] or [21]. Our presentation will highlight the differences between q and d. As is evident from two-phase conductors, these quantities depend on conductivities of individual phases and more importantly on the microstrucure (e-g: their volume proportions and other statistical measures). Motivated by applications in Optimal Design Problems, the goal of the theory is to bring out the dependence on the microstructure in its entirety at various intermediate scales mentioned in the previous section. Our results are mainly in the form of bounds on these macro quantities valid independent of microstructure or at best depending on volume proportions of phases.

5.1. Spatial Representation

For (q, d), we already have a representation in the (energy-momentum) space:

$$q = \frac{1}{2!}\lambda_1^{(2)}(0), \quad d = \frac{1}{4!}\lambda_1^{(4)}(0).$$

In order to bring out their dependence on the microstructure, it is essential that we seek spatial representation for them. In the case of q, we saw such a representation, namely a spatial average in Section 2. Let us recall it here:

$$q\eta = \left\langle a(y)\nabla(\chi^{(1)} + \eta \cdot y)\right\rangle$$

which can be used to deduce an expression for the quadratic form associated with q :

$$q\eta^2 = q\eta \cdot \eta = \left\langle a(y)\nabla(\chi^{(1)} + \eta \cdot y) \cdot \nabla(\chi^{(1)} + \eta \cdot y)\right\rangle \quad \forall\, \eta \in \mathbb{R}^N,$$

from which bounds below follow at once

$$\min\{|a(y)|; y \in Y\} \leq q \leq \max\{|a(y)|; y \in Y\}.$$

In particular, q is positive definite. These results are classical and well-known in the literature. What is perhaps not well-known is that an analogous representation exists for the tensor d. In fact, the corresponding 4th order form is given by [22]:

$$d\eta^4 = -\langle a(y)\nabla(\chi^{(2)} - \tfrac{1}{2}(\chi^{(1)})^2) \cdot \nabla(\chi^{(2)} - \tfrac{1}{2}(\chi^{(1)})^2)\rangle \quad \forall\, \eta \in \mathbb{R}^N.$$

Here appears a new cell test function $\chi^{(2)}$ capturing more information on the microstructure, and it is defined as follows: $\chi^{(2)}$ is Y-periodic and satisfies

$$-\text{div}(a(y)\nabla\chi^{(2)}) = (a(y) - q)\eta^2 + a(y)\nabla\chi^{(1)} \cdot \eta + \text{div}(a(y)\eta\chi^{(1)}).$$

As one may expect, $\chi^{(2)}$ is related to the second order derivatives of $\phi_1(y, \eta)$ with respect to η at $\eta = 0$. An immediate consequence of above representation is the following universal sign for d independent of microstructure:

$$d\eta^4 \leq 0, \quad \forall \eta \in \mathbb{R}^N.$$

This result is somewhat surprising because one does not expect the tensor d to have a sign at this generality. Secondly, the sign of d is opposite to that of q. One may ask whether the next order macro tensor $\frac{1}{6!}\lambda_1^{(6)}(0)$ keeps a non-negative sign. Unfortunately, this is not so [23].

Another observation on the above representation is the following: while q depends on the microstructure implicitly through $\chi^{(1)}$, the dependence of d is through both $(\chi^{(1)}, \chi^{(2)})$. Further, the representation for d contains some nonlinear terms signifying interaction between the scales including a self interaction term $(\chi^{(1)})^2$. This is a new feature which is absent in q.

5.2. One Dimensional Results

Let us consider the case of conductors with two-phases $\{\alpha_0, \alpha_1\}$ in proportions $\{1-\gamma, \gamma\}$. In this case, it is well-known that the value of the homogenized coefficient (which is a scalar) remains fixed at the Harmonic Mean of $\{\alpha_0, \alpha_1\}$ as microstructure varies:

$$\frac{1}{q} = \frac{1-\gamma}{\alpha_0} + \frac{\gamma}{\alpha_1}.$$

In contrast, as we shall see, (the scalar) d shows rich variation with the microstructure. In such a case, one would like to exploit the negativity of d in optimization problems involving microstructures. This begs the following questions: how negative d can be when microstructure varies? To answer this, we must minimize d over microstructures. Is the minimizing microstructure classical or generalized/relaxed one? What is the minimum value? In the present one dimensional case, we provide complete answers to these questions. In multi-dimensional problems, these questions remain open, in general. Let us state our precise results.

Theorem ([25]) Under the above situation, we have

(i) The minimum value of d is given by

$$\min d = -\frac{1}{12}|Y|^2 q^3 \gamma^2 (1-\gamma)^2 \left(\frac{1}{\alpha_1} - \frac{1}{\alpha_0}\right)^2.$$

(ii) The maximum value of d is zero.

(iii) The minimizer is unique upto translations and is given by the following classical microstructure: the phases $\{\alpha_0, \alpha_1\}$ occupy intervals of lengths $\{(1-\gamma)|Y|, \gamma|Y|\}$ inside Y.

(iv) The maximizer is also unique and is given by the following generalized/relaxed microstructure: the phases $\{\alpha_0, \alpha_1\}$ are present everywhere with uniform local proportions $\{1-\gamma, \gamma\}$ at every point y.

(v) The values of d over classical microstructures range over the entire interval $[\min d, \max d[$.

(vi) d has the monotonicity property: more interfaces between the two phases in the micrsotructure, the higher the value of d.

Let us make several remarks on the above Theorem. One general lesson learnt from the above Theorem is the following: many complicated microstructures may be associated with a particular value of q; however when d is considered along with q, a particularly simple microstructure is chosen among them as minimizer for d. This remarkable property is stated in (iii) above in the one dimensional case. The minimum value d is a curious combination of $(|Y|, \gamma, \alpha_0, \alpha_1)$. Unlike the Harmonic Mean, it is not clear how this is related to the standard statistical quantities associated with the microstructure. In optimization problems involving microstructures, the optimizers are, in general, generalized/relaxed microstructures. It is exceptional here that one finds the minimizer to be a classical microstructure. This is due to the monotonicity property of d stated in (vi): At its minimum value, d has just two interfaces between the phases $\{\alpha_0, \alpha_1\}$, whereas d has a continuum of interfaces at its maximum value. These properties could be exploited in high contrast materials to produce ones with negative index, particularly at scales where d is significant. Finally, it is surprising to see continuous variation of d over periodic classical microstructures: d assumes all intermediate values between its minimum and its maximum. This is very different from the behaviour of q. In one-dimension, its value does not change at all and remains fixed at the Harmonic Mean. In higher dimensions, the eigenvalues of q over all periodic microstructures are merely dense in the corresponding set over arbitrary microstructures [20].

Let us say a few words about the proof of the above result. First of all, two simplifications occur in the physical space representation formula for d, given in Subsection 5.1. Dependence of

d on the microstructure through the cell test function $\chi^{(2)}$ can be taken into account explicitly and thus $\chi^{(2)}$ can be eliminated. Similarly, the self-interaction term also disappears from the scene yielding a simple representation formula for d ([22]) :

$$d = -q\left\langle v^2\right\rangle,$$

where v is defined more directly in terms of microstructure via

$$-\frac{dv}{dy} = 1 - q\left(\frac{\chi_T(y)}{\alpha_1} + \frac{1-\chi_T(y)}{\alpha_0}\right) \text{ in } \mathbb{R},$$

$$v \text{ is } Y \text{ - periodic.}$$

Observe that the minimization problem for d has the structure of Optimal Control Problem with microstructures being controls and the equation for v as the state equation and d as the cost functional. As usual, the main difficulty in minimization problems over microstructures is the non-convexity.

In the problem at hand, we see non-convexity at two places. One is the non-convexity of the constraint set on controls: convex combination of characteristic functions need not be a characteristic function. This can be rectified by replacing the original problem by its relaxed version in which characteristic functions are replaced by local density functions $0 \le \theta(y) \le 1$, which are weak limits of characteristic functions. This process called **relaxation** does not alter the minimum value of the problem. For a general description of relaxation procedures in optimal deign problems, the reader may refer to [24], for instance. The second source of non-convexity is in the functional to be minimized and this is more serious. The major technical point of our paper [25] is to show how to overcome this difficulty. Last but not least, one exploits the fact that one-dimensional minimal microstructures cannot display complicated patterns. For instance, any open set in $\mathbb{R}$ is a countable union of open intervals. Various covering theorems show that this is not the case in multi-dimensions.

In the proof of above theorem, we follow the usual procedure for solving Optimal Control Problems. We derive first order optimality conditions which are necessary at minimum. This is formulated in terms of the state and the adjoint state. This reduces greatly the set of possible minimizers. The task of finding the minimizer is reduced to comparing the values of the cost functional on the above set. An informed guess for the minimizer enables us to finish the proof.

5.3. Bounds on Laminates

Let us now consider the multi-dimensional case of a two-phase conductor with phases $\{\alpha_0, \alpha_1\}$ in given proportions $\{1-\gamma, \gamma\}$. A celebrated theorem describes the variation of the eigenvalues of the homogenized matrix q as the microstructure varies. More precisely, the eigenvalues $\{\mu_j\}$ of q are characterized by the following bounds (assuming $\alpha_0 > \alpha_1$) :

$$\mu_-(\gamma) \le \mu_j \le \mu_+(\gamma),$$

$$\sum_{j=1}^{N} \frac{1}{\mu_j - \alpha_1} \le \frac{1}{\mu_-(\gamma) - \alpha_1} + \frac{N-1}{\mu_+(\gamma) - \alpha_1},$$

$$\sum_{j=1}^{N} \frac{1}{\alpha_0 - \mu_j} \le \frac{1}{\alpha_0 - \mu_-(\gamma)} + \frac{N-1}{\alpha_0 - \mu_+(\gamma)}.$$

Here, we have used the following notations for the Arithmetic and Harmonic Means:

$$\mu_+(\gamma) = (1-\gamma)\alpha_0 + \gamma\alpha_1,$$

$$\frac{1}{\mu_-(\gamma)} = \frac{1-\gamma}{\alpha_0} + \frac{\gamma}{\alpha_1}.$$

Above result has a long history of developments. Many authors have made their contributions. K.Lurie and A.Cherkaev are some of the original contributors [26], [27], [28], [29]. The reader can also refer to [4] for proofs based on compensated compactness and H-measures. Obtaining such a characterization for the tensor d is a far away dream.

In the above description, laminates play a crucial role. Indeed, they are optimal micro-structures for q in the sense that the eigenvalues of q on laminates provide upper and lower bounds for the eigenvalues of q on arbitrary microstructures.

In this subsection, our attention turns to multi-dimensional aspects of macro coefficients by studying their variation on laminates. One of the first phenomena which comes to our mind is the direction-dependence of physical properties, i.e., lack of isotropy at the macro scale even when the materials at microstructure level are isotropic. In particular, we are aware of different longitudinal and transverse behaviour in laminates. Further, there is no interaction between longitudinal and transverse modes in these structures. Somewhat surprisingly, this is no more true when we decrease the scale, as seen below.

The first task in our endeavour is to find a replacement which will play the role of eigenvalues. In this context, let us recall that bounds on eigenvalues of a symmetric matrix such as q are equivalent to the bounds on the associated quadratic form $q\eta^2 = q_{kl}\eta_k\eta_l$. This motivates us to consider the quartic form $d\eta^4 \equiv d_{ijkl}\eta_i\eta_j\eta_k\eta_l$ associated to the 4th order tensor d and study its variation on laminated microstructures. Another reason for considering the quadratic or quartic form is to capture direction-dependent phenomena.

Periodic laminates, by definition, correspond to the following choice of the conductivity matrix : $a_{kl}(y) = a(y)\delta_{kl}$ with $a(y) = a(y_1)$ which is 2π-periodic function of a single variable. Further, since we are considering two-phase media, we have

$$a(y_1) = \alpha_0\chi_{Y\setminus T}(y_1) + \alpha_1\chi_T(y_1), \text{ with } \gamma = \frac{|T|}{|Y|}.$$

It is implicitly assumed that T as a subset of the basic reference cell Y has one-dimensional structure: $T = T_1\times]0,2\pi[^{N-1}$ where T_1 is an arbitrary measurable subset of $]0,2\pi[$. Examples include vertical strips and their unions. Note that there are two types of directions in this microstrucutre: parallel (longitudinal) and perpendicular (transverse) directions to lamination. Accordingly, we split the (momentum) variable $\eta = (\eta_1, \tilde{\eta})$ with $\eta_1 \in \mathbb{R}, \tilde{\eta} = (\eta_2, \eta_3 \dots \eta_N) \in \mathbb{R}^{N-1}$.

As before, the aim is to study the variation of (q,d) as the laminated microstructure T varies inside the basic cell Y, assuming that the volume proportion $\gamma = \frac{|T|}{|Y|}$ remains fixed. Here are our results:

Theorem Under the above situation, we have

(i) the value $q\eta^2 = q_{kl}\eta_k\eta_l$ remains fixed at

$$q\eta^2 = \mu_-(\gamma)\eta_1^2 + \mu_+(\gamma)|\tilde{\eta}|^2 \quad \forall \eta \in \mathbb{R}^N,$$

as microstructure varies.

(ii) The minimum value of $d\eta^4 = d_{klmn}\eta_k\eta_l\eta_m\eta_n$ is given by

$$\min d\eta^4 = -\frac{1}{12}|Y|^2\gamma^2(1-\gamma)^2(\mu_-(\gamma))^{-1}\left\{(\mu_-(\gamma))^2\eta_1^2\left(\frac{1}{\alpha_1}-\frac{1}{\alpha_0}\right)+|\tilde{\eta}|^2(\alpha_1-\alpha_0)\right\}^2.$$

(iii) The maximum value of $d\eta^4$ is zero.

(iv) As microstructure varies among classical laminates, $d\eta^4$ fills up the entire interval $[\min d\eta^4, \max d\eta^4[$.

(v) Minimizer for $d\eta^4$ exists and is unique upto translations. It is the classical laminate consisting of a single vertical strip inside Y.

(vi) Maximizer for $d\eta^4$ exists and is unique. It is however a relaxed laminated microstructure defined by the local density $a(y_1) \equiv \gamma \quad \forall\, y_1 \in]0, 2\pi[$.

A few remarks about the previous Theorem are in order. It is a natural generalization of our one-dimensional result in Section 5.2; indeed, if $\tilde{\eta} = 0$, the above result is reduced to the result in one dimension. Thus all our remarks on one-dimensional results of Subsection 5.2 hold good. So we comment only on additional features here. The statement (i) of the above Theorem is rephrasing the well-known fact, namely the conductivities in the transverse and the longitudinal directions are given by the Harmonic and the Arithmetic Means respectively. Further, there is no interaction between the two modes. Because of the presence of the cross term in statement (ii), there is interaction between the longitudinal and the transverse modes in the case of higher order macro coefficient d. This is a new phenomenon occurring because of the decrease of scale. The origin of this macro phenomenon lies in the presence of interaction terms $(\chi^{(1)})^4, \chi^{(2)}(\chi^{(1)})^2$ etc. in the spatial representation formula for d. (Section 5.1).

Finally, taking successively one after the other $\tilde{\eta} = 0$ or $\eta_1 = 0$ in the statement (ii), and comparing it with the statement (i), we realize that the quantities

$$-\frac{1}{12}|Y|^2\gamma^2(1-\gamma)^2\,(\mu_-(\gamma))^3\left(\frac{1}{\alpha_1}-\frac{1}{\alpha_0}\right)^2,$$

$$-\frac{1}{12}|Y|^2\gamma^2(1-\gamma)^2\,(\mu_-(\gamma))^{-1}\,(\alpha_1-\alpha_0)^2,$$

represent the next order corrections to the Harmonic and the Arithmetic Means respectively as the scale is decreased. The details of the proof of the above results will appear elsewhere.

6. Conclusions and Perspectives

Probing the matter and the space-time at high energies and fine scales through experiments, observations and theoretical studies of models is part of modern day trend. One can hope to carry out similar activities to understand the behaviour of heterogeneous structures at fine scales and at high energies. This article has treated periodically heterogeneous structures with the above point of view. In this context, we have introduced higher order macro coefficients and studied their properties. Several new phenomena that appear with the decrease of scale or with the increase of energy are pointed out. Analogous treatment of arbitrary heterogeneous structures is a huge program for the future. Even among the restricted class of periodic structures with two-phases but with arbitrary micro-geometry, the analysis of behaviour of higher order macro coefficients is a challenging task and efforts are being taken to face upto it.

Acknowledgments

C.C thanks the MICDB for partial support through Grant ICM P05-001-F, Fondap-Basal-Conicyt and the Chilean & French Governments through Ecos-Conicyt Grant C07 E05. J.S.M was partially supported by Grant Fondecyt 1090239 and BASAL-CMM Project. L.S was partially supported by Grant RP-2, no. 6/01.07.2009 of CNCSIS-UEFISCSU. M.V thanks French CNRS and Institut de Mathématiques de Toulouse for hospitality and support for his stay in France during which the final stages of this work were completed. The authors thank the anonymous referee for critical observations and for pointing out the correct references.

References

[1] Bensoussan A, Lions J-L and Papanicolaou G 1978 *Asymptotic Analysis in Periodic Structures* (Amsterdam: North-Holland)
[2] Jikov V.V, Kozlov V.M and Oleinik O 1994 *Homogenization of Differential Operators and Integral Functionals* (Berlin: Springer-Verlag)
[3] Sanchez-Palencia E 1980 *Non-Homogeneous Media and Vibration Theory* (Berlin: Springer-Verlag)
[4] Tartar L 2009 *The General Theory of Homogenization A Personalized Introduction* (Berlin:Springer)
[5] Vanninathan M 2009 in *Current Trends in Science, Platinum Jubilee Special ed. Mukunda M* (Bangalore: Indian Academy of Sciences) 603
[6] Conca C, Orive R and Vanninathan M 2002 *SIAM J. Math. Anal.* **33** 1166
[7] Spagnolo S 1968 *Ann. Sc. Norm. Sup. Pisa,* **22** 571
[8] Gelfand I.M 1950 *Dokl. Akad. Nauk. SSSR* **73** 1117
[9] Odeh F and Keller J.B 1964 *J.Math. Phys.* **5** 1499
[10] Wilcox C.H 1978 *J.Anal. Math.* **33** 146
[11] Khavin V P and Nikol' skij N K 1991 *Commutative Harmonic Analysis I* (Berlin: Springer-Verlag)
[12] Kato T. 1995 *Perturbation Theory for Linear Operators* (Berlin: Springer-Verlag)
[13] Reed M and Simon B 1972 *Methods of Modern Mathematical Physics, Vol.4, Analysis of Operators* (New York: Academic Press)
[14] Conca C and Vanninathan M 1997 *SIAM J. Math. Anal.* **57** 1639
[15] Allaire G, Capdeboscq Y, Piatnitski A, Seiss V and Vanninathan M 2004 *Arch. Ration. Mech. Anal.* **174** 179
[16] Allaire G and Vanninathan M 2006 *Discrete Contin. Dyn. Syst. Ser. B.* **6** 1
[17] Sivaji Ganesh S and Vanninathan M 2006 *ESAIM Control Optim. Calc. Var.* **11** 542
[18] Allaire G 1992 *SIAM J. Math. Anal.* **23** 1482
[19] Milton G 2002 *The Theory of Composites* (Cambridge: Cambridge University Press)
[20] Allaire G 2002 *Shape Optimization by the Homogenization Method* (New York: Springer-Verlag)
[21] Cherkaev A 2000 *Variational Methods for Structural Optimization* (New York: Springer-Verlag)
[22] Conca C, Orive R and Vanninathan M 2006 *J. Math. Phys.* **47** 032902
[23] Conca C, Orive R and Vanninathan M in preparation
[24] Kohn R.V and Strang G 1986 *Comm. Pure Appl. Math.* **39** 113
[25] Conca C, San Martín J, Smaranda L and Vanninathan M 2009 *Math. Models Methods Appl. Sci.* **19** 1743
[26] Lurie K and Cherkaev A 1984 *Proceed.Roy. Soc.Edinburgh, sect.A* **99** 71
[27] Lurie K and Cherkaev A 1982 Report 783, Physical Technical. Inst. Acad. Sci. USSR,1982,32p.(in Russian)
[28] Lurie K and Cherkaev A 1986*Proceed.Roy.Soc. Edinburgh, sect.A* **104** 21
[29] Lurie K and Cherkaev A Report 894, Physical Technical. Inst. Acad. Sci. USSR, 1986, 26 pp. (in Russian)

Journal of Physics: Conference Series **319** (2011) 012021 doi:10.1088/1742-6596/319/1/012021

Principles of Optimization of Structures Against an Impact

Andrej Cherkaev*, Elena Cherkaev†, Seubpong Leelavanichkul‡

September 15, 2011

Abstract

We describe a macro- and micro-behavior of damageable structures. Damage is understood as a process of irreversible phase transition from an initial to a damaged state. An unstructured material can absorb energy until it melts. However, a tiny portion of this energy yields to a construction's disintegration due to instability of the damage that leads to energy concentration. Appearance of concentrated damage zones, such as cracks or delaminating, destroys the construction. The remains of the disintegrated construction can still absorb large amount of energy. The goal of a rational design is to maximally use the energy absorption ability of a damageable material. We propose a concept of a replaceable protective structure design. Such structure should dissipate maximum energy, be able to spread energy of damage, and the damage process should be as stable as possible. The absorption rate and rate of damage propagation are times amplified by waiting-link structures that we describe below.

1 Preliminaries

1.1 Protective structures

We investigate a problem of propagation of structural damage, formulate principles of optimization of the impact-protecting structures, and present models of a protective structure with improved characteristics. The designed structures must sustain a sudden impact: they absorb the energy while keep their structural integrity. The goal is in contrast with the conventional optimal design of maximally stiff constructions that are designed to be loaded by a given static load. Commonly used protective devices include car bumpers, helmets, tempered glass, climbing screamers, and woven baskets of hot balloons, see Figure 1.

These very different constructions are all designed to be damaged or destroyed in a collision saving the protected object. These constructions might be used only once, they absorb the collision (impact) energy shielding the protected object and keeping the structural integrity. Both features are equally important. In a collision, a properly designed bumper will be badly damaged, thereby absorbing the energy of impact and saving the vehicle and its passengers. A strong and stiff bumper that stays undamaged while the car is ruined does not fit its purpose. Likewise, an easily damageable bumper does not fit it because it does not absorb enough energy. An optimally designed structure

*Department of Mathematics, University of Utah, Salt Lake City, Utah, USA 84112, cherk@math.utah.edu
†Department of Mathematics, University of Utah, Salt Lake City, Utah, USA 84112, cherk@math.utah.edu
‡Department of Mechanical Engineering, University of Utah, Salt Lake City, Utah, USA 84112, sleelava@eng.utah.edu

realizes a familiar recommendation of defensive driving instructions "Choose to hit something that will give way (such as bush or shrubs) rather than something hard".

The energy density determines the capability of a structure to withstand dynamical loading. In most structures, due to a highly uneven distribution of energy density, localized yielding occurs at a particular location while the total energy associated with fracture may be relatively small. To increase stability, a concept of a bistability has been introduced [1]. In bistable structures, yielding propagates in a larger volume, thus increasing the total fracture energy. A bistable structure may be created by forming a chain or lattice of bistable structural elements [2-4]. Multiple regions of such structure exhibit yielding prior to ultimate failure. Thus, the bistable structures allow for the delocalization of yielding. Based on results of our previous research dealing with the bistable structures and their applications, the present paper formulates optimization principles for constructions and structures designed to sustain a damaging impact.

1.2 Elements and features of an impact resistive structure

An amorphous or crystalline material with only micron-range order can absorb energy until it fails locally. The lack of longer-range microstructural order precludes distributing the damage away from a concentrated loading zone. Energy concentration leads to instabilities, such as cracking or delamination, which destroy the structural integrity of a protective structure made from such material. In all unstructured materials, the stress rate decreases with increasing strain due to development of imperfections, opening of micro-cracks, dislocation concentration, amongst other mechanisms. These phenomenona lead to stress concentrations that eventually destroy the material sample.

The stress concentration can be reduced by means of bistable structures. A simple bistable structure assembly consists of two roughly parallel brittle-elastic or elastic-plastic components, one of which is longer than the other [1, 2, 3]. When the shorter "sacrificial" element fails (indicated as "first failure" in Fig. 2), the load is assumed by the "waiting element" that was initially inactive. Unlike conventional structures, the bistable structure transforms the energy of an impact into directed traveling waves that dissipate energy throughout a larger domain. Delocalization of damage is achieved by allowing partial internal failure to precede ultimate failure. Due to their morphology, bistable structures become stiffer and stronger when damaged. Such structures would evenly distribute the stress and will be stable even when partially destroyed. Structures with inner instabilities, creating multiple inner breakages that lead to stable structural transformation, transport damage away from impacted zone. As a result, the structure can sustain large deformation

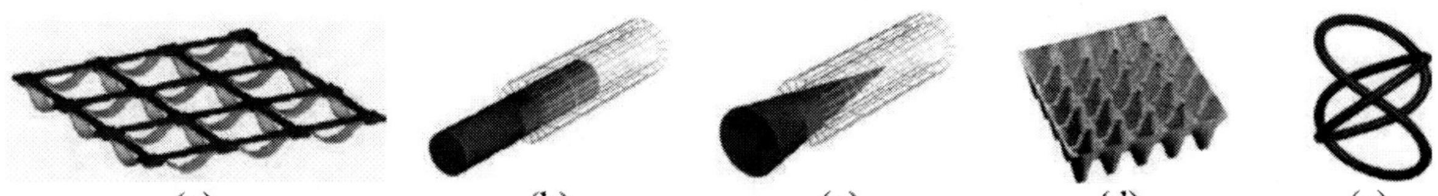

(a) (b) (c) (d) (e)

Figure 1: Structures that absorb energy and become partially damaged, but maintain structural integrity. (a) Lattice with redundant load paths. (b) Tube and cylinder. (c) Tube and cone. (d) Multilayer structure. (e) Structure with helices that mimics protein.

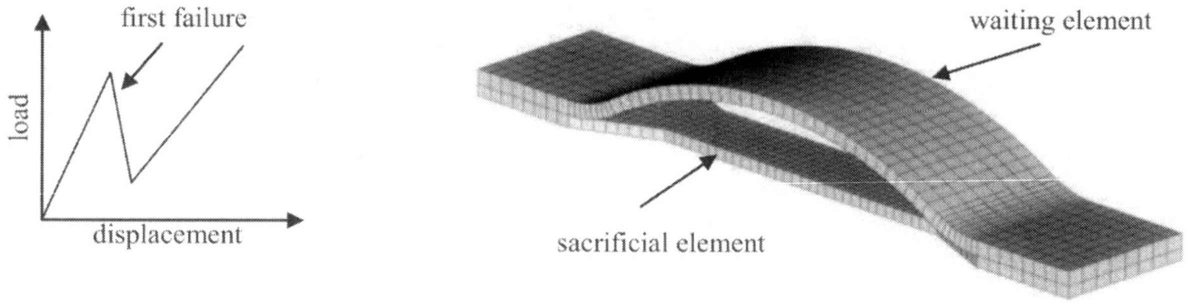

Figure 2: A force-displacement response and a conceptual visualization of a bistable link assembly

without total destruction.

1.3 Lattices: Stages of structural damage

We study damage of lattice structures that consist of a number of connected rods. A simple frame in Figure 3 illustrates a cascade of breakages. We assume that the rods in the frame are brittle-elastic: the elongation growths with the stress and when it reaches a threshold, the rod fails. Unlike chains with sequentially joined links, lattices are not broken when one or several links fail. We assume that a frame from five breakable rods A, B, C, D, and E is attached to a vertical support and is loaded by a force F that is applied to the end of rod E, as shown in Figure 3.

The failure of the structure passes through the following stages:

1. Initially, all rods are stressed by a loading F.
2. When the magnitude of F increases, rod A is breaks.Then, rod B assumes the load; rods C and D become inactive, they form a *waiting link.* The equilibrium remains locally stable but the stiffness of the structure decreases.
3. When rod B fails next, the structure loses its equilibrium and moves to a new equilibrium position that corresponds to straight aligning of rods C and D, which now become active. In the new equilibrium position, the stiffness may increase.
4. Next, one of rods C or D (say, rod C) breaks and the structure moves to another equilibrium position.
5. Finally, rod E breaks and the force is no longer supported by a structure.

Notice, that all rods but D are broken before the load is relieved. The percentage of broken rods shows effectiveness of the support because each breakage absorbs some energy.

Generally, a sequence of breakage of a loaded structure includes: (a) breaks within an equilibrium position, (b) transition from one equilibrium to another one with some waiting links being activated, and (c) a complete breakage when the load is no longer supported by a structure. Step (a) correspond to damage within a stiff structure. An equilibrium of a loaded structure or a substructure is called stiff if it corresponds to a local minimum of the energy, that is no node can move without elongating of at least one link and change the energy balance. The number of degrees of

Journal of Physics: Conference Series **319** (2011) 012021 doi:10.1088/1742-6596/319/1/012021

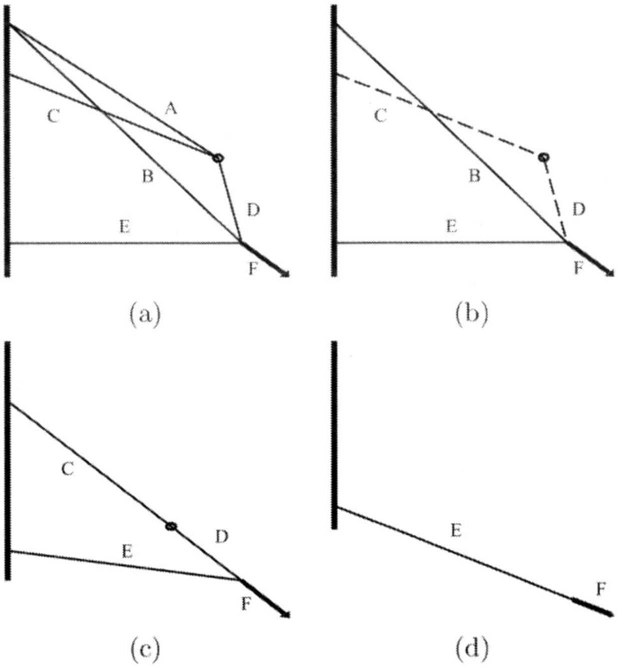

Figure 3: Stages of structural failure

freedom of the nodes of a n-nodes structure is equal to $3n$, out of which 6 degrees correspond to motion of the structure as a solid. The number of active (stressed) links may exceed or be equal to the number of $3n - 6$ of degrees of freedom of a deformation. In this case, the equilibrium is stiff; moreover, if several active links are removed (broken), the structure remains stiff. When the damaged links are removed, remaining active links redistribute the load, and inactive links are not activated. This process can be analyzed by linearized equilibrium equations. When a certain number of links is removed, the structure loses the stiffness (Step (b)) and moves from the equilibrium point. The motion stops when some previously inactive waiting links are activated and the structure assumes a new equilibrium position. The process repeats, until the structure fails and stops to resist the load. The analysis of this process requires consideration of dynamics since the transitions between equilibria correspond to a significant dynamic load factor that could result in further damages that are accompanied by intensive wave generation.

Any addition of new rods, straight or curved, increases the resistance. Additional curved rods may either join the neighboring nodes, as in chains with waiting links, or they may join any two nodes in the structure. The problem of optimal position of additional rods is open.

1.4 Dynamics of bistable lattices

Strong elastic waves and waves of damage propagate throughout the bistable structures, reflecting from its boundaries and interfering with each other. This process leads to a high concentration of stresses that destroys the construction. On the other hand, the elastic waves can be redirected, thereby sending energy away from the zone of contact as it is discussed for the case of lattice shown below in Fig. 10 in Section 3. Initiation and redirection of waves can be effectively done by bistable structures. Because of their local instabilities, the structures are able to absorb the kinetic energy of an impact by exciting intensive waves which more evenly distribute the *partial damage* and relax the stress by allowing for large local deformations of the elements. Stressed bistable cellular lattices naturally excite and transmit waves of damage, acting as a domino train.

Waves in locally unstable systems are controllable. The possibility to redirect the transition wave in elongated strip-like structures is demonstrated in [5]. Depending on structural parameters, the damage may either start at the impacted end and propagate toward the opposite end where the structure is fasten, or it may start at the fastened end and propagate toward the end where the force is applied, or two transition waves may be initiated at both ends simultaneously [4]. The process is controlled by parameters of the structure; optimally designed structures generate waves along the protective plate orthogonal to the direction of the impact.

An optimized structure should trigger stress waves propagation and thereby spread the stress more evenly; it must also contain multiple absorbers. This can be achieved by the use of a composite with a special structure made of a strong frame, and an absorbing filler material. The optimization strategy requires a proper choice of design parameters. They are: the shape of the whole protective construction, location of beams or frames and waiting links, and the choice of materials.

1.5 Analytic solutions for dynamics of bistable chains and lattices

Analytic solution of the problem of dynamics of bistable chains and lattices was derived based on ideas developed in [5, 6] for modeling of propagation of the cracks. Bistable-bond chain model was analyzed in [4, 7, 8, 9, 10, 11, 12, 13, 14, 15, 16]. In a bistable chain, the transition wave is accompanied by a significant dissipation due to the excitation of structure-associated high-frequency waves. The analytical derivation of the solution is presented in detail in [9, 15]. The work [15] also deals with a more complicated case of non-parallel branches of the two-branched strain-stress constitutive relation. In this case, the Wiener-Hopf technique is exploited to derive the solution. Analytically calculated speed of the transition wave [15] for a cellular structure with bonds characterized by a bistable constitutive model, was compared with numerically computed speed estimates. This comparison is shown in Figure 4(b). A more complex problem of breakable two-dimensional structure with waiting links was numerically investigated in [17, 18]. Bistable chain from elastic-plastic material was studied in [3]. Analytical solution of the problem of propagation of plane and crack-like transition waves in two-dimensional lattices was developed in [2].

Partial failure of a cellular structure propagates as a cascade of phase transitions. The dynamics of discrete mass-spring system with "phase transition" of damage is modeled by a system with radial forces:

$$m_k \ddot{u}_k = \sum_{j \in J_k} F_{kj}(\|u_k - u_j\|) e_{kj}, \quad e_{kj} = \frac{u_k - u_j}{\|u_k - u_j\|} \tag{1}$$

Breakage is modeled by assumption that springs instantly change their stiffness and equilibrium lengths when the elongations reach given thresholds,

$$F_{kj}(z) = \begin{cases} C_{kj}(z - z_{eq}) & \text{if sactificial link is unbroken} \\ C'_{kj}(z - z_{eq} - z_{slack}) & \text{if sacrificial link is broken} \end{cases} \tag{2}$$

Here, we assume that waiting links are not broken. The solution can be found if in addition a certain character of transition is assumed, such as periodicity (in reversible transition), a constant speed of the transition wave, or a randomness. Formally, we may replace $F_{kj}(z)$ by a time-dependent function

$$\tilde{F}_{kj}(z, t, t_{kj}^{trans}) = \begin{cases} C_{kj}(z - z_{eq}) & \text{if } t \le t_{kj}^{trans} \\ C'_{kj}(z - z_{eq} - z_{slack}) & \text{if } t > t_{kj}^{trans} \end{cases} \tag{3}$$

where t_{kj}^{trans} is the time of transition that depends on assumed transitional parameters such as the wave's speed. We assign these parameters, and run the simulations adjusting them to the transition conditions

$$t_{kj}^{trans} : \ \|u_k(t_{kj}^{trans}) - u_j(t_{kj}^{trans})\| = z_{threshold} \tag{4}$$

For example, the dynamics of a mass-spring chain with damageable links is modeled [16] under assumption that springs instantly change their equilibrium lengths when the elongations reach given thresholds. Between these instances, the springs are linear and a motion of masses is described by a system of linear differential equations. Because of similarity in motion of all masses,

$$u_k(t) = u_{k-1}(t - \tau) + \eta \quad \forall k,$$

where η is an equilibrium distance, it is enough to consider position u of mass m_0 in the period $t \in (0, \tau)$:

$$m_0 \ddot{u}(t) = C\left[u(t + \tau) - 2u(t) + u(t - \tau)\right] - L\left[H(t) - H\left(t - \frac{\tau}{2}\right)\right]. \tag{5}$$

where C is stiffness and L is the relaxation parameter. This system can be solved by Fourier transform. To close the system, it remains to solve turn-on turn off condition to determine the wave's speed

$$\tau_i : \ \|u_1(\tau_i) - u_0(\tau_i)\| = z_{threshold} \tag{6}$$

In [15], the problem is analyzed and the speed of the transitional wave is computed. It depends on the load applied at an infinitely distant point, see Figure 4(b). The calculation was verified by simulation of the transitional wave in a spring-mass system and measuring its speed, see Figure 4(a). The points obtained by numerical simulation are marked at the analytically calculated speed curve in Figure 4(b).

2 Model of an impact resistive bistable structure

Transmission and redirection of stress waves can be achieved and optimized through use of the bistable design. The local instabilities excite intensive waves that evenly distribute the "partial damage" to allow large local deformations without fully damaging the global protective system (see Fig. 5).

2.1 One dimensional constitutive material model

The constitutive relation for a one-dimensional elastic-brittle breakable specimen is modeled as

$$\sigma = (1 - c)D\varepsilon, \quad \varepsilon = \ln\left(\frac{\Delta L}{L}\right) \tag{7}$$

where D is the material modulus, c is a damage parameter that takes the value of zero (undamaged specimen) or one (damaged specimen), L is the undeformed length, and ΔL is the elongation.

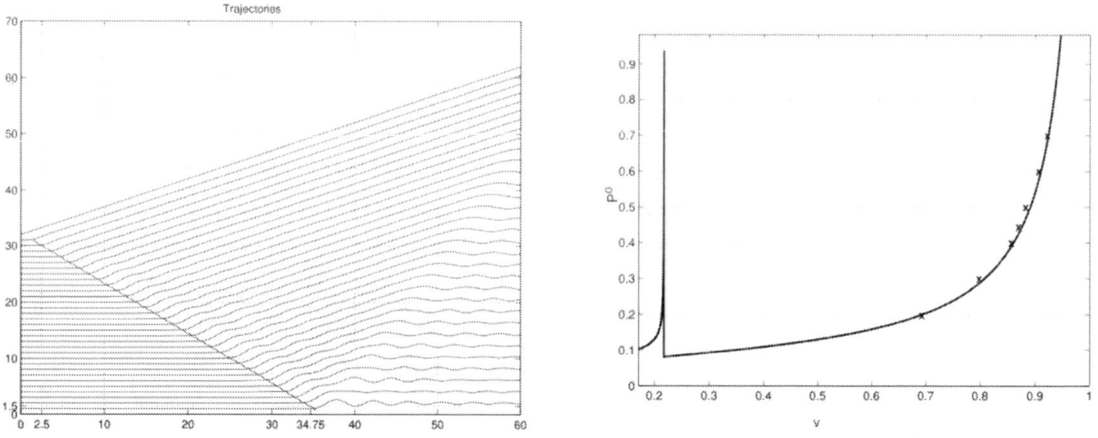

(a) Numerical simulation (b) Speed of transition wave

Figure 4: (a) Numerically simulated wave of transition in a 32-cell chain. (b) Analytically calculated speed of the transition wave compared with numerical estimates shown by stars.

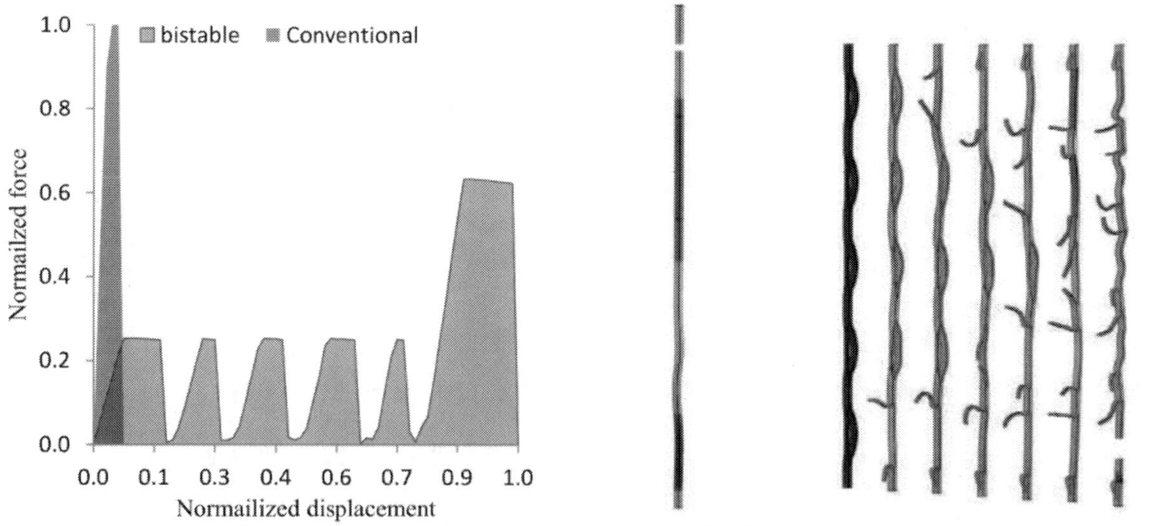

Regular bar (failed at time 0.07T)

Breaking process of a chain with 5 bistable links (failed at time T)

Figure 5: Finite element simulations of a bistable versus a conventional chain having identical mass and subjected to the same loading (displacement control). The simulation used basic elastic-plastic material model. The area under the curve represents the amount of energy absorbed by the structure. This particular example shows that the bistable chain absorbed approximately 4.5 times the amount of energy the conventional chain absorbed before failure.

For numerical simulation, we use the discretized model. The axial stress at time step $n+1$ is computed through stress σ^n at the previous time step n.

$$\sigma^{n+1} = \sigma^n + D\left[(1-c^{n+1})\varepsilon^{n+1} - (1-c^n)\varepsilon^n\right], \tag{8}$$

where c^n control the status of the links at time step n; if $c^n = 0$, the link is undamaged, if $c^n = 1$, it is broken, and $0 < c^n < 1$ corresponds to a partially damaged link. Evolution of c^n is described by the equation

$$c^{n+1} = \begin{cases} 0 & \text{if } \varepsilon_{\text{total}} \leq \varepsilon_c \text{ and } c^n \neq 1 \\ c^n + \dot{c}\Delta t & \text{if } \varepsilon_{\text{total}} \geq \varepsilon_c \text{ and } c^n < 1 \\ 1 & \text{otherwise,} \end{cases} \tag{9}$$

where ε_c is the critical strain and $\dot{c}$ is the speed of damage.

In a bistable assembly, the stress is expressed as a sum of the stresses in the sacrificial and waiting elements,

$$\sigma^{n+1} = \sigma_s^{n+1} + \sigma_w^{n+1}, \tag{10}$$

where subscripts s and w denote the sacrificial and waiting links, respectively. To obtain the the bistable response shown in Fig. 2, the stress in the sacrificial element is evaluated using Eq. (8). The waiting element is initially inactive and is activated after the strain in the bistable assembly reaches a predefined value, $\varepsilon_{\text{init}}$. The stress in the waiting element is computed using

$$\sigma^{n+1} = \left[\sigma^n + D\left((1-c^{n+1})\varepsilon^{n+1} - (1-c^n)\varepsilon^n\right)\right] \times b(\varepsilon_{\text{init}}). \tag{11}$$

where

$$b(\varepsilon_{\text{init}})^{n+1} = \begin{cases} 0 & \text{if } \varepsilon_w \leq \varepsilon_{\text{init}} \text{ and } b(\varepsilon_{\text{init}})^n \neq 1 \\ 1 & \text{otherwise,} \end{cases} \tag{12}$$

Before failure of the sacrificial element, damage parameter c_s is zero, therefore $\sigma_w = 0$ and $\sigma_s > 0$. When the sacrificial element fails, σ_s becomes zero (see Eq. (8)) and the waiting element picks up the load (see Eq. (11)) $\sigma_w > 0$. When the deformation reaches a critical value ε_c, waiting element fails as well. This material model is used in the finite element simulations below. The damage speed $\dot{c}$ plays an important role, the effectiveness of the bistable assembly depends on this value relative to the speed of the impact.

3 Numerical examples

The numerical simulations are carried out using LS-Dyna [19] with the custom material model presented above. The structure is assembled using truss elements.

3.1 Example 1: A rigid ball impacting a net

Using the model in Eqs. 7 to 12, we have performed 3-D simulations of a rigid body impacting two-dimensional lattices, one assembled from conventional links and another one assembled from

Table 1: Sacrificial element's parameters

Parameter	Value
Young modulus, E_s	120 GPa
Poisson's ratio, ν_{12s}	0.36
Length, L_s	1.0 m
Cross-sectional area, A_s	0.01 m^2
Critical strain, ε_{cs}	0.008

Table 2: Waiting element's parameters

Parameter	Value
Young modulus, E_w	120 GPa
Poisson's ratio, ν_{12w}	0.36
Length, L_w	1.008 m
Cross-sectional area, A_w	0.0198 m^2
Critical strain, ε_{cw} (after fully stretched)	0.008

bistable links. These two lattice systems were given identical areal densities. Under various loading conditions, the exit velocity, the kinetic energy of the impacting ball, and distribution of damage and stress in the conventional and bistable nets are compared. It is assumed that the ball impacts the lattice perpendicularly. The model parameters are given in Table 1 and 2 for the simulation time of $t = 0.5$ s. The parameters are chosen so that the transition to the waiting element is sufficiently fast to simulate the response in Fig. 2.

A 100mx100m net was constructed and constrained in all direction along the edges. The net is constructed using 23539 truss elements, and contains 8404 nodes. Here we use the general three-dimensional segment-based penalty contact algorithm [19] with the lattice containing the slave nodes and the ball containing master nodes. No contact friction is accounted for. The results are compared with a similar net from conventional links of the same mass as the sacrificial and waiting elements together; they are made from the same material.

The comparison of the impact resistance is illustrated in Fig. 6. Moreover, as the rigid body penetrates the lattice, cracks are formed in the absence of bistable links, but are avoided when using bistable links. Based on this observation, further investigation will help us to understand the mechanism of damage waves as well as to develop a technique that allows to control these waves in multilevel lattice structures. For example, the path of the damage propagation, or branching of the damage zone in a desired direction, can be controlled through introduction of weak zones or by optimizing the orientation or density of the bistable links. Thus, it is possible to design a "topological trap" for the arrest of the damage propagation.

The results demonstrated in Fig. 6 are based on the constitutive model described in Eqs. 7 to 12, which crudely accounts for damage and eliminates the need of sacrificial element deletion. The model does not have the capability to replace or rearrange the elements when the structure is subjected to different loading conditions during its life cycle. However, the presented results already demonstrate the potential of the bistable link concept.

Effectiveness To evaluate the effectiveness, the effective value R is the ratio of the initial impact velocity, $v(t_0)$, to the exit velocity, $v(t_{\mathrm{final}})$ [17]:

$$R = \frac{v(t_{\mathrm{final}})}{v(t_0)}, \tag{13}$$

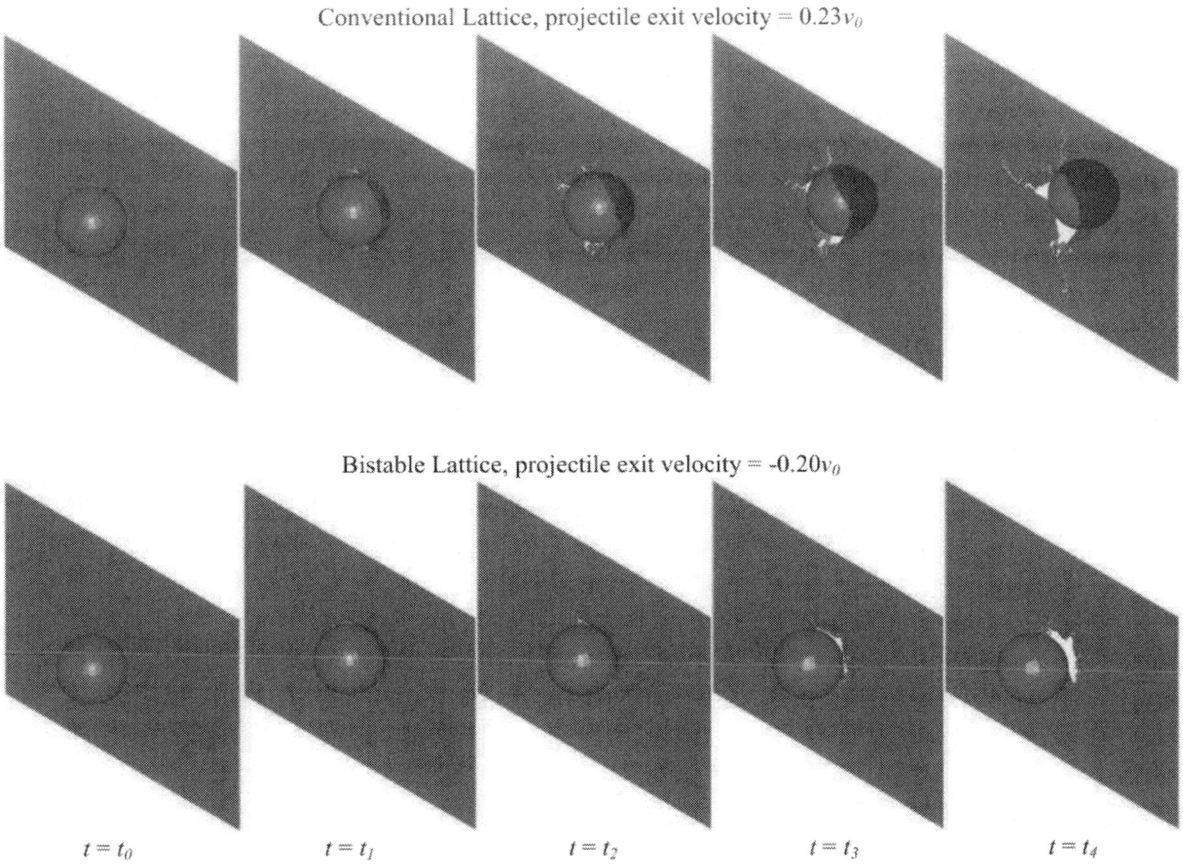

Figure 6: Finite element simulations of lattices impacted by a rigid body. The conventional (top) and the bistable (bottom) lattices have the same total mass, are made of the same material, and are subjected to the same boundary conditions and impact loading. The simulations show that, under the same conditions, the bistable lattice captures and rejects the impacting ball, while the conventional lattice fails to resist the impact.

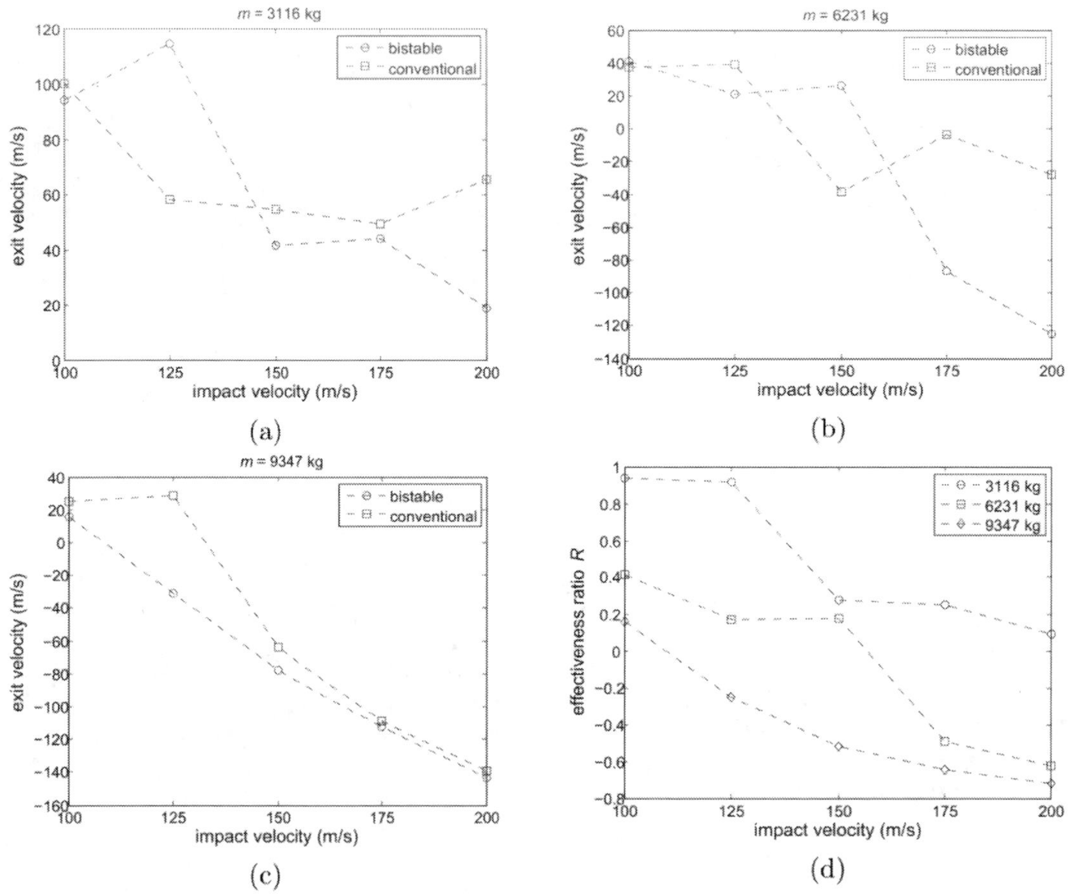

Figure 7: (a)-(c) Comparison of the ball velocity at $t = 0.5$ s. The positive velocity implies that the ball bounces off the net, while the negative velocity indicates that the ball continues to penetrate the net. (d) Effectiveness ratio of the bistable net under an impact of various masses. The net is most effective against impact when the $m \leq 6231$ kg and $v_{\text{impact}} \leq 150$ m/s.

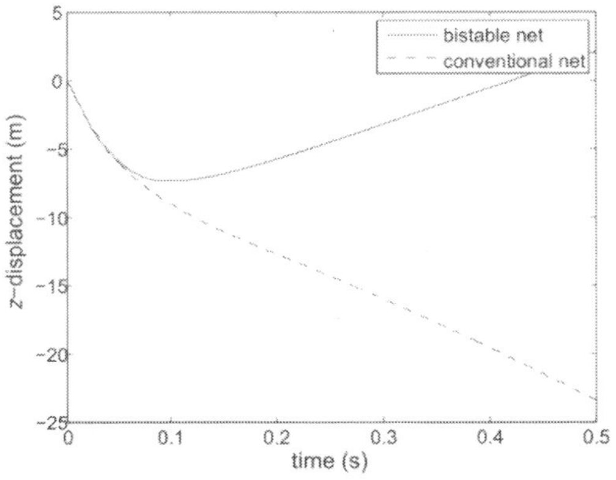

(a)

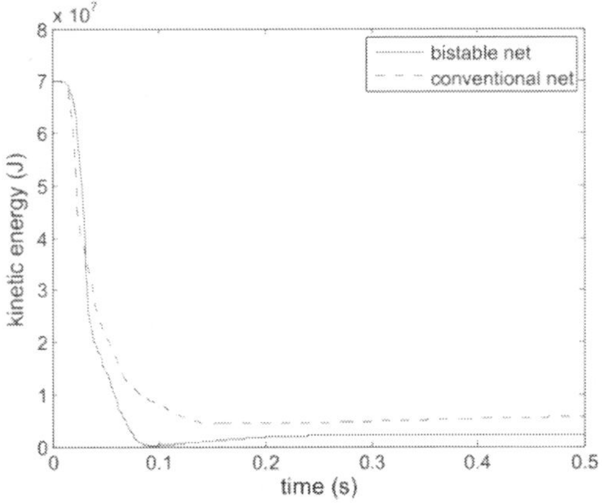

(b)

Figure 8: (a) Kinetic energy. Mass = 6231 kg and impact velocity = 150 m/s. (b) Displacement along the z-direction; mass = 6231 kg and impact velocity = 150 m/s.

where t_0 and t_{final} are the initial and the final moments of the observation, respectively. The variation of impulse of the ball R shows how much of it is transformed to the motion of structural elements. Parameter R evaluates the structure's performance using the impacting ball as a measuring device without considering the energy dissipated in each element of the structure; it does not vary when the ball is not in contact with the structure.

For the chosen configuration, at lower mass (3116 kg), the ball is rejected and neither conventional nor bistable design is damaged. When the speed of the impact is moderate (see Fig. 7), the bistable structure is more effective. It stops heavier balls or decreases the speed of penetrating ones. The bistable net has the most advantage over the conventional net when the initial velocity of the ball is 150 m/s and the mass of the ball is 6231 kg. At this combination, the ball is bounced off the bistable net, while the conventional net is penetrated, see Fig. 7d. In addition, Fig. 8 shows that more energy is dissipated by the bistable net than the conventional net. However, at high speed both structures behave similarly. For example, when the ball's mass is 9347 kg and its speed is above 175 m/s, the ball penetrates both nets and have similar exit velocity.

Damage distribution We observe that a conventional net is destroyed sooner than the one with waiting elements, and it requires less energy to be penetrated. The larger energy dissipation is achieved by delocalization of damage. The damage in the conventional net (see Fig. 9) concentrates in the impact area. The damage in the bistable net occurs by the failure of sacrificial elements and it is more distributed (see Fig. 10). Notice that some sacrificial elements located far away from the impact zone are destroyed while others are not. The damaged links are clustered in small regions scattered throughout the net.

The dynamics of the damage and failure of the net can be viewed as a series of phase transitions. After each break, the network changes its elastic properties and its equilibrium position. In contrast with the conventional structure, the bistable structure may become stronger after sacrificial

(a) Conventional net

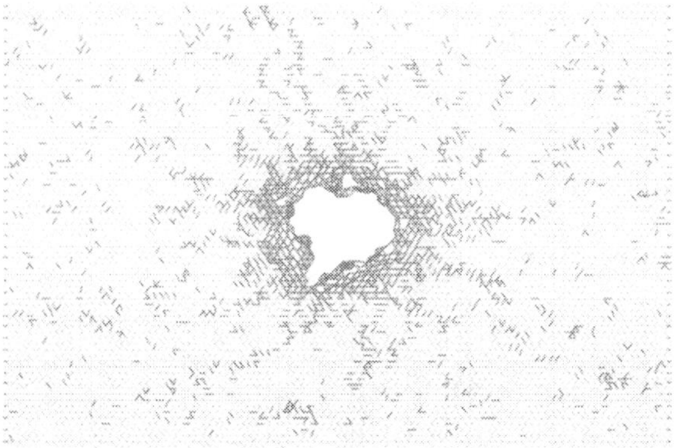

(b) Bistable net

Figure 9: Damage at $t = 0.5$ s.: (a) The completely destroyed sacrificial elements are shown in red (2435 damaged links), while the completely damaged waiting element are hidden from view. There are no partially damaged links in this example because the damage speed $\dot{c}$ is set to the same value as the time step size used in the simulation. (b) The completely destroyed links (284 links) are hidden from view.

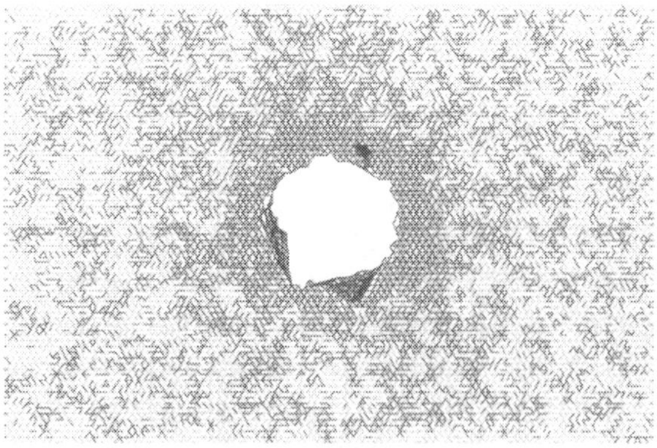

(a) Impact speed = 50 m/s 1700 damaged sacrificial elements

(b) Impact speed = 250 m/s, 9161 damaged sacrificial elements

Figure 10: Damage at $t = 1.0$ s.: (a) These figures illustrate the effect of the waiting elements when the net is subjected to slower and faster impacts having initial kinetic energy of 70×10^6 N·m (similar to the case shown in Fig. 9. The number of the partially damaged links increases as the speed of the projectile decreases.

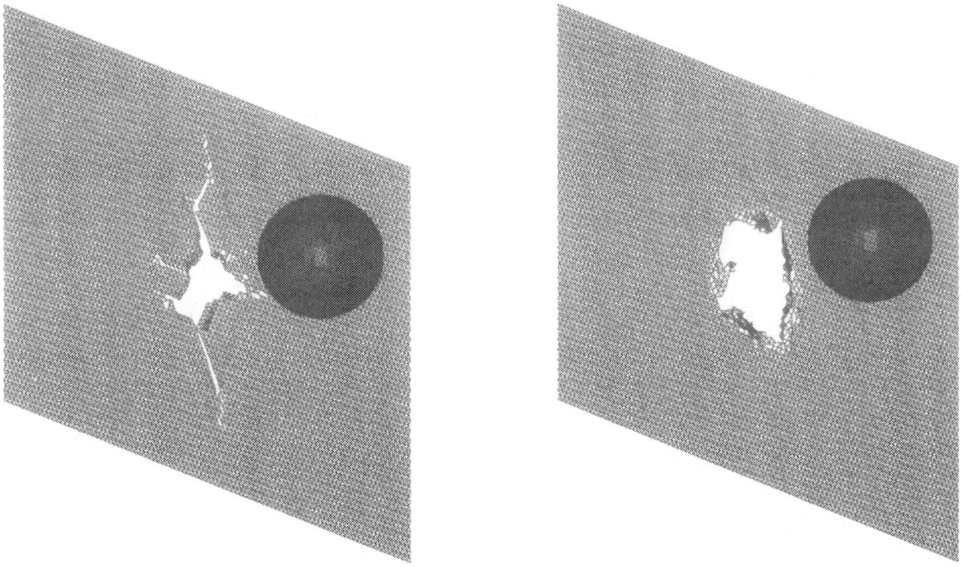

(a) Conventional net (b) Bistable net

Figure 11: Penetration of a 9347 kg rigid ball with initial velocity of 175 m/s, $t = 0.3s$. The phase transition leads to a *pseudo-plastic* response of the net.

elements break. After a sacrificial element breaks, the slackness of the waiting element causes the elastic wave to travel away which helps to spread the damage across the structure.

As the ball penetrates through the net, cracks are formed in the conventional net (see Fig. 11), while no crack formation is observed in the bistable net. Crack is stopped in the bistable net because bistability leads to stress delocalization. Notice that bistable nets, made from an elastic-brittle material, show large but stable *pseudo-plastic* strains and thus increase the resistivity.

4 Damage tracking and optimization of bistable structures

The resistance against fracture and yielding of a protective lattice of bistable links can be optimized by material redistributing, i.e., reinforcing the overloaded links at the expense of underloaded ones. An optimization algorithm can be formulated similar to the approach such as those in [20, 21]. The main challenge of the optimization problem is the time dependent loading and structural variation due to damage, which leads to high sensitivity of the design variables as illustrated in the simulations of the rigid body impacting a bistable lattice discussed earlier (Fig. 10). These figures show that the effectiveness of the design (number of fully damaged sacrificial links) at a given geometry depends greatly on the speed of the impact.

In a chain, the number of partially damaged links is related to the absorbed energy. The load carrying capacity of the chain is lost as soon as one link in the chain is broken. On the contrary, the triangular lattice can still resist concentrated loads even if some of the links are completely broken. We use two basic criteria to compare the state of the structure before and after the collision: (1) The percentage of partially damaged links, and (2) The percentage of destroyed links. The first

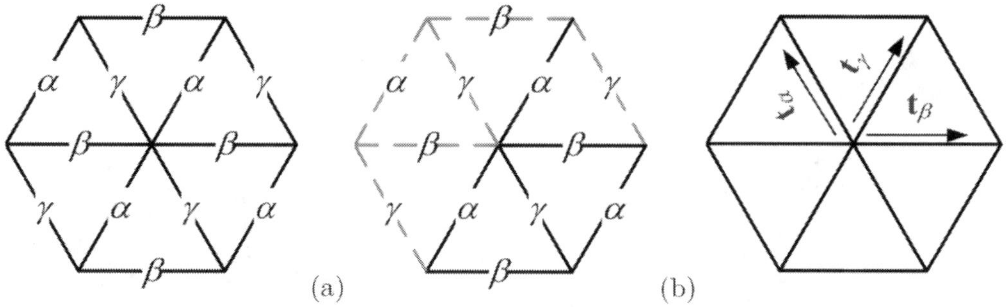

Figure 12: (a) The lattice consists of three families of parallel links α, β, and γ; (b) Example of lattice with damaged links (dashed lines); and (c) Tangent vectors to the links' families.

number shows how effectively the partial damage is spread, and the second one shows how badly the structure is damaged. Ideally, we wish to have a structure in which all links are partially damaged, but none is completely destroyed. The number of destroyed links is a rough overall quality criterion, but it ignores a significant factor - the distribution and orientation of the broken links. To describe various configurations of broken links, we introduce a "damage tensor" that can distinguish partial damage and quantify damage texture based on which links are broken [22, 23]. For example, consider an equilateral triangular lattice that consists of three families of parallel links α, β, and γ with tangent vectors $\mathbf{t}_\alpha, \mathbf{t}_\beta, \mathbf{t}_\gamma$, respectively (see Fig. 12). In this case, we can construct the damage tensor $\mathbf{D}$ as follows:

1. In a reference configuration, select a domain Ω that is larger than distance between nodes (i.e., a domain containing many nodes),
2. Count the number of links l_α, l_β and l_γ that contain at least one node in Ω
3. Count the number of broken links, b_α, b_β and b_γ in (Fig. 12b) and introduce the relative directional damage, p_α, p_β and p_γ

$$p_\alpha = \frac{b_\alpha}{l_\alpha}, \quad p_\beta = \frac{b_\beta}{l_\beta}, \quad p_\gamma = \frac{b_\gamma}{l_\gamma}. \tag{14}$$

4. Finally, the damage tensor $\mathbf{D}(\Omega)$ can be formulated using

$$\mathbf{D}(\Omega) = p_\alpha \mathbf{t}_\alpha \otimes \mathbf{t}_\alpha + p_\beta \mathbf{t}_\beta \otimes \mathbf{t}_\beta + p_\gamma \mathbf{t}_\gamma \otimes \mathbf{t}_\gamma. \tag{15}$$

Likewise, damage tensors can be constructed for the sacrificial and the waiting elements. The dimensionless damage tensor $\mathbf{D}(\Omega)$ allows the characterization of an average damage in a region Ω and hence the average irreversible deformation in that region. It can be used for an intermediate scale description of the damage, which in turn would be useful in optimization of link morphology as well as visualization using any standard methods for displaying tensor fields [24].

5 Acknowledgment

Support of this work by the United States Army Research Office (ARO) is gratefully acknowledged.

References

[1] Cherkaev A and Slepyan L 1995. *Int. J. of Damage Mech,* **4**(1), pp 58–82.

[2] Slepyan L and Ayzenberg-Stepanenko M 2004. *J. Mech. Phys. Solids,* **52**(7), pp 1447–1479.

[3] Vinogradov V *et al* 2006. *Mech. of Materials,* **38**, pp 748–756.

[4] Cherkaev A *et al* 2005. *J. Mech. Phys. Solids,* **53**, pp 383–405.

[5] Slepyan L and Troyankina L 1984. *J. Appl. Mech. Techn. Phys.,* **25**(6), pp 921–927.

[6] Slepyan, L., and Troyankina, L., 1988. *In: 'Strength and Visco-plasticity', Nauka,* pp 301–305. in Russian.

[7] Slepyan L 2000. *J. Mech. Phys. Solids,* **48**, pp 931–964.

[8] Slepyan L 2001. *J. Mech. Phys. Solids,* **49**, pp 513–550.

[9] Slepyan L 2002. *Models and Phenomena in Fracture Mechanics.* Springer. Berlin.

[10] Ngan S-C and Truskinovsky L 1999. *J. Mech. Phys. Solids,* **47**, pp 141–172.

[11] Puglisi G and Truskinovsky L 2000. *J. Mech. Phys. Solids,* **48**, pp 1–27.

[12] Balk A *et al* 2001. *J. Mech. Phys. Solids,* **49**, pp 131–148.

[13] Balk A *et al* 2001. *J. Mech. Phys. Solids,* **49**, pp 149–171.

[14] Charlotte M and Truskinovsky L 2002. *J. Mech. Phys. Solids,* **50**, pp 217–251.

[15] Slepyan L *et al* 2005. *J. Mech. Phys. Solids,* **53**, pp 407–436.

[16] Cherkaev A *et al* 2006. *Vibration Problems ICOVP 2005, Springer Proc in Phys* , **111** pp 111–122.

[17] Cherkaev A and Zhornitskaya L 2004. *Asymptotics, Singularities and Homogenisation in Problems of Mechanics,* pp 273–284.

[18] Cherkaev A and Zhornitskaya L 2004. *Multibody System Dyn,* **13**(1), pp 63 – 67.

[19] Hallquist J O 2006. *LS-DYNA Theory Manual.* Livermore Software Techn. Corp.

[20] Bensoe, M. P., and Sigmund, O., 2004. *Topology Optimization.* Springer, New York.

[21] Cherkaev E and Cherkaev A 2008. *Computers and Structures,* **86**, pp 1426–1435.

[22] Betten J 1983. *J. Mec. Theor. et Appl.,* **2**(1), pp 13–32.

[23] Skrzypek J and Ganczarski A 2010. Springer, New York.

[24] Wilson A T and Brannon R M 2005. *Proceeding of IEEE Visualization 2005,* pp 11–18

AUTHOR INDEX

Institute of Physics
Dirac House, Temple Back
Bristol BS1 6BE UK

CURRAN ASSOCIATES INC
proceedings.com